NONLINEAR TIME SERIES ANALYSIS

This book represents a modern approach to time series analysis which is based on the theory of dynamical systems. It starts from a sound outline of the underlying theory to arrive at very practical issues, which are illustrated using a large number of empirical data sets taken from various fields. This book will hence be highly useful for scientists and engineers from all disciplines who study time variable signals, including the earth, life and social sciences.

The paradigm of deterministic chaos has influenced thinking in many fields of science. Chaotic systems show rich and surprising mathematical structures. In the applied sciences, deterministic chaos provides a striking explanation for irregular temporal behaviour and anomalies in systems which do not seem to be inherently stochastic. The most direct link between chaos theory and the real world is the analysis of time series from real systems in terms of nonlinear dynamics. Experimental technique and data analysis have seen such dramatic progress that, by now, most fundamental properties of nonlinear dynamical systems have been observed in the laboratory. Great efforts are being made to exploit ideas from chaos theory wherever the data display more structure than can be captured by traditional methods. Problems of this kind are typical in biology and physiology but also in geophysics, economics and many other sciences.

This revised edition has been significantly rewritten an expanded, including several new chapters. In view of applications, the most relevant novelties will be the treatment of non-stationary data sets and of nonlinear stochastic processes inside the framework of a state space reconstruction by the method of delays. Hence, non-linear time series analysis has left the rather narrow niche of strictly deterministic systems. Moreover, the analysis of multivariate data sets has gained more attention. For a direct application of the methods of this book to the reader's own data sets, this book closely refers to the publicly available software package TISEAN. The availability of this software will facilitate the solution of the exercises, so that readers now can easily gain their own experience with the analysis of data sets.

HOLGER KANTZ, born in November 1960, received his diploma in physics from the University of Wuppertal in January 1986 with a thesis on transient chaos. In January 1989 he obtained his Ph.D. in theoretical physics from the same place, having worked under the supervision of Peter Grassberger on Hamiltonian many-particle dynamics. During his postdoctoral time, he spent one year on a Marie Curie fellowship of the European Union at the physics department of the University of

Florence in Italy. In January 1995 he took up an appointment at the newly founded Max Planck Institute for the Physics of Complex Systems in Dresden, where he established the research group 'Nonlinear Dynamics and Time Series Analysis'.

In 1996 he received his venia legendi and in 2002 he became adjunct professor in theoretical physics at Wuppertal University. In addition to time series analysis, he works on low- and high-dimensional nonlinear dynamics and its applications. More recently, he has been trying to bridge the gap between dynamics and statistical physics. He has (co-)authored more than 75 peer-reviewed articles in scientific journals and holds two international patents. For up-to-date information see http://www.mpipks-dresden.mpg.de/mpi-doc/kantzgruppe.html.

THOMAS SCHREIBER, born 1963, did his diploma work with Peter Grassberger at Wuppertal University on phase transitions and information transport in spatio-temporal chaos. He joined the chaos group of Predrag Cvitanović at the Niels Bohr Institute in Copenhagen to study periodic orbit theory of diffusion and anomalous transport. There he also developed a strong interest in real-world applications of chaos theory, leading to his Ph.D. thesis on nonlinear time series analysis (University of Wuppertal, 1994). As a research assistant at Wuppertal University and during several extended appointments at the Max Planck Institute for the Physics of Complex Systems in Dresden he published numerous research articles on time series methods and applications ranging from physiology to the stock market. His habilitation thesis (University of Wuppertal) appeared as a review in *Physics Reports* in 1999. Thomas Schreiber has extensive experience teaching nonlinear dynamics to students and experts from various fields and at all levels. Recently, he has left academia to undertake industrial research.

NONLINEAR TIME SERIES ANALYSIS

HOLGER KANTZ AND THOMAS SCHREIBER

Max Planck Institute for the Physics of Complex Systems, Dresden

CAMBRIDGE
UNIVERSITY PRESS

CAMBRIDGE UNIVERSITY PRESS
Cambridge, New York, Melbourne, Madrid, Cape Town, Singapore, São Paulo

Cambridge University Press
The Edinburgh Building, Cambridge CB2 2RU, UK

Published in the United States of America by Cambridge University Press, New York

www.cambridge.org
Information on this title: www.cambridge.org/9780521821506

First published 2000
Second edition published 2004
Reprinted 2005

A catalogue record for this publication is available from the British Library

Library of Congress Cataloguing in Publication data
Kantz, Holger, 1960–
Nonlinear time series analysis/Holger Kantz and Thomas Schreiber. – [2nd ed.].
p.   cm.
Includes bibliographical references and index.
ISBN 0 521 82150 9 – ISBN 0 521 52902 6 (paperback)
1. Time-series analysis. 2. Nonlinear theories.   I. Schreiber, Thomas, 1963–   II. Title

QA280.K355 2003
519.5′5 – dc21   2003044031

ISBN-13 978-0-521-82150-6 hardback
ISBN-10 0-521-82150-9 hardback

ISBN-13 978-0-521-52902-0 paperback
ISBN-10 0-521-52902-6 paperback

Transferred to digital printing 2006

Contents

Contents

Preface to the first edition

The paradigm of deterministic chaos has influenced thinking in many fields of science. As mathematical objects, chaotic systems show rich and surprising structures. Most appealing for researchers in the applied sciences is the fact that deterministic chaos provides a striking explanation for irregular behaviour and anomalies in systems which do not seem to be inherently stochastic.

The most direct link between chaos theory and the real world is the analysis of time series from real systems in terms of nonlinear dynamics. On the one hand, experimental technique and data analysis have seen such dramatic progress that, by now, most fundamental properties of nonlinear dynamical systems have been observed in the laboratory. On the other hand, great efforts are being made to exploit ideas from chaos theory in cases where the system is not necessarily deterministic but the data displays more structure than can be captured by traditional methods. Problems of this kind are typical in biology and physiology but also in geophysics, economics, and many other sciences.

In all these fields, even simple models, be they microscopic or phenomenological, can create extremely complicated dynamics. How can one verify that one's model is a good counterpart to the equally complicated signal that one receives from nature? Very often, good models are lacking and one has to study the system just from the observations made in a single time series, which is the case for most non-laboratory systems in particular. The theory of nonlinear dynamical systems provides new tools and quantities for the characterisation of irregular time series data. The scope of these methods ranges from invariants such as Lyapunov exponents and dimensions which yield an accurate description of the structure of a system (provided the data are of high quality) to statistical techniques which allow for classification and diagnosis even in situations where determinism is almost lacking.

This book provides the experimental researcher in nonlinear dynamics with methods for processing, enhancing, and analysing the measured signals. The theorist will be offered discussions about the practical applicability of mathematical results. The time series analyst in economics, meteorology, and other fields will find inspiration for the development of new prediction algorithms. Some of the techniques presented here have also been considered as possible diagnostic tools in clinical research. We will adopt a critical but constructive point of view, pointing out ways of obtaining more meaningful results with limited data. We hope that everybody who has a time series problem which cannot be solved by traditional, linear methods will find inspiring material in this book.

Dresden and Wuppertal
November 1996

Preface to the second edition

In a field as dynamic as nonlinear science, new ideas, methods and experiments emerge constantly and the focus of interest shifts accordingly. There is a continuous stream of new results, and existing knowledge is seen from a different angle after very few years. Five years after the first edition of "Nonlinear Time Series Analysis" we feel that the field has matured in a way that deserves being reflected in a second edition.

The modification that is most immediately visible is that the program listings have been be replaced by a thorough discussion of the publicly available software TISEAN. Already a few months after the first edition appeared, it became clear that most users would need something more convenient to use than the bare library routines printed in the book. Thus, together with Rainer Hegger we prepared stand-alone routines based on the book but with input/output functionality and advanced features. The first public release was made available in 1998 and subsequent releases are in widespread use now. Today, TISEAN is a mature piece of software that covers much more than the programs we gave in the first edition. Now, readers can immediately apply most methods studied in the book on their own data using TISEAN programs. By replacing the somewhat terse program listings by minute instructions of the proper use of the TISEAN routines, the link between book and software is strengthened, supposedly to the benefit of the readers and users. Hence we recommend a download and installation of the package, such that the exercises can be readily done by help of these ready-to-use routines.

The current edition has be extended in view of enlarging the class of data sets to be treated. The core idea of phase space reconstruction was inspired by the analysis of deterministic chaotic data. In contrast to many expectations, purely deterministic and low-dimensional data are rare, and most data from field measurements are evidently of different nature. Hence, it was an effort of our scientific work over the past years, and it was a guiding concept for the revision of this book, to explore the possibilities to treat other than purely deterministic data sets.

There is a whole new chapter on non-stationary time series. While detecting non-stationarity is still briefly discussed early on in the book, methods to deal with manifestly non-stationary sequences are described in some detail in the second part. As an illustration, a data source of lasting interest, human speech, is used. Also, a new chapter deals with concepts of synchrony between systems, linear and nonlinear correlations, information transfer, and phase synchronisation.

Recent attempts on modelling nonlinear stochastic processes are discussed in Chapter 12. The theoretical framework for fitting Fokker–Planck equations to data will be reviewed and evaluated. While Chapter 9 presents some progress that has been made in modelling input–output systems with stochastic but observed input and on the embedding of time delayed feedback systems, the chapter on modelling considers a data driven phase space approach towards Markov chains. Wind speed measurements are used as data which are best considered to be of nonlinear stochastic nature despite the fact that a physically adequate mathematical model is the deterministic Navier–Stokes equation.

In the chapter on invariant quantities, new material on entropy has been included, mainly on the ϵ- and continuous entropies. Estimation problems for stochastic versus deterministic data and data with multiple length and time scales are discussed.

Since more and more experiments now yield good multivariate data, alternatives to time delay embedding using multiple probe measurements are considered at various places in the text. This new development is also reflected in the functionality of the TISEAN programs. A new multivariate data set from a nonlinear semiconductor electronic circuit is introduced and used in several places. In particular, a differential equation has been successfully established for this system by analysing the data set.

Among other smaller rearrangements, the material from the former chapter "Other selected topics", has been relocated to places in the text where a connection can be made more naturally. High dimensional and spatio-temporal data is now discussed in the context of embedding. We discuss multi-scale and self-similar signals now in a more appropriate way right after fractal sets, and include recent techniques to analyse power law correlations, for example detrended fluctuation analysis.

Of course, many new publications have appeared since 1997 which are potentially relevant to the scope of this book. At least two new monographs are concerned with the same topic and a number of review articles. The bibliography has been updated but remains a selection not unaffected by personal preferences.

We hope that the extended book will prove its usefulness in many applications of the methods and further stimulate the field of time series analysis.

Dresden
December 2002

Acknowledgements

If there is any feature of this book that we are proud of, it is the fact that almost all the methods are illustrated with real, experimental data. However, this is anything but our own achievement – we exploited other people's work. Thus we are deeply indebted to the experimental groups who supplied data sets and granted permission to use them in this book. The production of every one of these data sets required skills, experience, and equipment that we ourselves do not have, not forgetting the hours and hours of work spent in the laboratory. We appreciate the generosity of the following experimental groups:

NMR laser. Our contact persons at the Institute for Physics at Zürich University were Leci Flepp and Joe Simonet; the head of the experimental group is E. Brun. (See Appendix B.2.)

Vibrating string. Data were provided by Tim Molteno and Nick Tufillaro, Otago University, Dunedin, New Zealand. (See Appendix B.3.)

Taylor–Couette flow. The experiment was carried out at the Institute for Applied Physics at Kiel University by Thorsten Buzug and Gerd Pfister. (See Appendix B.4.)

Atrial fibrillation. This data set is taken from the MIT-BIH Arrhythmia Database, collected by G. B. Moody and R. Mark at Beth Israel Hospital in Boston. (See Appendix B.6.)

Human ECG. The ECG recordings we used were taken by Petr Saparin at Saratov State University. (See Appendix B.7.)

Foetal ECG. We used noninvasively recorded (human) foetal ECGs taken by John F. Hofmeister as the Department of Obstetrics and Gynecology, University of Colorado, Denver CO. (See Appendix B.7.)

Phonation data. This data set was made available by Hanspeter Herzel at the Technical University in Berlin. (See Appendix B.8.)

Human posture data. The time series was provided by Steven Boker and Bennett Bertenthal at the Department of Psychology, University of Virginia, Charlottesville VA. (See Appendix B.9.)

Autonomous CO_2 laser with feedback. The data were taken by Riccardo Meucci and Marco Ciofini at the INO in Firenze, Italy. (See Appendix B.10.)

Nonlinear electric resonance circuit. The experiment was designed and operated by M. Diestelhorst at the University of Halle, Germany. (See Appendix B.11.)

Nd:YAG laser. The data we use were recorded in the University of Oldenburg, where we wish to thank Achim Kittel, Falk Lange, Tobias Letz, and Jürgen Parisi. (See Appendix B.12.)

We used the following data sets published for the Santa Fe Institute Time Series Contest, which was organised by Neil Gershenfeld and Andreas Weigend in 1991:

NH_3 laser. We used data set A and its continuation, which was published after the contest was closed. The data was supplied by U. Hübner, N. B. Abraham, and C. O. Weiss. (See Appendix B.1.)

Human breath rate. The data we used is part of data set B of the contest. It was submitted by Ari Goldberger and coworkers. (See Appendix B.5.)

During the composition of the text we asked various people to read all or part of the manuscript. The responses ranged from general encouragement to detailed technical comments. In particular we thank Peter Grassberger, James Theiler, Daniel Kaplan, Ulrich Parlitz, and Martin Wiesenfeld for their helpful remarks. Members of our research groups who either contributed by joint work to our experience and knowledge or who volunteered to check the correctness of the text are Rainer Hegger, Andreas Schmitz, Marcus Richter, Mario Ragwitz, Frank Schmüser, Rathinaswamy Bhavanan Govindan, and Sharon Sessions. We have also considerably profited from comments and remarks of the readers of the first edition of the book. Their effort in writing to us is gratefully appreciated.

Last but not least we acknowledge the encouragement and support by Simon Capelin from Cambridge University Press and the excellent help in questions of style and English grammar by Sheila Shepherd.

Part I

Basic topics

Chapter 1

Introduction: why nonlinear methods?

You are probably reading this book because you have an interesting source of data and you suspect it is not a linear one. Either you positively know it is nonlinear because you have some idea of what is going on in the piece of world that you are observing or you are led to suspect that it is because you have tried linear data analysis and you are unsatisfied with its results.[1]

Linear methods interpret all regular structure in a data set, such as a dominant frequency, through linear correlations (to be defined in Chapter 2 below). This means, in brief, that the intrinsic dynamics of the system are governed by the linear paradigm that small causes lead to small effects. Since linear equations can only lead to exponentially decaying (or growing) or (damped) periodically oscillating solutions, all irregular behaviour of the system has to be attributed to some random external input to the system. Now, chaos theory has taught us that random input is not the only possible source of irregularity in a system's output: nonlinear, chaotic systems can produce very irregular data with purely deterministic equations of motion in an autonomous way, i.e., without time dependent inputs. Of course, a system which has both, nonlinearity *and* random input, will most likely produce irregular data as well.

Although we have not yet introduced the tools we need to make quantitative statements, let us look at a few examples of real data sets. They represent very different problems of data analysis where one could profit from reading this book since a treatment with linear methods alone would be inappropriate.

Example 1.1 (NMR laser data). In a laboratory experiment carried out in 1995 by Flepp, Simonet & Brun at the Physics Department of the University of Zürich, a **N**uclear **M**agnetic **R**esonance laser is operated under such conditions that the amplitude of the (radio frequency) laser output varies irregularly over time. From

---

[1] Of course you are also welcome to read this book if you are not working on a particular data set.

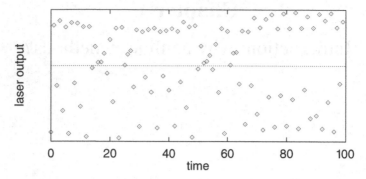

Figure 1.1 100 successive measurements of the laser intensity of an NMR laser. The time unit here is set to the measurement interval.

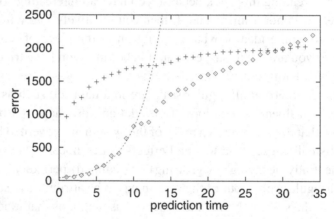

Figure 1.2 The average prediction error (in units of the data) for a longer sample of the NMR laser output as a function of the prediction time. For an explanation of the different symbols see the text of Example 1.1.

the set-up of the experiment it is clear that the system is highly nonlinear and random input noise is known to be of very small amplitude compared to the amplitude of the signal. Thus it is not assumed that the irregularity of the signal is just due to input noise. In fact, it has been possible to model the system by a set of differential equations which does not involve any random components at all; see Flepp *et al.* (1991). Appendix B.2 contains more details about this data set.

Successive values of the signal appear to be very erratic, as can be seen in Fig. 1.1. Nevertheless, as we shall see later, it is possible to make accurate forecasts of future values of the signal using a nonlinear prediction algorithm. Figure 1.2 shows the mean prediction error depending on how far into the future the forecasts are made. Quite intuitively, the further into the future the forecasts are made, the larger

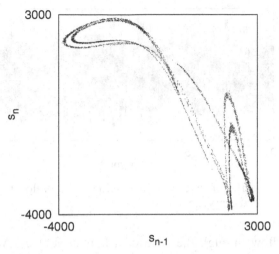

Figure 1.3 Phase portrait of the NMR laser data in a stroboscopic view. The data are the same as in Fig. 1.1 but all 38 000 points available are shown.

will the uncertainty be. After about 35 time steps the prediction becomes worse than when just using the mean value as a prediction (horizontal line). On short prediction horizons, the growth of the prediction error can be well approximated by an exponential with an exponent of 0.3, which is indicated as a dashed line. We used the simple prediction method which will be described in Section 4.2. For comparison we also show the result for the best *linear* predictor that we could fit to the data (crosses). We observe that the predictability due to the linear correlations in the data is much weaker than the one due to the deterministic structure, in particular for short prediction times. The predictability of the signal can be taken as a signature of the deterministic nature of the system. See Section 2.5 for details on the linear prediction method used. The nonlinear structure which leads to the short-term predictability in the data set is not apparent in a representation such as Fig. 1.1. We can, however, make it visible by plotting each data point versus its predecessor, as has been done in Fig. 1.3. Such a plot is called a *phase portrait*. This representation is a particularly simple application of the *time delay embedding*, which is a basic tool which will often be used in nonlinear time series analysis. This concept will be formally introduced in Section 3.2. In the present case we just need a data representation which is printable in two dimensions. □

Example 1.2 (Human breath rate). One of the data sets used for the Santa Fe Institute time series competition in 1991–92 [Weigend & Gershenfeld (1993)] was provided by A. Goldberger from Beth Israel Hospital in Boston [Rigney *et al.* (1993); see also Appendix B.5]. Out of several channels we selected a 16 min

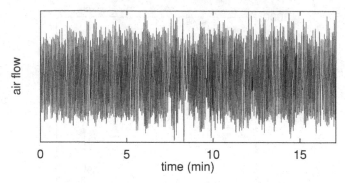

Figure 1.4 A time series of about 16 min duration of the air flow through the nose of a human, measured every 0.5 s.

record of the air flow through the nose of a human subject. A plot of the data segment we used is shown in Fig. 1.4.

In this case only very little is known about the origin of the fluctuations of the breath rate. The only hint that nonlinear behaviour plays a role comes from the data itself: the signal is not compatible with the assumption that it is created by a Gaussian random process with only linear correlations (possibly distorted by a nonlinear measurement function). This we show by creating an artificial data set which has exactly the same *linear* properties but has no further determinism built in. This data set consists of random numbers which have been rescaled to the distribution of the values of the original (thus also mean and variance are identical) and filtered so that the power spectrum is the same. (How this is done, and further aspects, are discussed in Chapter 4 and Section 7.1.) If the measured data are properly described by a linear process we should not find any significant differences from the artificial ones.

Let us again use a *time delay embedding* to view and compare the original and the artificial time series. We simply plot each time series value against the value taken a *delay time τ* earlier. We find that the resulting two *phase portraits* look qualitatively different. Part of the structure present in the original data (Fig. 1.5, left hand panel) is not reproduced by the artificial series (Fig. 1.5, right hand panel). Since both series have the same linear properties, the difference is most likely due to nonlinearity in the system. Most significantly, the original data set is statistically asymmetric under time reversal, which is reflected in the fact that Fig. 1.5, left hand panel, is non-symmetric under reflection with respect to the diagonal. This observation makes it very unlikely that the data represent a noisy *harmonic oscillation*.   □

Example 1.3 (Vibrating string data). Nick Tufillaro and Timothy Molteno at the Physics Department, University of Otago, Dunedin, New Zealand, provided a couple of time series (see Appendix B.3) from a vibrating string in a magnetic field.

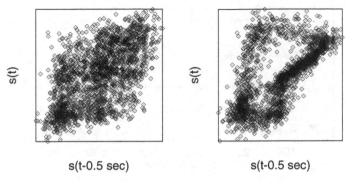

Figure 1.5 Phase portraits of the data shown in Fig. 1.4 (left) and of an artificial data set consisting of random numbers with the same linear statistical properties (right).

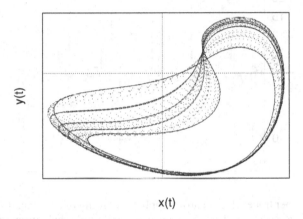

Figure 1.6 Envelope of the elongation of a vibrating string. Line: the first 1000 measurements. Dots: the following 4000 measurements.

The envelope of the elongation of the wire undergoes oscillations which may be chaotic, depending on the parameters chosen. One of these data sets is dominated by a period-five cycle, which means that after every fifth oscillation the recording (approximately) retraces itself. In Fig. 1.6 we plot the simultaneously recorded x- and y-elongations. We draw a solid line through the first 1000 measurements and place a dot at each of the next 4000. We see that the period-five cycle does not remain perfect during the measurement interval. This becomes more evident by plotting one point every cycle versus the cycle count. (These points are obtained in a systematic way called the *Poincaré section*; see Section 3.5.) We see in Fig. 1.7 that the period-five cycle is interrupted occasionally. A natural explanation could be that there are perhaps external influences on the equipment which cause the system to leave the periodic state. However, a much more appropriate explanation

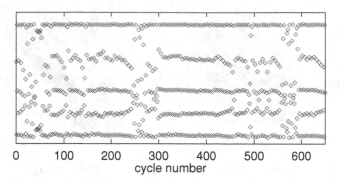

Figure 1.7 Data of Fig. 1.6 represented by one point per cycle. The period-five cycle is interrupted irregularly.

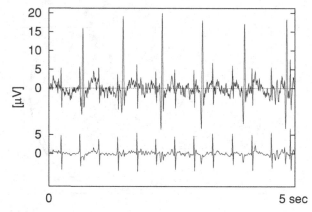

Figure 1.8 Upper trace: electrocardiographic recording of a pregnant woman. The foetal heart causes the small spikes. Lower trace: the foetal electrocardiogram has been extracted by a nonlinear projection technique.

of the data set is given in Example 8.3 in Section 8.4. The irregular episodes are in fact due to *intermittency*, a typical phenomenon found in nonlinear dynamical systems. □

Example 1.4 (Foetal electrocardiogram). Let us consider the signal processing problem of extracting the tiny foetal electrocardiogram component from the (small) electrical potentials on the abdomen of a pregnant woman (upper trace in Fig. 1.8). The faster and smaller foetal signal cannot be distinguished by classical linear techniques because both the maternal and foetal electrocardiograms have broad band power spectra. However, using a nonlinear phase space projection technique that was originally developed for noise reduction in chaotic data (Section 10.3.2), it is possible to perform the separation in an automated way. The lower trace in Fig. 1.8 shows the extracted foetal component. More explanations are given in

Section 10.4. Note that we do not need to assume (nor do we actually expect) that the heart is a deterministic chaotic system. □

There is a broad range of questions that we have to address when talking about nonlinear time series analysis. How can we get the most precise and meaningful results for a clean and clearly deterministic data set like the one in the first example? What modifications are necessary if the case is less clear? What can still be done if, as in the breath rate example, all we have are some hints that the data are not properly described by a linear model with Gaussian inputs? Depending on the data sets and the analysis task we have in mind, we will choose different approaches. However, there are a number of general ideas that one should be acquainted with no matter what one's data look like. The most important concepts analysing complex data will be presented in Part One of the book. Issues which are either theoretically more advanced, require higher-quality data, or are of less general interest will be found in Part Two.

Obviously, in such a diverse field as nonlinear time series analysis, any selection of topics for a single volume must be incomplete. It would be naive to claim that all the choices we have made are exclusively based on objective criteria. There is a strong bias towards methods that we have found either conceptually interesting or useful in practical work, or both. Which methods are most useful depends on the type of data to be analysed and thus part of our bias has been determined by our contacts with experimental groups. Finally, we are no exception to the rule that people prefer the methods that they have been involved in developing.

While we believe that some approaches presented in the literature are indeed useless for scientific work (such as determining Lyapunov exponents from as few as 16 data points), we want to stress that other methods we mention only briefly or not at all, may very well be useful for a given time series problem. Below we list some major omissions (apart from those neglected as a result of our ignorance). Nonlinear generalisations of ARMA models and other methods popular among statisticians are presented in Tong (1990). Generalising the usual two-point autocorrelation function leads to the *bispectrum*. This and related time series models are discussed in Subba Rao & Gabr (1984). Within the theory of dynamical systems, the analysis of unstable periodic orbits plays a prominent role. On the one hand, periodic orbits are important for the study of the topology of an attractor. Very interesting results on the template structure, winding numbers, etc., have been obtained, but the approach is limited to those attractors of dimension two or more which can be embedded in three dimensional phase space. (One dimensional manifolds, e.g. trajectories of dynamical systems, do not form knots in more than three dimensional space.) We refer the interested reader to Tufillaro *et al.* (1992). On the other hand, periodic orbit expansions constitute a powerful way of computing characteristic quantities which are defined as averages over the natural measure, such as dimensions, entropies, and Lyapunov exponents. In the case where a system is *hyperbolic*, i.e., expanding and contracting directions are nowhere tangent to each other, exponentially converging expressions for such quantities can be derived. However, when hyperbolicity is lacking (the generic case), very large numbers of periodic orbits are necessary for the cycle expansions. So far it has not been demonstrated that this kind of analysis is feasible based on experimental time series data. The theory is explained in Artuso *et al.* (1990), and in "The Webbook" [Cvitanović *et al.* (2001)].

After a brief review of the basic concepts of linear time series analysis in Chapter 2, the most fundamental ideas of the nonlinear dynamics approach will be introduced in the chapters on phase space (Chapter 3) and on predictability (Chapter 4). These two chapters are essential for the understanding of the remaining text; in fact, the concept of a phase space representation rather than a time or frequency domain approach is the hallmark of nonlinear dynamical time series analysis. Another fundamental concept of nonlinear dynamics is the sensitivity of chaotic systems to changes in the initial conditions, which is discussed in the chapter about dynamical instability and the Lyapunov exponent (Chapter 5). In order to be bounded and unstable at the same time, a trajectory of a dissipative dynamical system has to live on a set with unusual geometric properties. How these are studied from a time series is discussed in the chapter on attractor geometry and fractal dimensions (Chapter 6). Each of the latter two chapters contains the basics about their topic, while additional material will be provided later in Chapter 11. We will relax the requirement of determinism in Chapter 7 (and later in Chapter 12). We propose rather general methods to inspect and study complex data, including visual and symbolic approaches. Furthermore, we will establish statistical methods to characterise data that lack strong evidence of determinism such as scaling or self-similarity. We will put our considerations into the broader context of the theory of nonlinear dynamical systems in Chapter 8.

The second part of the book will contain advanced material which may be worth studying when one of the more basic algorithms has been successfully applied; obviously there is no point in estimating the whole spectrum of Lyapunov exponents when the largest one has not been determined reliably. As a general rule, refer to Part One to get first results. Then consult Part Two for *optimal* results. This rule applies to embeddings, Chapter 9, noise treatment, Chapter 10, as well as modelling and prediction, Chapter 12. Here, methods to deal with nonlinear processes with stochastic driving have been included. An advanced mathematical level is necessary for the study of invariant quantities such as Lyapunov spectra and generalised dimensions, Chapter 11. Recent advancements in the treatment of non-stationary signals are presented in Chapter 13. There will also be a brief account of chaotic synchronisation, Chapter 14.

Chaos control is a huge field in itself, but since it is closely related to time series analysis it will be discussed to conclude the text in Chapter 15.

Throughout the text we will make reference to the specific implementations of the methods discussed which are included in the TISEAN software package.[2] Background information about the implementation and suggestions for the use of the programs will be given in Appendix A. The TISEAN package contains a large number of routines, some of which are implementations of standard techniques or included for convenience (spectral analysis, random numbers, etc.) and will not be further discussed here. Rather, we will focus on those routines which implement essential ideas of nonlinear time series analysis. For further information, please refer to the documentation distributed with the software.

Nonlinear time series analysis is not as well established and is far less well understood than its linear counterpart. Although we will make every effort to explain the perspectives and limitations of the methods we will introduce, it will be necessary for you to familiarise yourself with the algorithms with the help of artificial data where the correct results are known. While we have almost exclusively used experimental data in the examples to illustrate the concepts discussed in the text, we will introduce a number of numerical models in the exercises. We urge the reader to solve some of the problems and to use the artificial data before actually analysing measured time series. It is a bad idea to apply unfamiliar algorithms to data with unknown properties; better to practise on some data with appropriate known properties, such as the number of data points and sampling rate, number of degrees of freedom, amount and nature of the noise, etc. To give a choice of popular models we introduce the logistic map (Exercise 2.2), the Hénon map (Exercise 3.1), the Ikeda map (Exercise 6.3), a three dimensional map by Baier and Klein (Exercise 5.1), the Lorenz equations (Exercise 3.2) and the Mackey–Glass delay differential equation (Exercise 7.2). Various noise models can be considered for comparison, including *moving average* (Exercise 2.1) and *Brownian motion* models (Exercise 5.3). These are only the references to the exercises where the models are introduced. Use the index to find further material.

Further reading

Obviously, someone who wants to profit from time series methods based on chaos theory will improve his/her understanding of the results by learning more about chaos theory. There is a nice little book about chance and chaos by Ruelle (1991). A readable introduction without much mathematical burden is the book by Kaplan & Glass (1995). Other textbooks on the topic include Bergé *et al.* (1986), Schuster

---

[2] The software and documentation is available at www.mpipks-dresden.mpg.de/~tisean.

(1988), and Ott (1993). A very nice presentation is that by Alligood *et al.* (1997). More advanced material is contained in Eckmann & Ruelle (1985) and in Katok & Hasselblatt (1996). Many articles relevant for the practical aspects of chaos are reproduced in Ott *et al.* (1994), augmented by a general introduction. Theoretical aspects of chaotic dynamics relevant for time series analysis are discussed in Ruelle (1989).

There are several books that cover chaotic time series analysis, each one with different emphasis. Abarbanel (1996) discusses aspects of time series analysis with methods from chaos theory, with a certain emphasis on the work of his group on false neighbours techniques. The small volume by Diks (1999) focuses on a number of particular aspects, including the effect of noise on chaotic data, which are discussed quite thoroughly. There are some monographs which are connected to specific experimental situations but contain material of more general interest: Pawelzik (1991), Buzug (1994) (both in German), Tufillaro *et al.* (1992) and Galka (2000). Many interesting articles about nonlinear methods of time series analysis can be found in the following conference proceedings volumes: Mayer-Kress (1986), Casdagli & Eubank (1992), Drazin & King (1992), and Gershenfeld & Weigend (1993). Interesting articles about nonlinear dynamics and physiology in particular are contained in Bélair *et al.* (1995). Review articles relevant to the field but taking slightly different points of view are Grassberger *et al.* (1991), Casdagli *et al.* (1991), Abarbanel *et al.* (1993), Gershenfeld & Weigend (1993), and Schreiber (1999). The TISEAN software package is accompanied by an article by Hegger *et al.* (1999), which can also be used as an overview of time series methods.

For a statistician's view of nonlinear time series analysis, see the books by Priestley (1988) or by Tong (1990). There have been a number of workshops that brought statisticians and dynamical systems people together. Relevant proceedings volumes include Tong (1993), Tong (1995) and Mees (2000).

Chapter 2
Linear tools and general considerations

2.1 Stationarity and sampling

Quite generally, a scientific measurement of any kind is in principle more useful the more it is reproducible. We need to know that the numbers we measure correspond to properties of the studied object, up to some measurement error. In the case of time series measurements, reproducibility is closely connected to two different notions of stationarity.

The weakest but most evident form of *stationarity* requires that all parameters that are relevant for a system's dynamics have to be fixed and constant during the measurement period (and these parameters should be the same when the experiment is reproduced). This is a requirement to be fulfilled not only by the experimental set-up but also by the process taking place in this fixed environment. For the moment this might be puzzling since one usually expects that constant external parameters induce a stationary process, but in fact we will confront you in several places in this book with situations where this is not true. If the process under observation is a probabilistic one, it will be characterised by probability distributions for the variables involved. For a stationary process, these probabilities may not depend on time. The same holds if the process is specified by a set of *transition probabilities* between different states. If there are deterministic rules governing the dynamics, these rules must not change during the time covered by a time series.

In some cases, we can handle a simple change of a parameter once this change is noticed. If the calibration of the measurement apparatus drifts, for example, we can try to rescale the data continuously in order to keep the mean and variance constant. This can be dangerous, though, unless we are *sure* that only the measurement scale, and not the dynamics, is drifting. A strictly periodic modulation of a parameter can be interpreted as a dynamical variable rather than a parameter and does not necessarily destroy stationarity.

13

Unfortunately, in most cases we do not have direct access to the system which produces a signal and we cannot establish evidence that its parameters are indeed constant. Thus we have to formulate a second concept of *stationarity* which is based on the available data itself. This concept has to be different since there are many processes which are formally stationary when the limit of infinitely long observation times can be taken but which behave effectively like non-stationary processes when studied over finite times. A prominent phenomenon belonging to this class is called *intermittency* and will be discussed in Section 8.4.

A time series, as any other measurement, has to provide enough information to determine the quantity of interest unambiguously. All algorithms given below will have some minimal requirements as to how long and how precise the time series must be, and how frequently measurements should be taken in order to obtain a meaningful result. If this cannot be achieved, a smaller sub-phenomenon has to be considered for study.

Example 2.1 Suppose you want to study how the temperature changes on the roof of your house. How many single measurements will you need? The answer is: it depends. When you intend to study what happens in the course of a day, you may climb up to the roof to take a measurement once every hour throughout one week to obtain a fair average of the temperature profile. If you are interested in the seasonal fluctuations, you will perhaps only climb up at a fixed time once every day, but now will do this throughout the year. If you want to make a statement about how the temperature fluctuates from year to year you will have to take a couple of measurements every year. And finally, if you are aiming for a complete description of the phenomenon, you will need to climb up to the roof once an hour every day of the year for a couple of years. Note that the requirement that the phenomenon under study has to be sampled sufficiently is not peculiar to any method of time series analysis, but, rather, is of a general nature.   □

In a more formal way, a signal is called stationary if all joint probabilities of finding the system at some time in one state and at some later time in another state are independent of time *within the observation period*, i.e. when calculated from the data. This includes the constancy of relevant parameters, but it also requires that phenomena belonging to the dynamics are contained in the time series sufficiently frequently, so that the probabilities or other rules can be inferred properly. If the observed signal is quite regular almost all of the time, but contains one very irregular burst every so often, then the time series has to be considered to be non-stationary for our purposes, even in the case where all parameters remained exactly constant but the signal is intermittent. Only if the rare events (e.g. the irregular bursts mentioned before) also appear several times in the time series can we speak of an effective independence of the observed joint probabilities and thus of stationarity.

Non-stationary signals are very common in particular when observing natural or cultural phenomena. Such signals can be worth studying and you can even make money by predicting stock prices in situations which have no parallel in history. In this book we adopt the point of view that a time series is just a special kind of scientific (i.e. reproducible) measurement which is produced to improve our knowledge about a certain phenomenon. Be aware that almost all the methods and results on time series analysis given below assume the validity of both conditions: that the parameters of the system remain constant and that the phenomenon is sufficiently sampled. As we will discuss in Chapter 13, the requirement of stationarity is a consequence of the fact that we try to approach a dynamical phenomenon through a single finite time series, and hence is a requirement of almost all statistical tools for time series data, including the linear ones. Clearly, since time series analysis methods in the end give rise to algorithms which just compress time series data into a set of a few numbers, you can apply them to any sequence of data, also to non-stationary data. The results, however, cannot be assumed to characterise the underlying system. Hence, we strongly discourage you to simply ignore known non-stationarity. One way out can be segmentation of the time series into almost stationary segments (see below for methods). We are well aware of the problem that many (perhaps even the most) interesting signals are non-stationary. Therefore, in Chapter 13 we will come back to the problem of non-stationarity on a more advanced level and also propose extensions of the concept of phase space reconstruction for deterministic and stochastic processes with slowly drifting parameters.

We should mention that we do not consider the notion of *weak stationarity*, which can be found in the literature on linear time series analysis. Weak stationarity only requires statistical quantities up to second order to be constant. In a nonlinear setting, this is certainly inadequate.

2.2 Testing for stationarity

After such emphasis has been put on the stationarity problem we have to address the question how non-stationarity can be detected for a given data set – obviously stationarity is a property which can never be positively established. It turns out that the matter is even more complicated because the stationarity requirement differs depending on the application. Thus in this section we will only give some general concepts and discuss particular techniques in the appropriate places in the context of the applications. In Chapter 13 we will discuss some ways to deal with time series from certain specific non-stationary processes. See also "Further reading" at the end of this chapter for references.

As a first requirement, the time series should cover a stretch of time which is much longer than the longest characteristic time scale that is relevant for the evolution of the system. For instance, the concentration of sugar in the blood of a human is driven by the consumption of food and thus roughly follows a 24 hour cycle. If this quantity is recorded over 24 hours or less, the process must be considered non-stationary no matter how many data points have been taken during that time. Quantitative information can be gained from the power spectrum to be introduced below. The longest relevant time scale can be estimated as the inverse of the lowest frequency which still contains a significant fraction of the total power of the signal. A time series can be considered stationary only on much larger time scales. See also Section 2.3.1.

Strong violations of the basic requirement that the dynamical properties of the system underlying a signal must not change during the observation period can be checked simply by measuring such properties for several segments of the data set. Transition probabilities, correlations, etc., computed, for example for the first and second half of the data available, must not differ beyond their statistical fluctuations. Characteristics with known or negligible statistical fluctuations are preferable for this purpose. The statistically most stable quantities are the mean and the variance. More subtle quantities such as spectral components, correlations or nonlinear statistics may be needed to detect less obvious non-stationarity.

Example 2.2 (Running variance of voice data). We perform a very simple test of stationarity on the phonation of a patient suffering from acute *laryngitis*. We computed the sample mean and variance for 50 consecutive segments of the data set, each containing 2048 data points. Within each segment, the standard error of the estimated mean $\langle s \rangle = \sum_{n=1}^{N} s_n / N$ is given by[1]

$$\sqrt{\frac{\sum_{n=1}^{N}(s - \langle s \rangle)^2}{N(N-1)}}, \tag{2.1}$$

neglecting the fact that the samples in the time series are not independent. The observed fluctuations of the running mean of the 50 segments were within these errors. The running variance, however, showed variability beyond the statistical fluctuations, as can be seen in Fig. 2.1. It is in general almost impossible to obtain stationary recordings from living beings. Although one should try to keep the experimental conditions as constant as possible, temporal changes of the spontaneous dynamics are unavoidable and, after all, very natural. This recording was done in order to study disorders of the phonatory system. Since in this context the (instantaneous) amplitude is not the most interesting quantity, one might decide that some slow change is of no particular concern. It is then necessary to use methods

---

[1] The standard deviation divided by an extra $\sqrt{N}$ is known to be the error when estimating the mean value of Gaussian distributed uncorrelated numbers, whereas the standard deviation itself describes the typical spread of the individual numbers from their mean.

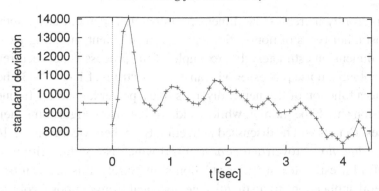

Figure 2.1 Running standard deviation for non-overlapping segments of length 93 ms (2048 points) of a phonation data set [Herzel & Reuter (1996); see Appendix B.8 for more information on the data]. The short horizontal line denotes the total amplitude of the data set. The slow variations indicate non-stationarity, probably due to slight changes in the air supply.

of time series analysis which are robust against changes in amplitude, such as the power spectrum. However, it can be preferable to do a time dependent analysis in order to avoid artefacts and to exploit the information on the system contained in these changes. Later in this chapter we will revisit this example and compute a *spectrogram*, that is, a running *power spectrum*, which gives much more interesting information about the changes going on in the dominant frequencies. □

In experimental chaotic systems, it is not uncommon for a parameter drift to result in no visible drift in the mean or the distribution of values. Linear correlations and the spectrum may also be unaffected. Only the nonlinear dynamical relations and transition probabilities change appreciably. This can be detected by comparing prediction errors with respect to a nonlinear model for different sections of the data set. This possibility is discussed in the context of nonlinear predictions, Section 4.4.

Whether the data set is a sufficient sample for a particular application, such as the estimate of a characteristic quantity, may be tested by observing the convergence of that quantity when larger and larger fractions of the available data are used for its computation. An attractor dimension obtained from the first half of the data must not differ substantially from the value determined using the second half, and should agree with the value computed for the whole data set within the estimated statistical errors. An example of this strategy is given in Example 7.6 in Section 7.2. Note that this test is a very crude one since the convergence of nonlinear statistics can be very slow and, indeed, not much is known about its rate.

A quantity which suffers considerably from non-stationarity in the data is the *correlation dimension* (to be introduced in Chapter 6). While a simple drift of the

calibration usually increases the dimension (since it wipes out any fractal structures), most other types of non-stationarity and insufficient sampling yield spuriously low dimension estimates. For example, if the process is not recurrent, such
as certain coloured noise processes with an $f^{-\alpha}$-spectrum of $1 < \alpha < 3$ (the power
spectrum as a function of frequency diverges like a power law at low frequencies),
the dimension should be infinite, while standard dimension algorithms yield small
finite values. Details will be discussed in Section 6.5, where we introduce the *space
time separation plot* [Provenzale *et al.* (1992)] which is a stationarity test particularly useful for estimates of the correlation sum. Surely, this test can be used in
more general applications to distinguish geometrical from temporal correlations.

2.3 Linear correlations and the power spectrum

Linear statistical inference is an extremely well-developed field. Besides the definition of several statistical tools to characterise data, its main power lies in the
fact that concepts can be derived in a rigorous way. For instance, it is relevant to
distinguish between a quantity such as a mean value, and the way one derives a
number for it from a finite sample. This is called an *estimate*, and there can exist
different estimates for the same quantity in the same data set, depending on the
assumptions made. Thus we will distinguish estimates of any quantity by a "hat"
above the symbol. Unlike in the linear theory, the estimates of nonlinear quantities
that we will make later cannot usually be studied sufficiently to rigorously prove
their correctness and convergence, for instance. In this book, we introduce only the
concepts of linear statistical processes and of the power spectrum and correlation
functions; any further aspects of linear time series methods will not be considered
explicitly.

The solutions of stable linear deterministic equations of motions with constant coefficients are linear superpositions of exponentially damped harmonic oscillations
(stability here implies that solutions remain bounded, otherwise also exponentially
growing solutions are possible). Hence, linear systems always need irregular inputs
to produce irregular signals. The most simple system which produces non-periodic
signals is a *linear stochastic* process. For this reason we will momentarily assume
the observable to be a random variable. However, this point of view will also be
reasonable if we speak about deterministic systems, for several reasons. First of
all, measurement noise always introduces some randomness. Secondly, the unknown initial condition of the system turns the current observation into a random
event. Finally, the instability of chaotic motion suggests a probabilistic description
in terms of invariant measures (ergodic theory). Hence, a measurement s_n of the
state at time n of a stochastic process can be regarded as drawn from an underlying probability distribution for observing different values or sequences of values.

A stochastic process is called *stationary* if these probabilities are constant over time: $p(s_n) = p(s)$.

Since it is not known beforehand from which probability distribution $p(s)$ the measurements are drawn, information about $p(s)$ can only be inferred from the time series. The *mean* of the probability distribution $p(s)$, defined as

$$\langle s \rangle := \int_{-\infty}^{\infty} ds' s' p(s'), \tag{2.2}$$

can be estimated by the mean of the finite time series

$$\widehat{\langle s \rangle}_n = \frac{1}{N} \sum_{n=1}^{N} s_n, \tag{2.3}$$

where $\langle \cdot \rangle_n$ denotes the average over time[2] and N is the total number of measurements in the time series. The *variance* of the probability distribution will be estimated by the variance of the time series:

$$\sigma^2 := \langle (s - \langle s \rangle)^2 \rangle = \int_{-\infty}^{\infty} ds' (s' - \langle s \rangle)^2 p(s'),$$

$$\hat{\sigma}^2 = \frac{1}{N-1} \sum_{n=1}^{N} (s_n - \widehat{\langle s \rangle})^2 = \frac{1}{N-1} \left(\sum_{n=1}^{N} s_n^2 - N \widehat{\langle s \rangle}^2 \right). \tag{2.4}$$

We will refer to σ as the standard deviation or the rms (**root mean squared**) amplitude as opposed to the *variance* σ^2.

For these two quantities the time ordering of the measurements is irrelevant and thus they cannot give any information about the time evolution of a system. Such information can be obtained from the *autocorrelations* of a signal. The autocorrelation at *lag* τ is given by

$$c_\tau = \frac{1}{\sigma^2} \langle (s_n - \langle s \rangle)(s_{n-\tau} - \langle s \rangle) \rangle = \frac{\langle s_n s_{n-\tau} \rangle - \langle s \rangle^2}{\sigma^2}. \tag{2.5}$$

The estimation of the autocorrelations from a time series is straightforward as long as the lag τ is small compared to the total length of the time series. Therefore, estimates $\hat{c}_\tau$ are only reasonable for $\tau \ll N$. If we plot values s_n versus the corresponding values a fixed lag τ earlier, $s_{n-\tau}$, the autocorrelation c_τ quantifies how these points are distributed. If they spread out evenly over the plane, then $c_\tau = 0$. If they tend to crowd along the diagonal $s_n = s_{n-\tau}$, then $c_\tau > 0$, and if they are closer to the line $s_n = -s_{n-\tau}$ we have $c_\tau < 0$. The latter two cases reflect some tendency of s_n and $s_{n-\tau}$ to be proportional to each other, which makes it plausible that the autocorrelation function reflects only linear correlations. If the signal is observed

---

[2] Henceforth we will suppress the index when it is clear that an average is taken with respect to time.

over continuous time, one can introduce the *autocorrelation function* $c(\tau)$, and the correlations of Eq. (2.5) are estimates of $c(\tau = \tau \Delta t)$. Obviously, $c_\tau = c_{-\tau}$ and $c_0 = 1$.

A particular case is given if the measurements are drawn from a Gaussian distribution,

$$p(s) = \frac{1}{\sigma \sqrt{2\pi}} e^{-(s-\langle s \rangle)^2/2\sigma^2} , \qquad (2.6)$$

and, moreover, the joint distributions of multiple measurements are also Gaussian. As a consequence, their second moments are given by the autocorrelations. Since a multivariate Gaussian distribution is completely specified by its first and second moments, such a random process is fully described statistically by the mean, the variance σ^2 and the autocorrelation function c_τ, and is called a *linear Gaussian random process*.

Obviously, if a signal is periodic in time, then the autocorrelation function is periodic in the lag τ. Stochastic processes have decaying autocorrelations but the rate of decay depends on the properties of the process. Autocorrelations of signals from deterministic chaotic systems typically also decay exponentially with increasing lag. Autocorrelations are not characteristic enough to distinguish random from deterministic chaotic signals.

Instead of describing the statistical properties of a signal in real space one can ask about its properties in Fourier space. The *Fourier transform* establishes a one-to-one correspondence between the signal at certain times (*time domain*) and how certain frequencies contribute to the signal, and how the phases of the oscillations are related to the phases of other oscillations (*frequency domain*). Apart from different conventions concerning the normalisation, the Fourier transform of a function $s(t)$ is given by

$$\tilde{s}(f) := \frac{1}{\sqrt{2\pi}} \int_{-\infty}^{\infty} s(t) e^{2\pi i f t} dt \qquad (2.7)$$

and that of a finite, discrete time series by

$$\tilde{s}_k = \frac{1}{\sqrt{N}} \sum_{n=1}^{N} s_n e^{2\pi i k n/N} . \qquad (2.8)$$

Here, the frequencies in physical units are $f_k = k/N\Delta t$, where $k = -N/2, \ldots, N/2$ and Δt is the sampling interval. The normalisation of both transforms is such that the inverse transform only amounts to a change of the sign in the exponential and an exchange of f (or, respectively, k) and t (respectively n). Both operations are thus invertible and are linear transformations.

The *power spectrum* of a process is defined to be the squared modulus of the continuous Fourier transform, $S(f) = |\tilde{s}(f)|^2$. It is the square of the amplitude, by which the frequency f contributes to the signal. If we want to estimate it from a finite, discrete series, we can use the *periodogram*, $S_k = |\tilde{s}_k|^2$. This is not the most useful estimator for $S(f)$ for two reasons. First, the finite frequency resolution of the discrete Fourier transform leads to *leakage* into adjacent frequency bins. For example, in the limit of a pure harmonic signal, where the continuous transform yields a δ distribution, the discrete transform yields only a finite peak which becomes higher for longer time series. Second, its statistical fluctuations are of the same order as $S(f)$ itself. This can be remedied either by averaging over adjacent frequency bins or, more efficiently, by averaging over running windows in the time domain. If you do not use a packaged routine for the estimation of the power spectrum anyway, you should consult the *Numerical Recipes* by Press *et al.* (1992), or similar texts for further details of the numerical estimation of the power spectrum and windowing techniques (and routines). The TISEAN software package contains a relatively non-sophisticated routine for power spectral estimation, as well as other linear tools. For serious linear analysis of time series, one should use specialised software such as the Matlab signal processing toolbox.

The important relation between the power spectrum and the autocorrelation function is known as the *Wiener–Khinchin theorem* and states that the Fourier transform of the autocorrelation function equals the power spectrum. The total power increases linearly with the length of the time series and can be computed both in the time and frequency domain, via $S = \sum_{n=1}^{N} |s_n|^2 = \sum_{k=-N/2}^{N/2} |\tilde{s}_k|^2$ (*Parseval's theorem*).

The power spectrum is particularly useful for studying the oscillations of a system. There will be sharper or broader peaks at the dominant frequencies and at their integer multiples, the harmonics. Purely periodic or quasi-periodic signals show sharp spectral lines; measurement noise adds a continuous floor to the spectrum. Thus in the spectrum, signal and noise are readily distinguished. Deterministic chaotic signals may also have sharp spectral lines but even in the absence of noise there will be a continuous part of the spectrum. This is an immediate consequence of the exponentially decaying autocorrelation function. Without additional information it is impossible to infer from the spectrum whether the continuous part is due to noise on top of a (quasi-)periodic signal or to chaoticity. An example where this additional information is available can be found in Gollub & Swinney (1975). In a Taylor–Couette flow experiment, the continuous floor in the power spectrum occurred suddenly when a system parameter was changed. Since it is unlikely that a parameter change has such a dramatic influence on the noise in the system, the change was attributed to the transition from quasi-periodic motion to chaos.

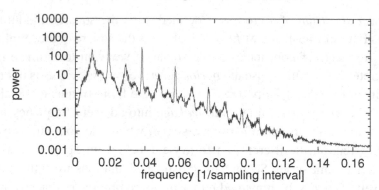

Figure 2.2 Lower part of the spectrum of a time series from the nonlinear feedback laser (Appendix B.10). The sharp peaks form a series of harmonics; the broad peak at low frequencies is a sub-harmonic.

Example 2.3 (Nonlinear feedback laser). Let us show a typical power spectrum of a continuous time chaotic system, the nonlinear optical feedback laser described in Appendix B.10. We estimated the power spectrum of these 128 000 points by sliding a square window of length 4096 over the data and averaging the squared moduli of the Fourier transforms of all the windows (Fig. 2.2).   □

When we study a signal in the frequency domain or by time averages such as the autocorrelations, we lose all time information. Sometimes, in particular when system properties can change over time (which leads to non-stationary signals), we want to preserve this information. In this case, spectral analysis can be done on consecutive segments of a longer time series to track temporal changes. This is usually called a *spectrogram*.

Example 2.4 (Spectrogram of voice data). Let us use the power spectrum for the analysis of the phonation data (Appendix B.8) which has been found to be non-stationary in Example 2.2. First we compute a power spectral estimate of the whole data set. There are rather sharp peaks at discrete frequencies, namely, two incommensurate pitches occur. This phenomenon is called *biphonation*; see Herzel & Reuter (1996). Then we compute power spectra for 100 segments of 2048 points each. The overlap of consecutive segments is now 1024 points. Such a time dependent power spectrum is usually called a *spectrogram* and is routinely used in speech research. In Fig. 2.3 for each of the segments we plot a symbol at the frequencies where the power assumes a relative maximum. To the left we also show the frequencies contributing to the total power spectrum, thus averaging over the entire series. Further we plot lines at the two fundamental frequencies, 306 Hz and 426 Hz. Towards the end of the recording, sub-harmonics of one-half or one-third of the lower fundamental frequency occur.   □

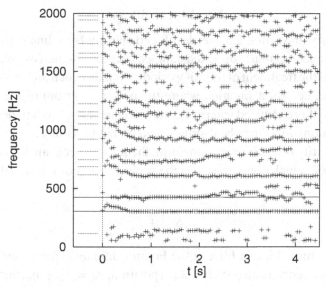

Figure 2.3 Spectrogram of phonation data. In this representation one can follow the appearance and disappearance of the higher harmonics and sub-harmonics of the two fundamental frequencies (horizontal lines).

2.3.1 Stationarity and the low-frequency component in the power spectrum

We said in Section 2.1 that effective non-stationarity of a finite time series is related to the fact that time scales of the dynamics are of the order of the total observation time. The power spectrum reflects the contribution of all possible periods, from the sampling interval up to the total time covered by the series, and thus should contain the relevant information. If there is too much power in the low frequencies, the time series must be considered non-stationary since the corresponding Fourier modes have only very few oscillations during the observation time. What is less clear is how much "too much" really is. Recall that white noise (or better, in discrete time, an *iid* process: independently identically distributed random numbers), which is a prototype for a stationary process, has equal power in all frequency bins! Moreover, the power for frequency $f = 0$ can be arbitrarily modified by a simple offset on the data, and is furthermore related to the integral over the autocorrelation function (a consequence of the Wiener–Khinchin theorem). Hence, we have to expect a real problem only if the power spectrum is diverging around zero, such as $f^{-\alpha}$ with positive α. In particular, Brownian motion has a f^{-2} spectrum and is known not to be recurrent in two and more dimensions. The decision whether data are stationary or not is much more difficult if the power spectrum approaches a nonzero constant for small frequencies which is much larger than the background. We can only suggest that this be interpreted as a warning.

2.4 Linear filters

While the theory of Section 2.3 was developed starting from linear stochastic processes which yield irregularly distorted damped oscillations, here we will reinterpret some aspects of the treatment of (quasi-)periodic signals which are distorted by measurement noise, i.e., for a situation where the power spectrum of the unperturbed signal consists of pronounced spikes at certain frequencies. We will describe how a *signal* and random *noise* may be distinguished on the basis of the power spectrum. Due to the linearity of the Fourier transform, the power spectrum $S_k^{(s)}$ of an additive superposition of signal x and noise η is the superposition of the signal $S_k^{(x)}$ and the noise $S_k^{(\eta)}$ power,

$$S_k^{(s)} = S_k^{(x)} + S_k^{(\eta)} . \tag{2.9}$$

It can be easily turned into a filter in the Fourier domain. That is, we assume that the observed sequence is also such a superposition, $s_n = x_n + \eta_n$, and we want to estimate the pure signal, $\hat{x}_n \approx x_n$, by filtering the measured sequence appropriately: $\hat{\tilde{x}}_k = \phi_k \tilde{s}_k$. If we require the mean squared distance e^2 between $\hat{x}_n$ and x_n to be minimal, the *optimal filter* can be constructed by solving a minimisation problem. In the Fourier domain we have to minimise:

$$
\begin{aligned}
e^2 &= \sum_{k=-N/2}^{N/2} |\hat{\tilde{x}}_k - \tilde{x}_k|^2 \\
&= \sum_{k=-N/2}^{N/2} |\phi_k(\tilde{x}_k + \tilde{\eta}_k) - \tilde{x}_k|^2 \\
&= \sum_{k=-N/2}^{N/2} (\phi_k - 1)^2 |\tilde{x}_k|^2 + \phi_k^2 |\tilde{\eta}_k|^2 ,
\end{aligned}
\tag{2.10}
$$

(assuming statistical independence of x and η) which we can do by setting the derivative with respect to ϕ_k equal to zero. The solution is

$$\phi_k = \frac{S_k^{(x)}}{S_k^{(s)}} . \tag{2.11}$$

This is called the *Wiener filter*. Obviously, we cannot compute $S_k^{(x)}$ directly but we can try to estimate it by inspection of the total power spectrum. This is the point where we need a criterion to distinguish signal and noise by their respective properties. As we have said, this is often possible for signals with dominant oscillations. For signals with intrinsic broad band spectra, like those from deterministic chaotic systems, this is usually impossible.

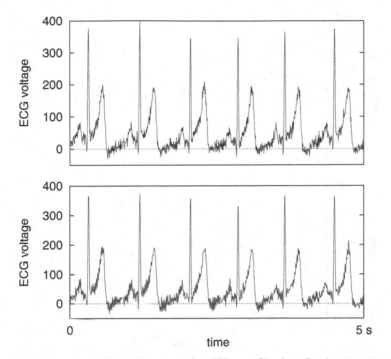

Figure 2.4 Human ECG before and after Wiener filtering. In the upper panel, considerable baseline noise is visible. After filtering (bottom), the situation has not improved. Presumably, some of the high-frequency noise is removed but at the same time, power from the sharp ventricular complexes has been spread in the time domain which leads to baseline contamination.

Example 2.5 (Failure to reduce noise in electrocardiograms using a Wiener filter). Fig. 2.4 shows a relatively noisy human electrocardiogram (ECG). In the upper and lower panels are plots before and after Wiener filtering, respectively. (See Appendix B.7 for more information about the data.) $S_k^{(x)}$ has been estimated by $S_k^{(s)} - \bar{S}^{(\eta)}$, where the noise power $S_k^{(\eta)}$ has been assumed to be k-independent (white noise) and identical to the noise floor $\bar{S}^{(\eta)}$, which persists up to high frequencies. We see that the Wiener filter is unable to clean the ECG, which is related to the fact that also the noise free ECG power spectrum is broad band due to the sharp spikes. See Schreiber & Kaplan (1996a, b) and Section 10.4 for a way of filtering ECGs with nonlinear techniques. □

There are situations, however, where we have some external information about the signal or the noise contributions to the power spectrum. One example is the 50 Hz (or 60 Hz in the US) AC contamination in signals recorded with electronic circuits. Here we know that most of the power at 50 Hz is due to this contamination and we

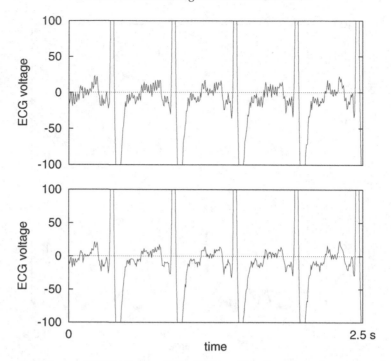

Figure 2.5 Human ECG before and after Wiener filtering. The enlargement clearly shows typical 50 Hz contamination on the baseline (upper panel). Of course it is also present during cardiac activity but it is less visible then. After filtering (bottom) only some random noise remains.

can construct a filter by interpolating between the adjacent unaffected frequency bins.

Example 2.6 (Removing 50 Hz AC contamination with a Wiener filter). An example where linear filters are fully adequate even for nonlinear signals is the 50 Hz (or 60 Hz) noise in the output of electrical amplifiers. The contamination is well localised in the Fourier domain and the signal spectrum can easily be estimated by interpolating over the spurious peak in the power spectrum. Figure 2.5 shows a human electrocardiogram (ECG) before and after filtering. (See Appendix B.7 for more information about the data.) A filter which basically removes a single-frequency component is called a *notch* filter in the literature. For technical issues please refer to the standard literature, e.g. Hamming (1983).   □

Let us mention that certain linear filters which are sometimes built into data acquisition hardware can add one deterministic degree of freedom and hence can increase the attractor dimension. The reason is that for online data smoothing one can use a filter with feedback which introduces another degree of freedom into the system. See Badii & Politi (1986), Badii *et al.* (1988), Mitschke *et al.* (1988), Mitschke (1990), Broomhead *et al.* (1992), Sauer & Yorke (1993), Theiler & Eubank (1993), and the footnote on page 231.

2.5 Linear predictions

Closely related to optimal filtering is the problem of prediction. For this discussion it will be more natural to work in the time domain. Suppose we have a sequence of measurements s_n, $n = 1, \ldots, N$ and we want to predict the outcome of the following measurement, s_{N+1}. One often wants to find the prediction $\hat{s}_{N+1}$, which minimises the expectation value of the squared prediction error $\langle (\hat{s}_{N+1} - s_{N+1})^2 \rangle$. When we assume stationarity, we can estimate this expectation value by its average over the available measured values. If we further restrict the minimisation to linear time series models which incorporate the m last measurements, we can express this by

$$\hat{s}_{n+1} = \sum_{j=1}^{m} a_j s_{n-m+j} \qquad (2.12)$$

and minimise

$$\sum_{n=m}^{N-1} (\hat{s}_{n+1} - s_{n+1})^2 \qquad (2.13)$$

with respect to the parameters a_j, $j = 1, \ldots, m$. Here we have assumed that the mean of the time series has already been subtracted from the measurements. By requiring that the derivatives with respect to all the a_js to be zero, we obtain the solution by solving the linear set of equations

$$\sum_{j=1}^{m} C_{ij} a_j = \sum_{n=m}^{N-1} s_{n+1} s_{n-m+i}, \qquad i = 1, \ldots, m. \qquad (2.14)$$

Here C_{ij} is the $m \times m$ auto-covariance matrix

$$C_{ij} = \sum_{n=m}^{N-1} s_{n-m+i} s_{n-m+j}. \qquad (2.15)$$

The linear relation, Eq. (2.12), is justified for harmonic motion but also for linear stochastic processes. Although we will postpone the discussion of the various linear time series models to Section 12.1.1, we want to note here that all these models require random inputs which we do not observe. Thus, the so-called *moving average* part of these models which describes the response to the external noise must be replaced here by the *average response* which is zero since the noise is assumed to have zero mean.

Further, note that the model of Eq. (2.12) cannot be expected to be stable under iteration. Performing more step predictions by iteration would mean that we use the coefficients a_j, which have been optimised for predictions on the (noisy) measurements, for predictions based on the best estimators $\hat{s}_n$. Their errors have different properties than the noise η_n. The easiest way around this problem is to modify Eq. (2.12) in order to make predictions further into the future in a single iteration. For an example of a linear prediction of time series data, refer to Example 1.1 where a comparison with a simple nonlinear method is also given. A second example will be presented in Chapter 4, Example 4.2. It is always important to report the achievements of a new approach in comparison to standard methods. Thus, if you manage to predict some time series with a nonlinear algorithm, you should make sure that you have not just picked up the linear autocorrelations in the data which can best be exploited by a linear predictor.

We have not said anything yet about how to choose the right *order m* of the linear relation, Eq. (2.12). Some general remarks will be provided in Section 12.5, however, we refer the reader to the specialised literature on linear time series analysis for more rigorous statements. For now let us only suggest that although the model has been built on the error averaged over the available data, we should not quote this error as a prediction error since it has been determined *in-sample*. Instead, we should hold back part of the data while we construct the model in order to be able to compute the relevant *out-of-sample* error once we have established a model.

Further reading

Practical material about linear time series analysis, filtering and prediction, including the essential programs, can be found in the famous *Numerical Recipes*, Press *et al.* (1992). There are many textbooks on spectral time series analysis; classical examples are Box & Jenkins (1976), Priestley (1988), Jenkins & Watts (1986), and Brockwell & Davis (1987). Digital filters are treated, for example, in Hamming (1983). Classical tests for stationarity are described in Priestley (1992). Chapter 13 deals with non-stationary time series and gives further references. Other references include Isliker & Kurths (1993), Kennel (1997), as well as Schreiber (1997).

Exercises

2.1   Create a sample $\{x_n\}$ of uncorrelated Gaussian random numbers. You may use a routine from a software package [e.g. *Numerical Recipes*, Press *et al.* (1992)] or the TISEAN routine `makenoise`. Now apply the *moving average filter* $s_n = 1/15 \sum_{j=-7}^{7} x_{n+j}$ to obtain 1024 correlated Gaussian variates. Obtain a routine from a software package (e.g., TISEAN) which estimates the power spectrum of a data set. Apply the routine to the sequences $\{x_n\}$ and $\{s_n\}$. What difference do you observe?

2.2 Create two artificial time series. The first, $\{\eta_n, \quad n = 1, \ldots, 4096\}$, contains uniformly distributed random numbers (use your favourite random number generator) in the interval [0, 1]. The second series, $\{s_n, \quad n = 1, \ldots, 4096\}$, is based on the deterministic evolution of x_n which follows the rules $x_0 = 0.1$ and $x_{n+1} = 1 - 2x_n^2$ (the *Ulam map*, a special case of the *logistic equation*). The values x_n are not measured directly but through the nonlinear observation function $s_n = \arccos(-x_n)/\pi$. Compare the mean, variance and the power spectra of the two time series.

Chapter 3

Phase space methods

3.1 Determinism: uniqueness in phase space

The nonlinear time series methods discussed in this book are motivated and based on the theory of *dynamical systems*; that is, the time evolution is defined in some phase space. Since such nonlinear systems can exhibit deterministic chaos, this is a natural starting point when irregularity is present in a signal. Eventually, one might think of incorporating a stochastic component into the description as well. So far, however, we have to assume that this stochastic component is small and essentially does not change the nonlinear properties. Thus all the successful approaches we are aware of either assume the nonlinearity to be a small perturbation of an essentially linear stochastic process, or they regard the stochastic element as a small contamination of an essentially deterministic, nonlinear process. If a given data set is supposed to stem from a genuinely *non-linear stochastic* processes, time series analysis tools are still very limited and their discussion will be postponed to Section 12.1.

Consider for a moment a purely deterministic system. Once its present state is fixed, the states at all future times are determined as well. Thus it will be important to establish a vector space (called a *state space* or *phase space*) for the system such that specifying a point in this space specifies the state of the system, and vice versa. Then we can study the dynamics of the system by studying the dynamics of the corresponding phase space points. In theory, dynamical systems are usually defined by a set of first-order ordinary differential equations (see below) acting on a phase space. The mathematical theory of ordinary differential equations ensures the existence and uniqueness of the trajectories, if certain conditions are met. We will not hold up any academic distinction between the state and the phase space, but we remark that except for mathematical dynamical models with given equations of motion, there will not be a unique choice of what the phase space of a system can be.

The concept of the *state of a system* is powerful even for nondeterministic systems. A large class of systems can be described by a (possibly infinite) set of states and some kind of transition rules which specify how the system may proceed from one state to the other. Prominent members of this category are the stochastic *Markov processes* for which the transition rules are given in the form of a set of transition probabilities and the future state is selected randomly according to these probabilities. The essential feature of these processes is their strictly finite memory: the transition probabilities to the future states may only depend on the present state, not on the past.[1] If you like you can regard a purely deterministic system as a limiting case of a Markov process on a continuum of states. The transition to the state specified by the deterministic rule occurs with probability 1 and every other transition has probability 0. We mention this approach because it treats uncertainties in the transition rule – what we call *dynamical noise* – as the generic case of a peaked distribution of transition probabilities. The noise-free – deterministic – case is only obtained in the limit of a delta peak. This is not the most useful formulation for the purpose of this book. Instead, we formulate the particular case of a deterministic system in its own right as a starting point which allows us to establish some interesting ways of understanding signals and systems. Let us keep in mind the fact that we can treat dynamical noise not only as an additional complication in an otherwise clean situation, but we can also regard strict determinism as a limiting case of a very general class of models.

Let us now introduce some notation for deterministic dynamical systems in phase space. Modifications necessary for not exactly deterministic systems will be discussed later. For simplicity we will restrict ourselves to the case where the phase space is a finite dimensional vector space $\mathbb{R}^m$ (partial differential equations such as the Navier–Stokes equation for hydrodynamic flow form highly interesting dynamical systems as well, living in infinite dimensional phase spaces). A state is specified by a vector $\mathbf{x} \in \mathbb{R}^m$. Then we can describe the dynamics either by an m dimensional map or by an explicit system of m first-order ordinary differential equations. In the first case, the time is a discrete variable:

$$\mathbf{x}_{n+1} = \mathbf{F}(\mathbf{x}_n), \qquad n \in \mathbb{Z}, \tag{3.1}$$

and in the second case it is a continuous one:

$$\frac{d}{dt}\mathbf{x}(t) = \mathbf{f}(\mathbf{x}(t)), \qquad t \in \mathbb{R}. \tag{3.2}$$

The second situation is usually referred to as a *flow*. The vector field $\mathbf{f}$ in Eq. (3.2) is defined not to depend explicitly on time, and thus is called *autonomous*.

---

[1] In a Markov chain of order m, the present *state* is represented by the values of the process during the last m discrete time steps.

If **f** contains an explicit time dependence, e.g., through some external driving term, the mathematical literature does not consider this a dynamical system any more since time translation invariance is broken. The state vector alone (i.e., without the information about the actual time t) does not define the evolution uniquely. In many cases such as periodic driving forces, the system can be made autonomous by the introduction of additional degrees of freedom (e.g., a sinusoidal driving can be generated by an additional autonomous harmonic oscillator with a unidirectional coupling). Then, one can typically define an extended phase space in which the time evolution is again a unique function of the state vectors, even without introducing auxiliary degrees of freedom; just by introducing, e.g., a phase angle of the driving force. When admitting arbitrary time dependences of **f**, however, this also includes the case of a noise driven stochastic system which is not a dynamical system in the sense of unique dependence of the future on some actual state vector.

In the autonomous case, the solution of the initial value problem of Eq. (3.2) is known to exist and to be unique if the vector field **f** is Lipshitz continuous. A sequence of points x_n or $x(t)$, solving the above equations is called a *trajectory* of the dynamical system, with x_0 or $x(0)$, respectively, the *initial condition*. Typical trajectories will either run away to infinity as time proceeds or stay in a bounded area forever, which is the case we are interested in here.[2] The observed behaviour depends both on the form of **F** (or, respectively, **f**) and on the initial condition; many systems allow for both types of solution. The set of initial conditions leading to the same asymptotic behaviour of the trajectory is called the *basin of attraction* for this particular motion.

On many occasions we will find the discrete time formulation more convenient. The formal solution of the differential equation Eq. (3.2) relating an initial condition to the end of the trajectory one unit of time later is sometimes called the *time one map* of **f**. We will sometimes refer to it, since, after all, the time series we will have to deal with are only given at discrete time steps. Also the numerical integration of the differential equations, Eq. (3.2), with a finite time step Δt yields a map. For example, the Euler integration scheme yields

$$x(t + \Delta t) \approx x(t) + \Delta t \mathbf{f}(x(t)). \tag{3.3}$$

When the time step Δt is small, the difference between consecutive values $x(t), x(t + \Delta t)$ is small as well, which is characteristic for the particular kind of map which arises from differential equations. We will refer to such time series as

---

[2] In this book we do not discuss scattering problems, where trajectories approach an interaction region coming from infinity, and after being scattered they disappear towards infinity. However, there exists a class of very interesting phenomena called *chaotic scattering*. Deterministic but chaotic dynamics lead to irregular dependence of the scattering angle on the impact parameter.

"flow-like". The fundamental difference between flow-like and map-like data will be discussed below and in Section 3.5.

Example 3.1 The two linear differential equations

$$dx/dt = -\omega y, \qquad dy/dt = \omega x, \qquad (3.4)$$

form a dynamical system with the periodic solution $x(t) = a \cos \omega(t - t_0)$, $y(t) = a \sin \omega(t - t_0)$. The solution will obviously stay finite forever. $\square$

Example 3.2 (Hénon map). The map given by Hénon (1976):

$$x_{n+1} = a - x_n^2 + by_n, \qquad y_{n+1} = x_n, \qquad (3.5)$$

yields irregular solutions for many choices of a and b. For $|b| \leq 1$ there exist initial conditions for which trajectories stay in a bounded region but, for example, when $a = 1.4$ and $b = 0.3$, a typical sequence of x_n will not be periodic but *chaotic*. We urge the reader to verify this claim by doing Exercise 3.1. $\square$

The dynamical systems used as examples in this book do not only have bounded solutions but they are usually also *dissipative*, which means that on average a phase space volume containing initial conditions is contracted under the dynamics. Then we have on average $|\det \mathbf{J_F}| < 1$ or $\operatorname{div} \mathbf{f} < 0$ respectively. ($\mathbf{J_F}$ is the Jacobian matrix of derivatives of $\mathbf{F}$: $(\mathbf{J_F})_{ij} = \partial/\partial x^{(j)} F_i(\mathbf{x})$.) For such systems, a set of initial conditions of positive measure will, after some transient time, be attracted to some sub-set of phase space. This set itself is invariant under the dynamical evolution and is called the *attractor* of the system. Simple examples of (non-chaotic) attractors are fixed points (after the transient time the system settles to a stationary state) and limit cycles (the system approaches a periodic motion). See Fig. 3.1. For an autonomous system with two degrees of freedom and continuous time, these (together with homoclinic and heteroclinic orbits connecting fixed points which are not discussed here) are the only possibilities (Poincaré–Bendixon theorem). More interesting attractors can occur for flows in at least three dimensions (see Fig. 3.2), where the mechanism of stretching and folding can produce *chaos*.

Example 3.3 The divergence of the flow introduced in Example 3.1 is zero, such that the dynamics are area preserving. The Hénon map, Example 3.2, is dissipative if the time independent determinant of the Jacobian, $|\det \mathbf{J}| = |b|$, is taken to be smaller than unity. The Lorenz system (see Exercise 3.2) is strongly dissipative, since its divergence, $-b - \sigma - 1$, for the usual choice of parameters is about -10. This means that on average a volume will shrink to e^{-10} of its original volume for each time unit. $\square$

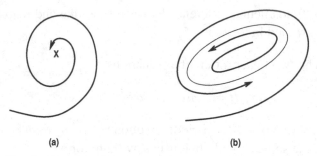

(a)   (b)

Figure 3.1 Fixed point (a) and limit cycle (b) attractors in the plane. Depending on the initial condition, a trajectory approaches the limit cycle (b) either from within or from without.

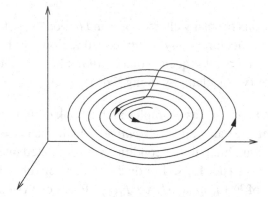

Figure 3.2 In three dimensional phase space a trajectory can be re-injected without violating determinism. Thus more complicated attractors are possible.

Characteristic for chaotic systems is that the corresponding attractors are complicated geometrical objects, typically exhibiting *fractal* structure. They are also called *strange attractors*. We will explain in Chapter 6, and in particular in Section 6.1, what we mean by a strange or fractal attractor.

So far we have illustrated that it is natural to describe a deterministic dynamical system as an object in phase space since this is the optimal way of studying its dynamical and geometrical properties. Since the dynamical equations (or the equations of motion) are defined in phase space, it is also most natural to use a phase space description for approximations to these equations. Such approximate dynamics will be important for predictions (Chapters 4 and 12), the determination of Lyapunov exponents (Chapters 5 and 11), noise filtering (Chapter 10) and most other applications (Chapter 15). For a deeper understanding of the nature of the underlying system, knowledge of the attractor geometry is desirable. How closely dynamics and geometry are related is expressed by theoretical results which relate Lyapunov exponents (dynamics) and dimensions (geometry). But practical algorithms can also sometimes be formulated equivalently in terms of geometry or dynamics. (See the discussion about noise reduction methods, Section 10.3.)

3.2 Delay reconstruction

Having stressed the importance of phase space for the study of systems with deterministic properties, we have to face the first problem: what we observe in an experiment is not a phase space object but a time series, most likely only a sequence of scalar measurements. We therefore have to convert the observations into state vectors. This is the important problem of *phase space reconstruction* which is technically solved by the method of delays (or related constructions).

Most commonly, the time series is a sequence of scalar measurements of some quantity which depends on the current state of the system, taken at multiples of a fixed sampling time:

$$s_n = s(\mathbf{x}(n\Delta t)) + \eta_n. \tag{3.6}$$

That is, we look at the system through some measurement function s and make observations only up to some random fluctuations η_n, the *measurement noise*. Let us neglect the effect of noise at this level of presentation. (We will discuss its effect later in Section 10.2.)

A *delay reconstruction* in m dimensions is then formed by the vectors $\mathbf{s}_n$, given as

$$\mathbf{s}_n = (s_{n-(m-1)\tau}, s_{n-(m-2)\tau}, \ldots, s_{n-\tau}, s_n). \tag{3.7}$$

The time difference in number of samples τ (or in time units, $\tau\Delta t$) between adjacent components of the delay vectors is referred to as the *lag* or *delay time*. Note that for $\tau > 1$, only the time window covered by each vector is increased, while the number of vectors constructed from the scalar time series remains roughly the same. This is because we create a vector for every scalar observation, s_n, with $n > (m - 1)\tau$. A number of embedding theorems are concerned with the question under which circumstances and to what extent the geometrical object formed by the vectors $\mathbf{s}_n$ is equivalent[3] to the original trajectory $\mathbf{x}_n$. In fact, under quite general circumstances the attractor formed by $\mathbf{s}_n$ is equivalent to the attractor in the unknown space in which the original system is living if the dimension m of the delay coordinate space is sufficiently large. To be precise, this is guaranteed if m is larger than twice the *box counting dimension* D_F of the attractor, i.e. roughly speaking, larger than twice the number of *active* degrees of freedom, regardless of how high the dimensionality of the true state space is. Depending on the application, even smaller

---

[3] In the sense that they can be mapped onto each other by a uniquely invertible smooth map. See Section 9.1.

m values satisfying $m > D_F$ can be sufficient. For theoretical details see Ding *et al.* (1993). One can hope to reconstruct the motion on attractors in systems such as hydrodynamic flows or in lasers, where the number of microscopic particles is huge, if only a few dominant degrees of freedom eventually remain as a result of some collective behaviour. In Chapter 9, we shall discuss the background of embedding and the theorems in more detail.

3.3 Finding a good embedding

When we start to analyse a scalar time series, we neither know the *box counting dimension*,[4] which is formally necessary to compute *m*, nor do we have any idea of how to choose the time lag τ. How we proceed depends on the underlying dynamics in the data, and on the kind of analysis intended. Most importantly, the embedding theorems *guarantee* that for ideal noise-free data, there exists a dimension *m* such that the vectors s_n are equivalent to phase space vectors. We will use this knowledge for the design of methods for the determination of the *dimension* of the attractor (Chapter 6), the *maximal Lyapunov exponent* (Chapter 5), and the *entropy* (Chapter 11). Generally, there are two different approaches for optimising the embedding parameters *m* and τ: either, one exploits specific statistical tools for their determination and uses the optimised values for further analysis, or one starts with the intended analysis right away and optimises the results with respect to *m* and τ. For example, dimension and Lyapunov estimates will be carried out by increasing the values of *m* until the typical behaviour of deterministic data appears.

For many practical purposes, the most important embedding parameter is the product $m\tau$ of the delay time and the embedding dimension, rather than the embedding dimension *m* or the delay time τ alone. The reason is that $m\tau$ is the time span represented by an embedding vector [see Kugiumtzis (1996) for a discussion]. For clarity, let us discuss the choice of *m* and τ separately. A precise knowledge of *m* is desirable since we want to exploit determinism with minimal computational effort. Of course, if an *m* dimensional embedding yields a faithful representation of the state space, every m' dimensional reconstruction with $m' > m$ does so as well. Choosing too large a value of *m* for chaotic data will add redundancy and thus degrade the performance of many algorithms, such as predictions and Lyapunov exponents. Due to the instability of chaotic motion, the first and last elements of a delay vector are the less related the larger their time difference. Therefore, taking a large value for *m* would not help much and it would risk "confusing" the algorithm.

---

[4] The box counting dimension is roughly the number of coordinates one needs to span the invariant subset on which the dynamics lives. We will be more specific in Chapter 6.

3.3.1 False neighbours

If we assume that the dynamics in phase space is represented by a smooth vector field, then neighbouring states should be subject to almost the same time evolution. Hence, after a short time interval into the future, the two trajectories emerging from them should be still close neighbours, even if chaos can introduce an exponential divergence of the two. This reasoning will be used very extensively in the next chapter on prediction. Here we want to refer to this property in order to discuss statistics which sometimes helps to identify whether a certain embedding dimension is sufficient for a reconstruction of a phase space.

The concept, called *false nearest neighbours*, was introduced by Kennel, Brown & Abarbanel (1992). We present it here with some minor modifications which avoid certain spurious results for noise [Hegger & Kantz (1999)].

The basic idea is to search for points in the data set which are neighbours in embedding space, but which should not be neighbours since their future temporal evolution is too different. Imagine that the correct embedding dimension for some data set is m_0. Now study the same data in a lower dimensional embedding $m < m_0$. The transition from m_0 to m is a projection, eliminating certain axes from the coordinate system. Hence, points whose coordinates which are eliminated by the projection differ strongly, can become "false neighbours" in the m dimensional space. The statistics to study is now obvious: for each point of the time series, take its closest neighbour in m dimensions, and compute the ratio of the distances between these two points in $m + 1$ dimensions and in m dimensions. If this ratio is larger than a threshold r, the neighbour was false. This threshold has to be large enough to allow for exponential divergence due to deterministic chaos.

Now, when we denote the standard deviation of the data by σ and use the maximum norm, the statistics to compute is

$$X_{\text{fnn}}(r) = \frac{\sum_{n=1}^{N-m-1} \Theta\left(\frac{|s_n^{(m+1)} - s_{k(n)}^{(m+1)}|}{|s_n^{(m)} - s_{k(n)}^{(m)}|} - r\right) \Theta\left(\frac{\sigma}{r} - |s_n^{(m)} - s_{k(n)}^{(m)}|\right)}{\sum_{n=1}^{N-m-1} \Theta\left(\frac{\sigma}{r} - |s_n^{(m)} - s_{k(n)}^{(m)}|\right)} \tag{3.8}$$

where $s_{k(n)}^{(m)}$ is the closest neighbour to s_n in m dimensions, i.e., $k(n)$ is the index of the time series element k different from n for which $|s_n - s_k| = \min$. The first step function in the numerator is unity, if the closest neighbour is false, i.e., if the distance increases by a factor of more than r when increasing the embedding dimension by unity, whereas the second step function suppresses all those pairs, whose initial distance was already larger than σ/r. Pairs whose initial distance is larger than σ/r by definition cannot be false neighbours, since, on average, there is not enough space to depart farther than σ. Hence, these are invalid candidates for the method and should not be counted, which is also reflected in the normalisation. There can be some false nearest neighbours even when working in the correct embedding dimension. Paradoxically, due to measurement noise there can be more false neighbours if more data is given. With more data, the closest neighbour is typically closer (there are more opportunities for a good recurrence on the attractor), whereas the chance for the distance of the $m + 1$st components to be at least of the order of the noise level remains the same. For data with a very coarse discretisation (say, 8 bit), there can even be *identical* delay vectors in m dimensions, which are not identical in $m + 1$ dimensions, so that the ratio of distances diverges. These appear to be false neighbours for any choice of r. Nonetheless, it is reasonable to study the false nearest neighbour ratio X_{fnn} as a function of r. The results of the false nearest neighbours analysis may depend on the time lag τ. Hence, if one wants to use this statistics for the distinction of chaos and noise, it is indispensable to verify the results with a surrogate data test (see Chapter 7).

Example 3.4 (False nearest neighbours of resonance circuit data). The electric resonance circuit described in Appendix B.11 creates rather noise free data on an attractor with a dimension D_F slightly above two. The false nearest neighbour statistics shown in Fig. 3.3 suggests that a 5 dimensional embedding is clearly enough (and this is also granted by the inequality $m > 2D_F$ from the embedding theorem). It is also evident that the few false neighbours found in 4 dimensions (for $N = 10000$ these are about 3 in absolute value) are hard to interpret. In Hegger *et al.* (1998) it was possible to do rather precise predictions and modelling in a four dimensional embedding, so that it seems that the false neighbours are noise artefacts.   □

3.3.2   The time lag

A good estimate of the lag time $\tau = \tau \Delta t$ is even more difficult to obtain. The lag is not the subject of the embedding theorems, since they consider data with infinite precision. Embeddings with the same m but different τ are equivalent in the mathematical sense for noise-free data, but in reality a good choice of τ facilitates the analysis. If τ is small compared to the internal time scales of the system, successive elements of the delay vectors are strongly correlated. All vectors s_n are then clustered around the diagonal in the $\mathbb{R}^m$, unless m is very large. If τ is very large, successive elements are already almost independent, and the points fill a large cloud in the $\mathbb{R}^m$, where the deterministic structures are confined to the very small scales. The first zero of the autocorrelation function Eq. (2.5) of the signal often yields a good trade-off between these extrema. A more refined concept is called

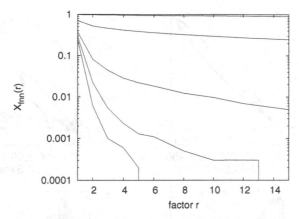

Figure 3.3 The relative number of false nearest neighbours as defined by Eq.(3.8) for electric resonance circuit data. The embedding dimension is $m = 1$ to 5 from top to bottom, with time lag 4.

mutual information and will be presented in Section 9.2. At this point we just give two recipes. The first is to study Example 3.5 in order to get a feeling for how a reasonable choice of τ can be verified visually. The second is that for a signal with a strong (almost-)periodic component, a time lag identical to one quarter of the period is a good first guess. All the nonlinear statistics in the next three chapters which rely on scaling behaviour suffer from reduced scaling ranges when τ has been chosen inappropriately, i.e., if τ is either unreasonably large or small.

Example 3.5 (Human ECG). Although the human electrocardiogram (ECG, see Appendix B.7) is probably not a deterministic signal, it can be interesting to view such signals in delay coordinates. Let us compare such representations for different delay times $\tau = \tau \Delta t$, where Δt is the sampling time, illustrated in Fig. 3.4. □

3.4 Visual inspection of data

Although this book contains a lot of refined methods for the characterisation and analysis of data, the first thing that we should do with a new data set is to look at it in several different ways. A plot of the signal as a function of time already gives the first hints of possible stationarity problems such as drifts, systematically varying amplitudes or time scales, and rare events. It allows us to select parts of the series which look more stationary. Quite often experimental data contain some faults which can be detected by visual inspection.

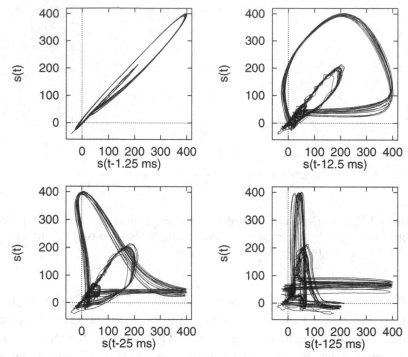

Figure 3.4 Delay representations of a human ECG signal, taken by Petr Saparin (1995). Upper left: $\tau = 1.25$ ms. All the data are close to the diagonal since consecutive values are very similar. Right: $\tau = 12.5$ ms. At this delay the large loop corresponding to the QRS complex (see Fig. B.4 for explanation) is now well unfolded. Lower left: $\tau = 25$ ms. The slower features, P- and T-waves, are better represented, although somewhat hidden by the QRS complex. Right: $\tau = 125$ ms. Larger delay times lead to unnecessarily complicated graphs.

The next step would be a two dimensional representation such as that shown in Fig. 1.5. Even if no clear structures are visible (as in the right hand panel), this gives a feeling about which time lag may be reasonable for an embedding of the data set.

Visual inspection can also reveal symmetries in the data or can guide us to a more useful representation of the data. An example of the latter is a measurement of the output power of a laser. It contains a nonlinear distortion of the physically more relevant variables, the electric and magnetic field strength inside the laser cavity: the power is proportional to the *square* of these quantities. Therefore, the data exhibit sharp maxima and smooth minima, and the signal changes much faster around the maxima than around the minima. Taking the square root of the data renders them more convenient for data analysis.

Sometimes one finds exact symmetries in the data, e.g., under change of sign of the observable. In this case, one can enlarge the data base for a purely geometric analysis by just replicating every data point by the symmetry operation. For an analysis of the dynamics one can apply the symmetry operation to the time series as a whole. Since almost all nonlinear methods exploit local neighbourhood relationships, both tricks double the data base. When modelling the dynamics of the system the class of functions chosen should respect the symmetry.

3.5 Poincaré surface of section

If we consider the phase space of a system of m autonomous differential equations, we find that, locally, the direction tangential to the flow does not carry much interesting information. The position of the phase space point along this direction can be changed by re-parametrising time. It has no relationship to the *geometry* of the attractor and does not provide any further information about the dynamics. We can use this observation to reduce the phase space dimensionality by one, at the same time turning the continuous time flow into a discrete time map.

The method, called the *Poincaré section*, is the following. First form a suitable oriented surface in phase space (hatched area in Fig. 3.5). We can construct an invertible map on this surface by following a trajectory of the flow. The iterates of the map are given by the points where the trajectory intersects the surface in a specified direction (from above in Fig. 3.5). Note that the discrete "time" n of this map is the intersection count and is usually not simply proportional to the original time t of the flow. The time a trajectory spends between two successive intersection points will vary, depending both on the actual path in the reconstructed state space and on the surface of section chosen. Sometimes, after applying this technique, a disappointingly small number of points remain. Experimentalists tend to adjust the sampling rate such that about 10–100 observations are made per typical cycle of the signal. Each cycle yields at most one point in the Poincaré map (if there are

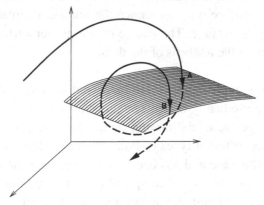

Figure 3.5 Poincaré section of a flow in three dimensions. The successive intersections A, B, . . . of the trajectory with the surface of sections define iterates of an invertible two dimensional map.

considerably less intersections, the surface is inappropriate). However, if the data base in the Poincaré map appears to be poor, keep in mind the fact that the flow data themselves are not much better since the additional information they contain is largely redundant with respect to the chaotic properties of the signal.

Apart from the construction of intersections we can also collect all minima or all maxima of a scalar time series. As we shall argue in Section 9.3, the (numerical) time derivative of the signal is a "legal" coordinate in a reconstructed state space. Hence, in the hypothetical reconstructed space spanned by vectors $(s(t), \dot{s}(t), \ddot{s}(t), \ldots)$, intersections of the trajectory with the surface given by $\dot{s}(t) = 0$ are precisely given by the minima (or maxima) of the time series. The minima (or maxima) are interpreted as the special measurement function which projects onto the first component of a vector applied to the state vectors inside this surface. Hence, they have to be embedded with a time delay embedding with lag unity themselves in order to form the invariant set of this particular Poincaré map. The TISEAN package contains a utility to form Poincaré sections (called `poincare`) and one that finds extremal points, called `extrema`. Finding good sections often requires some experimentation with parameter settings.

The simplest class of non-autonomous systems that we can handle properly using methods from nonlinear dynamical systems consists of those which are driven periodically. Their phase space is the extended space containing the phase of the driving force as an additional variable. The most natural surfaces of section are those of constant phase. The resulting Poincaré map is also called a *stroboscopic map*. In this case the system always has the same time span between intersections, which is a very useful feature for quantitative analysis.

Let us note in passing that whenever we study periodic solutions of a time continuous system (limit cycles), the discrete period of a periodic orbit of the corresponding Poincaré map depends on the details of the surface of section chosen, since by moving the surface one can reduce the number of intersection points of the limit cycle with the surface. The choice of the proper surface of section will then be a crucial step in the analysis of the data.

Example 3.6 (Phase portrait of NMR laser data). These data comprise the chaotic output of a periodically driven system (Appendix B.2). The sampling interval is $\frac{1}{15}$th of the period of the driving force, $\Delta t = T_{\text{force}}/15$. Let us consider three out of the many different types of Poincaré maps. The stroboscopic view is the most natural in this case and has been used to produce the data shown in the introduction, Fig. 1.3. Apart from a complication to be explained in Example 9.4, $s_n = s(t = nT_{\text{force}} + t_0)$. Figure. 3.6 shows a section through a three dimensional delay embedding space using linear interpolation (left) across a diagonal plane,

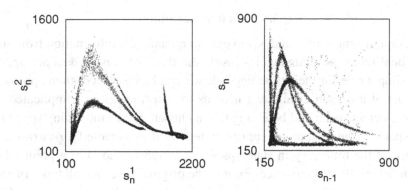

Figure 3.6 Phase portraits of the NMR laser data (Example 3.6). The left hand panel was obtained by a section through embedding space, see text. Right hand panel: successive minima of the parabolically interpolated series.

and the collection of all minima of the continuous time series after local parabolic interpolation (right). For the former, the section was taken such that $s_n^{(1)} = s(t_=)$, where $t_=$ is defined by $s(t_=) = s(t_= + \tau)$. The delay was taken to be $\tau = 4$. The second coordinate is $s_n^{(2)} = s(t_= - 2\tau)$. Thus the new time series is a series of two dimensional vectors. Alternatively, one could have used $s_n^{(1)}$ as a scalar time series and plot it as a two dimensional delay representation. The surface we selected is not optimal since in the upper left hand part of the attractor the noise is blown up considerably. In this part the surface of section cuts the attractor almost tangentially, such that the precise positions of the intersection points are strongly affected by noise and by the interpolation. The average time distance between intersection points is about 14. The deviation from the expected 15 is due to false crossings caused by the noise. The right hand panel uses a section at $\dot{s} = 0$, $\ddot{s} \geq 0$, i.e. every minimum of the parabolically interpolated series. The plot shows a time delay embedding of the time series given by the minima of these local parabolae. The average time distance between successive elements of the new series is about 15 and thus this section is as suitable as the stroboscopic view. Noise on the data causes some spurious extrema (the dots scattered outside the main structures).

The attractor looks different in all three cases since its geometrical details depend on the precise properties of the Poincaré section. Nevertheless, all three attractors are equivalent in the sense that they are characterised by identical values of the dimensions, Lyapunov exponents, and entropies. Numerical estimates of these quantities from finite and noisy data may differ however. □

3.6 Recurrence plots

The attempt of time series analysis to extract meaningful information from data to
a good deal relies on redundancies inside the data. At least if data are aperiodic
and no simple rule for their time dependence can be discerned, then approximate
repetitions of a certain events can help us to construct more complicated rules.
Assuming determinism and believing that a chosen delay embedding space forms
a state space of the system, an approximate repetition is called a *recurrence*, i.e.,
the return of the trajectory in state space to a neighbourhood of a point where it
has been before. Such recurrences exist in the original space for all types of motion
which are not transient. A system on a fixed point is trivially recurrent for all times.
In a system on a limit cycle, each point returns exactly to itself after one revolution.
A system on a chaotic attractor returns to an arbitrarily small neighbourhood of any
of its points. This is guaranteed by the invariance of the set which forms the support
of the attractor. If, however, the system never returns to all points which we find in
the initial part of the time series, then this indicates that this was a transient – the
initial condition was outside the invariant set and the trajectory relaxes towards this
set. Also non-stationarities through time dependent system parameters can cause a
lack of recurrence.

A very simple method for visualising recurrences is called a *recurrence plot* and
has been introduced by Eckmann *et al.* (1987): Compute the matrix

$$M_{ij} = \Theta(\epsilon - |\mathbf{s}_i - \mathbf{s}_j|) , \tag{3.9}$$

where $\Theta(.)$ is the Heaviside step function, ϵ is a tolerance parameter to be chosen,
and $\mathbf{s}_i$ are delay vectors of some embedding dimension. This matrix is symmetric
by construction. If the trajectory in the reconstructed space returns at time i into the
ϵ-neighbourhood of where it was at time j then $M_{ij} = 1$, otherwise $M_{ij} = 0$. One
can plot M_{ij} by black and white dots in the plane of indices for visual inspection.
This is the recurrence plot. Numerical schemes for its quantitative characterisation
have also been proposed. These are similar to many other statistical tools based
on neighbourhoods in embedding space and hence will not be discussed here. See
Casdagli (1997) for details.

So what can we learn from a visual inspection of the matrix M_{ij} by a recurrence
plot? It gives hints on the time series and on the embedding space in which we
are working. In a deterministic system, two points which are close should have
images under the dynamics which are also close (even if they are not as close due to
dynamical instability). Hence, one expects that black dots typically appear as short
line segments parallel to the diagonal. If in addition there are many isolated dots,
these rather indicate coincidental closeness and hence a strong noise component
on the data, or an insufficient embedding dimension. If there are mostly scattered

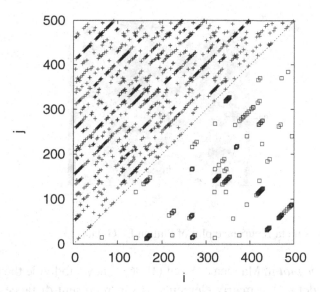

Figure 3.7 A recurrence plot of a short segment of the data from the electric resonance circuit (Appendix B.11). The upper left triangle represents embedding dimension two, the lower right is for dimension three. In two dimensions, this slightly more than two dimensional attractor intersects itself, hence there are lots of false neighbours. Many of them create isolated points. Due to the periodic forcing of this system, recurrences in the full phase space can only occur at temporal distances $|i - j|$ which are integer multiples of the driving period of about 25 samples.

dots, the deterministic component is absent or at least weak. Black dots in high dimensions are also black in lower dimensional embeddings. Hence, the relative number of lines increases when increasing the embedding dimension, since isolated dots disappear.

Example 3.7 (Recurrence plot of data from an electric resonance circuit). In Fig. 3.7 we show a recurrence plot of the NMR laser data in embedding dimensions 2 and 4. □

The irregularity of the arrangement of the line segments indicates chaos. If a signal was periodic, theses should form a periodic pattern. Stationarity of the whole time series requires that the density of line segments is uniform in the $i - j$-plane. See Example 13.8 for a non-stationary example. The temporal spacing between parallel lines can also give hints to the existence of unstable periodic orbits inside the chaotic attractor.

The essential drawback of this nice visual tool lies in the fact that for a time series of length N the matrix M_{ij} has N^2 elements, so that a huge number of pixels must be drawn. A compressed version of the recurrence plot was baptised

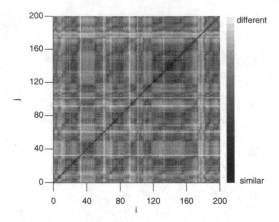

Figure 3.8 A meta-recurrence plot of human ECG data.

meta-recurrence plot in Manuca & Savit (1996). They subdivide the time series in segments and define the matrix elements M_{ij} to represent distances between the time series segments i and j. One possible distance measure is the *cross prediction error*. See the discussion in Schreiber (1999).

Example 3.8 (Meta-recurrence plot of ECG data). For ECG data, one can align segments along the main heart beat and then take a Euclidean distance. We define a segment by taking the last 100 ms before and the 500 ms after the upward zero crossing in the QRS-complex (see Appendix B.7). The resulting distance matrix is plotted in grey scale in Fig. 3.8. The non-stationarity is clearly visible. More importantly, almost stationary episodes can be recognised as light blocks on the diagonal. □

Further reading

Original sources about phase space embeddings are Takens (1981), Casdagli *et al.* (1991), and Sauer *et al.* (1991). Discussions of the proper choice of embedding parameters are contained for example in Fraser & Swinney (1986), Liebert & Schuster (1989), Liebert *et al.* (1991), Buzug & Pfister (1992), Kennel *et al.* (1992), and in Kugiumtzis (1996). See also Grassberger *et al.* (1991). *Recurrence plots* have been introduced by Eckmann *et al.* (1987). There is a nice more recent account by Casdagli (1997).

Exercises

3.1 Create different two dimensional phase portraits of a time series of the Hénon map (Example 3.2) using delay times $\tau = 1, 2, \ldots$ (here the sampling time is $\Delta t = 1$).

Which picture gives the clearest information? Rewrite the map in delay coordinates with unit delay.

3.2 Create a scalar series of flow data numerically by integration of the Lorenz system [Lorenz (1963)]:

$$\dot{x} = \sigma(y - x) ,$$
$$\dot{y} = rx - y - xz ,$$
$$\dot{z} = -bz + xy ,$$

with the parameters $\sigma = 10$, $r = 28$, and $b = 8/3$. This model was designed to describe convective motion of the Rayleigh–Bénard type, where x is the velocity of the fluid, y is the temperature difference between ascending and descending fluid, and z is the deviation of the temperature profile form linearity. For a laser system governed by the same equations and experimental data, see Appendix B.1. As an integrator, use, for example, a Runge–Kutta routine with a small step size. Sample the data at such a rate that you record on average about 25 scalar measurements of the x-coordinate during a single turn of the trajectory on one leaf of the attractor. Plot the attractor in two dimensional delay coordinates with different time lags. Convince yourself by visual inspection that the reconstruction is best when the lag is about one-quarter of the mean cycle time. Compute the autocorrelation function and the time delayed mutual information (to be defined in Section 9.2) to confirm this impression.

3.3 Numerically integrate the Lorenz system and perform a Poincaré section. Record (y, z) every time the x-coordinate equals zero and its derivative is negative.

3.4 Use the time series of the x-variable of Exercise 3.2 and collect all the maxima. Plot the series of the maxima with delay one and compare it to the attractor of Exercise 3.3.

Chapter 4
Determinism and predictability

In this chapter we will discuss the notion of the predictability of a system evolving over time or, strictly speaking, of a signal emitted by such a system. Forecasting future values of some quantity is a classical problem in time series analysis but the conceptual importance of the prediction problem is not limited to those who want to get rich by knowing tomorrow's exchange rates. Even if, instead, you are interested in describing, understanding or classifying signals, stay with us for a few pages.

In this book we are concerned with the detection and quantification of possibly complicated structures in a signal. We want to be able to convince others that the structures we find are real and not just fluctuations. The most convincing argument for the presence of some pattern is if it can be used to give an improved prediction. It is a necessary condition for a theory to be compatible with the known data but it is not sufficient. In order to become accepted, a theory must successfully predict something which can be verified subsequently. In time series analysis, we can take this requirement of predictive quality quite literally.

Most concepts, which we will introduce later in order to describe time series data, can be interpreted to some extent as indirect measures of predictability. Due to their indirect nature, some conclusions will remain controversial, especially if the structures are rather faint. The statistically significant ability to predict the signal better than other techniques do will then be a more convincing affirmation of nonlinear and deterministic structure than several dubious digits of the fractal dimension.

Readers whose primary interest is forecasting should see this chapter only as an introduction to concepts that are further elaborated in the chapter on modelling and forecasting, Chapter 12.

4.1 Sources of predictability

A signal which does not change is trivial to predict: the last observation is a perfect forecast for the next one. Even if the signal changes, this can be a reasonable method of forecasting,[1] and a signal for which this holds is called *persistent*. A system which changes periodically over time is also easy once you have observed one full cycle. Independent random numbers are easy as well: you do not have to work hard since working hard does not help anyway. The best prediction is just the mean value. Interesting signals are something in between; they are not periodic but they contain some kind of structure which can be exploited to obtain better predictions.

Before we mention the most common structures which can be exploited for predictions, let us state how we quantify the success of the predictions we make. The error measure that is most commonly used is the *root mean squared (rms) prediction error*. If we predict the outcome of measurements at times n to be $\hat{s}_n$, while the actual measurements are s_n, then the rms prediction error is

$$e = \sqrt{\langle (\hat{s}_n - s_n)^2 \rangle},\qquad (4.1)$$

where $\langle \cdot \rangle$ denotes the average over all the trials we have made. You can easily verify that for independent random numbers this error is minimised by $\hat{s}_n = const. = \langle s_n \rangle$. Other error measures are the *mean absolute error* $\langle |\hat{s}_n - s_n| \rangle$, or the *logarithmic error* $\langle \ln |\hat{s}_n - s_n| \rangle$. Error or cost functions can be much more complicated, in particular when money is involved. If you travelled to a country where credit cards are unknown and predicted the necessary amount of cash $100 short, then you would be in trouble: you would have to cable your bank for more cash, you would miss your flight home and end up in prison before the money was there. $100 too much in your pocket are harmless, the maximal cost being $100 if you lose it.

If for some reason you use a fancy cost function, the mean is not necessarily the best prediction even for independent quantities. Predictability is increased by knowledge of the distribution of the values that the predicted quantity takes.

The mean and the probability distribution are static characteristics of the data and do not involve any correlations in time. Depending on the strength and type of correlations, we can improve predictions considerably.

The most common and best-studied sources of predictability are linear correlations in time. Assume a fully random process with a slowly decaying autocorrelation function. The randomness prevents us from making any precise prediction, even with an absolute knowledge of the present state. The strong correlation assures us that the next observation will be given by a linear combination of the preceding observations, with an uncertainty of zero mean,

$$\hat{s}_{n+1} = \bar{s} + \sum_{j=1}^{m} a_j (s_{n-m+j} - \bar{s}),\qquad (4.2)$$

---

[1] It was claimed that weather forecasts were not much better than this until recently.

where the mean value $\bar{s}$ of the time series is conveniently split off. As we depicted in Section 2.3, the autocorrelation function and the power spectrum are intimately related by the *Wiener–Khinchin theorem* (see Section 2.3), which are prominent objects of linear statistical inference. Their information can be exhaustively exploited by linear predictors. However, due to the assumed random nature of the data, the underlying *auto-regressive models* and *moving average* models are stochastic.

Another kind of temporal correlation is present in nonlinear deterministic dynamical systems. Recall that pure dynamical systems are described either by discrete time maps,

$$\mathbf{x}_{n+1} = \mathbf{F}(\mathbf{x}_n), \qquad n \in \mathbb{Z}, \tag{4.3}$$

or by first-order ordinary differential equations,

$$\frac{d}{dt}\mathbf{x}(t) = \mathbf{f}(\mathbf{x}(t)), \qquad t \in \mathbb{R}. \tag{4.4}$$

Both variants are mathematical descriptions of the fact that all future states of such a system are *unambiguously* determined by specifying its present state at some time n (respectively t). This implies that there also exists a deterministic forecasting function. However, as we will see later, any inaccuracy in our knowledge of the present state will also evolve over time and, in the case of a chaotic system, will grow exponentially. In the latter case, we can no longer use the time evolution to calculate states arbitrarily far in the future since we can never know a physical quantity with zero uncertainty. However, even for a chaotic system the uncertainty is amplified only at a finite rate, and we can still hope to make reasonable short-term forecasts exploiting the deterministic evolution law. In contrast to correlations reflected by the autocorrelation function, those imposed by nonlinear deterministic dynamics may be visible only by using nonlinear statistics. They are usually called *nonlinear correlations*, and one has to employ new techniques in order to exploit them for predictions. There are of course lots of data which are neither deterministic chaos nor linear stochastic processes. They do not follow any of these two paradigms, and only for certain cases we will be able to do anything with them. See the discussion in Section 12.1.

4.2 Simple nonlinear prediction algorithm

Let us now be more specific and develop a simple prediction algorithm which exploits deterministic structures in the signal. As with most algorithms in this field, we will proceed by assuming that the data originates in an underlying dynamical system. Once we have established the algorithm, we will have to study its behaviour in more realistic cases.

A deterministic vector valued data set sampled at discrete times in its state space is perfectly described by Eq. (4.3). Unfortunately, only if we know the mapping $\mathbf{F}$ we can use this as a prediction scheme. Knowing $\mathbf{F}$ is an unrealistic expectation when working with real world data. With $\mathbf{F}$ unknown, we have to make some assumptions about its properties. Using the minimal assumption that the mapping $\mathbf{F}$ is continuous[2] we can construct a very simple prediction scheme. In order to predict the future state $\mathbf{x}_{N+1}$, given the present one $\mathbf{x}_N$, we search a list of all past states $\mathbf{x}_n$ with $n < N$ for the one closest to $\mathbf{x}_N$ with respect to some norm. If the state at time n_0 was similar to the present (and thus close in phase space), continuity of $\mathbf{F}$ guarantees that $\mathbf{x}_{n_0+1}$ will also be close to $\mathbf{x}_{N+1}$. Since by assumption $n_0 < N$, also $n_0 + 1 \leq N$, and hence we can read $\mathbf{x}_{n_0+1}$ from the data record.

If we have observed the system for a very long time, there will be states in the past which are arbitrarily close to the present state, and our prediction, $\hat{\mathbf{x}}_{N+1} = \mathbf{x}_{n_0+1}$, will be arbitrarily close to the truth. Thus, in theory we have a very nice prediction algorithm. It is usually referred to as the "Lorenz's method of analogues" because it was proposed as a forecasting method by Lorenz (1969).

Let us now face reality one step at a time. Even if the assumption of an underlying deterministic system is correct, we usually do not measure the actual states $\mathbf{x}_n$. Instead, we observe one (or a few) quantities which functionally depend on these states. Most commonly we have scalar measurements

$$s_n = s(\mathbf{x}_n), \qquad n = 1, \ldots, N, \tag{4.5}$$

in which, more often than not, the measurement function s is as unknown as $\mathbf{F}$. Obviously we cannot invert s but we can use a *delay reconstruction* (see Section 3.2) to obtain vectors equivalent to the original ones:

$$\mathbf{s}_n = (s_{n-(m-1)\tau}, s_{n-(m-2)\tau}, \ldots, s_{n-\tau}, s_n). \tag{4.6}$$

This procedure will introduce two adjustable parameters into the prediction method (which is in principle parameter free): the delay time τ and the embedding dimension m. The resulting method is still very simple. For all measurements $s_1, \ldots, s_N$ available so far, construct the corresponding delay vectors $\mathbf{s}_{(m-1)\tau+1}, \ldots, \mathbf{s}_N$. In order to predict a future measurement $s_{N+\Delta n}$, find the embedding vector $\mathbf{s}_{n_0}$ closest to $\mathbf{s}_N$ and use $s_{n_0+\Delta n}$ as a predictor. This scheme has been used in tests for determinism by Kennel & Isabelle (1992), for example. To our knowledge, the idea goes back to Pikovsky (1986). The method used by Sugihara & May (1990) is closely related.

As a next step towards reality, we no longer ignore the fact that every measurement of a continuous quantity is only valid up to some finite resolution. This resolution is

---

[2] We can usually allow the map to be composed of a few continuous pieces.

limited mainly by fluctuations in the measurement equipment, which are hopefully random, and by the fact that the results are eventually stored in some discretised form. In any case the finite resolution, call its typical size σ, implies that looking for the single closest state in the past is no longer the best we can do since inter-point distances are contaminated with an uncertainty of the order of σ. All points within a region in phase space of radius σ have to be considered to be equally good predictors *a priori*. Instead of choosing one of them arbitrarily, we propose to take the arithmetic mean of the individual predictions based on these values. The estimated resolution of the measurements introduces the third parameter into our prediction scheme.

Now we are ready to propose the following prediction algorithm. Given a scalar time series $s_1, \ldots, s_N$, choose a delay time τ and an embedding dimension m and form delay vectors $s_{(m-1)\tau+1}, \ldots, s_N$ in $\mathbb{R}^m$. In order to predict a time Δn ahead of N, choose the parameter ϵ of the order of the resolution of the measurements and form a neighbourhood $\mathcal{U}_\epsilon(s_N)$ of radius[3] ϵ around the point s_N. For all points $s_n \in \mathcal{U}_\epsilon(s_N)$, that is, all points closer than ϵ to s_N, look up the individual "predictions" $s_{n+\Delta n}$. The prediction is then the average of all these individual predictions:

$$\hat{s}_{N+\Delta n} = \frac{1}{|\mathcal{U}_\epsilon(s_N)|} \sum_{s_n \in \mathcal{U}_\epsilon(s_N)} s_{n+\Delta n}. \tag{4.7}$$

Here $|\mathcal{U}_\epsilon(s_N)|$ denotes the number of elements of the neighbourhood $\mathcal{U}_\epsilon(s_N)$. In the case where we do not find any neighbours closer than ϵ we might just increase the value of ϵ until we find some. A simple but effective implementation of this scheme is the program `zeroth` included in the TISEAN package and discussed in Section A.4.1. Later in this book we will describe more refined ways of prediction and modelling. In this more general setting, it will turn out that the scheme described here is based on a zeroth-order approximation of the dynamics, i.e. it is a locally constant predictor as opposed to a locally linear or even a global nonlinear model. Even more interestingly, it will be shown that for certain nonlinear stochastic processes, namely Markov chains, this is also the optimal predictor.

If more than a few predictions are needed, for instance because the average prediction error needs to be determined accurately, this implementation is very inefficient. For each prediction, *all* vectors in the embedding space are considered (excluding only those in the temporal vicinity of the one to be predicted in order to avoid causality conflicts). They are tested for closeness, and most of them are rejected as too far away. It is therefore wise to use an efficient method to find nearest neighbours in phase space, see Section A.1.1.

---

[3] For simplicity, we propose the use of the maximum norm: A point belongs to $\mathcal{U}_\epsilon(s_N)$ if none of its coordinates differs by more than ϵ from the corresponding coordinate of s_N.

4.3 Verification of successful prediction

If you made a prediction which came true, your success might have been just a coincidence. For significance, you have to study the errors of many predictions starting from independent actual states. This approach can be very tedious. Imagine that you can predict on 1 November whether there will be snow on Christmas Eve (let us say in Copenhagen – it is too easy to predict in Montreal). Only after a couple of years will it be statistically significant whether your prediction is useful. In practice, you would prefer to test your method on a record of the last one hundred years or so, but there is a problem: the "future" events you are predicting are already known, and you will have to make sure you do not use any existing information on the outcomes you want to predict. If you use the past to optimise your prediction method, you will eventually account for all the funny coincidences that happened over the last century. However, it is no longer appropriate to quote the performance of the predictor on this record as the prediction error. Such an error would be called an *in-sample* prediction error. If somebody tries to convince you of something by giving an impressive prediction error, always make sure it is an *out-of-sample* prediction error! (Otherwise one should rather speak of a "post-diction".) Out-of-sample error means that the given data set is divided into a *training set* and a *test set*. The parameters of the predictor (e.g., embedding dimension, time lag, neighbourhood size in Eq. (4.7)), are optimised by minimisation of the prediction error on the training set. The number which you use to access its performance should then be the prediction error on the test set. The predictor in Eq. (4.7) contains one ambiguity with this respect: Should the neighbourhood $\mathcal{U}_\epsilon(\mathbf{s}_n)$ be taken from the training set or from the test set? The concept of causality allows us to search for neighbours in the entire history of the current state $\mathbf{s}_N$, ignoring the splitting into training and test set. However, history means that also the "futures" of the neighbours, the elements $s_{n+\Delta n}$, have to be in the past of the current state, i.e., $n + \Delta n < N$. This is important for long term predictions of highly sampled data with large Δn.

The fact that in-sample errors often are much smaller than out-of-sample errors is more relevant the more parameters are adjusted to optimise the performance on the training set. The simple nonlinear predictor, Eq. (4.7), can almost be called a parameter free model, the few parameters being fixed initially. Nevertheless we will quote out-of-sample errors in the examples. They have been estimated from data which have been withheld while building the predictor. A notable exception where the use of in-sample prediction errors may be justified are tests for nonlinearity where the errors are not evaluated on their own, but are compared to *surrogate* data sets. A more thorough discussion on the statistical relevance of quantities such as prediction errors will be taken up in Chapter 7.

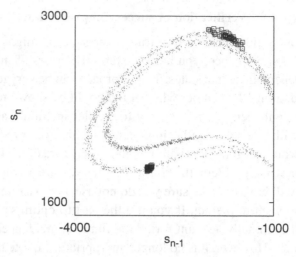

Figure 4.1 As an example for a simple nonlinear prediction, an enlarged region of a phase portrait of the NMR laser data is shown. (See Fig. 3.6 for a full portrait.) For an exemplary phase space point we highlighted all its neighbours closer than $\epsilon = 50$ (the cluster of squares in the lower part) and their images after two time steps. The cluster of the images is more spread out, but the centre of mass still yields a reasonable prediction.

Often, the data base is too small to be split into training and test sets. Then, a variant of this, called *leave-one-out statistics*, or, *complete cross-validation*, is employed: For every single prediction, the predictor uses information from the full data set, excluding only the information which is temporally correlated to the actual prediction to be made. In the simple predictor Eq. (4.7) this means that those delay vectors whose time indices obey $|n - N| < n_{corr}$ are excluded from the neighbourhood $\mathcal{U}_\epsilon(\mathbf{s}_N)$, where n_{corr} has to be fixed in accordance with the prediction horizon Δn and the correlations in the data set.

Example 4.1 (Prediction of NMR laser data). Let us test the proposed algorithm on an experimental but highly deterministic data set. In the introduction, we mentioned the NMR laser experiment carried out at the Institute for Physics at the University of Zürich. The data set we used to produce Fig. 1.2 consisted of 38 000 points, exactly one point per driving cycle (see Appendix B.2 for more information). We split the data set into two parts. The first 37 900 points form the data base for the prediction algorithm and the last 100 points form the test set. On each of the latter, we performed predictions for $\Delta n = 1, \ldots, 35$ time steps ahead. For each prediction horizon Δn, we computed the root mean squared prediction error $e_{\Delta n} = \sqrt{\langle (s_{n+\Delta n} - \hat{s}_{n+\Delta n})^2 \rangle}$, which measures the average deviation of the predictions from the subsequently measured actual values. This prediction error is shown as diamonds in Fig. 1.2 (see also Fig. 4.1). We used the TISEAN

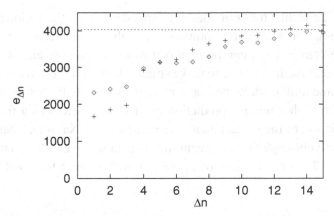

Figure 4.2 Prediction errors versus prediction time using a linear (diamonds) and the simple nonlinear (crosses) predictor for the human breath rate data from Example 1.2. For an explanation see Example 4.2.

program `zeroth` (see Appendix A.4.1) with embedding dimension $m = 3$, delay time $\tau = 1$ and neighbourhood size $\epsilon = 50$, which is about twice the amplitude of the measurement error in this data set. Note that for the small amplitude regime, the increase in the error is well described by an exponential $e_{\Delta n} \approx \epsilon e^{0.3\Delta n}$, which suggests that the prediction error is dominated by the chaotic divergence of initially close trajectories. Later we will (in Chapter 5) estimate the maximal Lyapunov exponent of this data set to be $\lambda = 0.3$. However, we do not recommend the use of the increase in prediction error as a method for estimating the Lyapunov exponent since the averaging is done in the wrong way. $\square$

Example 4.2 (Prediction of human breath rate). We can also try to predict data which are not as clearly deterministic as the NMR laser data in Example 4.1. However, we must be careful when it comes to interpreting the results. If linear correlations are present in the data, the nonlinear predictor will pick up this information to some extent. Thus, if an algorithm based on chaos theory is able to predict better than chance, this alone cannot be taken as a signature of deterministic chaos in the data. In Chapter 7, we will discuss how to make more significant statements in this respect. Now let us try to forecast future values of the human breath rate data set mentioned in the introduction (Example 1.2, Fig. 1.4; see also Appendix B.5). Again we split the data set into two parts, this time with a data base containing 1500 points. For each prediction horizon, the prediction error is estimated by an average of 100 trials. Figure 4.2 shows the root mean squared (rms) prediction error for the best linear predictor we could find (diamonds), and the simple nonlinear predictor introduced in this section (crosses). The linear predictor has five coefficients, the nonlinear one uses neighbourhoods of $\epsilon = 1000$ in two dimensions. The results for both does not depend much on the parameters chosen, and consequently no systematic optimisation of these parameters was performed. Thus, the prediction errors can be regarded as out-of-sample errors. We observe that the nonlinear method is able to make slightly better prediction over short times (up to three time units), while the linear method is marginally better from then on. Note the much higher relative prediction error (that is, error as a fraction of the standard deviation of the signal) than in the NMR laser example. There is a much stronger effectively random component in this data set. We will revisit this example in Chapter 7, where we will investigate it with respect to possible nonlinear structures. □

The quality of a nonlinear prediction can only be assessed by comparison with some benchmark given by standard techniques. Let us draw your attention to a common way of presenting the results of prediction which we consider not to be very instructive and sometimes misleading. You will often see plots showing a time series together with one- (or a few) step-ahead predictions. If the time series is autocorrelated (as is typical for a continuous time process), it is not particularly hard to predict a few steps ahead. As an extreme example, it is easy to predict the weather ten minutes from now! In analogy to this, we are not surprised that locally linear predictions in Abarbanel (1996) of the volume of the Great Salt Lake in Utah appear to be quite convincing. However, in this case we could verify that the prediction horizon compared to the autocorrelation time is so short that the one-step prediction error can easily be beaten by the linear model $s_{n+1} = 2s_n - s_{n-1}$, which just extrapolates the trend of the last two measurements.

More sophisticated prediction schemes and many issues related to modelling and prediction will be discussed in Chapter 12.

4.4 Cross-prediction errors: probing stationarity

It turns out that the locally constant prediction scheme described in Section 4.2 is one of the most robust indicators of nonlinear structure in a time series and is particularly sensitive to deterministic behaviour. We will thus use it later in Section 7.1 as a discriminating statistic in a test for nonlinearity. Indeed, this simple scheme works with much less data than, e.g., local linear predictors (see Section 12.3.2) and, unlike global nonlinear fits (see Section 12.4), no smart choice of basis functions for each new data set is necessary. These properties make the algorithm well suitable as a crude, but nevertheless sensitive, probe whenever we are looking for (nonlinear) predictability. As an example of the many situations where such a probe can be useful, let us use it to study non-stationarity in a long data set.

The idea is that, in a system where different dynamics are at work at different times, data records from different episodes are not equally useful as data bases for predictions. Let us break the available data into segments S_i, $i = 1, \ldots, I$, such that each segment is just long enough to make simple nonlinear predictions, say $N_s = N/I \approx 500$. Now for each two segments S_i and S_j, compute the rms prediction error using S_i as the training set (i.e. find neighbours in S_i) but use S_j as the test set (i.e. on which predictions are performed). The cross-prediction error as a function of i and j then reveals which segments differ in their dynamics. If a prediction error for a couple of segments S_i, S_j is larger than the average, the data in S_i obviously provide a bad model for the data in S_j. We expect the diagonal entries $i = j$ to be systematically smaller: they represent in-sample errors since training set and test set are identical. The use of mutual predictions as a test for stationarity is developed in Schreiber (1997). The TISEAN package has a program called $\texttt{nstatz}$ that does most of the work.

Example 4.3 (Non-stationarity of breath rate). The data set introduced in Appendix B.5 contains a recording of the breath rate of a human patient during almost a whole night (about 5 hr). Of course, conditions have probably changed during that time. As a first approach to the question of which episodes of the signal show similar dynamics, we split the recording into 34 non-overlapping segments S_i of 1000 points (500 s) each. Now we use every segment in turn as a data base for predictions with $\texttt{zeroth}$. We use unit delay and three dimensional embeddings. The size of the neighbourhoods is set to one quarter of the variance of the values in the data base. For each of the 34 data bases we compute 34 prediction errors, predicting values in segment S_j. In Fig. 4.3 we show the rms prediction error as a surface over the i–j-plane. Except for the expected lower errors for $i = j$, we see that there is a transition around the middle of the recording: segments from the first half are useless for predictions in the second half, and vice versa. □

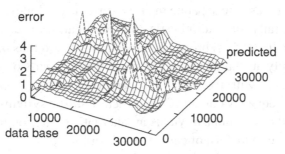

Figure 4.3 Mutual prediction errors for segments of a long, non-stationary recording of the breath rate of a human. See text of Example 4.3.

The idea of breaking a data set into parts and considering the fluctuations of a certain statistic computed on the single parts is widespread, and the only question is what statistic to use. We have chosen nonlinear predictability, since it is very appropriate in the case where data possess strong nonlinear correlations, and it yields meaningful results even for data from stochastic processes. The main advantage of our choice, however, lies in the fact that it allows the computation of cross-correlations as in Fig. 4.3. The variations of the cross-prediction errors are much larger and hence more significant than the fluctuations of the in-sample errors, which appear on the diagonal in Fig. 4.3. We will come back to this in Chapter 13.

4.5 Simple nonlinear noise reduction

A close relative of forecasting is noise reduction. Rather than predicting future values, we now want to replace noisy measurements by better values which contain less noise. For prediction, we had no information about the quantity to be forecasted other than the preceding measurements. For noise reduction, we are better off in this respect. We have a noisy measurement to start with, and, unless we have to process data items immediately in real time, we also have the future values. On the other hand this provides a benchmark: the replacements we propose have to contain errors which are on average less than the initial amplitude of the noise, otherwise the whole procedure is worthless.

Noise reduction means that one tries to decompose every single time series value into two components, one of which supposedly contains the signal and the other one contains random fluctuations. Thus, we always assume that the data can be thought of as an additive superposition of two different components which have to be distinguishable by some objective criterion. The classical statistical tool for obtaining this distinction is the power spectrum. Random noise has a flat, or at least a broad spectrum, whereas periodic or quasi-periodic signals have sharp spectral lines. After both components have been identified in the spectrum, a Wiener filter (Section 2.4) can be used to separate the time series accordingly.

This approach fails for deterministic chaotic dynamics because the output of such systems usually leads to broad band spectra itself and thus possesses spectral properties generally attributed to random noise. Even if parts of the spectrum can be clearly associated with the signal, a separation into signal and noise fails for most parts of the frequency domain. Chaotic deterministic systems are of particular interest because the determinism yields an alternative criterion to distinguish the signal and the noise (which is, of course, not deterministic).

The way we will exploit the deterministic structure will closely follow what we have done for nonlinear predictions. (Again we have to refer the reader who needs optimal results to a later chapter, Chapter 10 in Part Two.) Let us, for convenience, state the problem in delay coordinates right away. Let the time evolution of the signal be deterministic with a map f which is not known to us. All that we have knowledge of are noisy measurements of this signal:

$$s_n = x_n + \eta_n, \qquad x_n = f(x_{n-m}, \ldots, x_{n-1}). \qquad (4.8)$$

Here η_n is supposed to be random noise with at least fast decaying autocorrelations and no correlations with the signal x_n. We will refer to $\sigma = \sqrt{\langle \eta^2 \rangle}$ as the noise amplitude or absolute noise level.

How can we recover the signal x_n given only the noisy sequence s_n? After what we have learned about predictions, at a first glance the solution seems obvious: in order to clean a particular value of the time series, throw away the measurements and replace them by a prediction $\hat{x}_n$ based on previous measurements, say

$$\hat{x}_n = \hat{f}(s_{n-m}, \ldots, s_{n-1}). \qquad (4.9)$$

If you are a more conservative character, you may prefer to use some linear combination of the noisy measurements and their predictions. But even then, this scheme will fail for chaotic systems. Remember that the previous measurements are noisy as well, and the errors they contain will be amplified by the chaotic dynamics. The proposed replacement will on average be much worse than the noisy measurements we had. (How much worse depends on the average local expansion rate of the dynamics – expect a factor of $1.5 - 3.0$ for typical maps! See Section 10.2 for a more quantitative discussion.)

The next, smarter, suggestion would be to reverse time and "predict" the desired value from the subsequent measurements. The idea is that a reversed expansion is after all a contraction. Here, the catch is that, in bounded systems, not all errors can be expanded in all directions, and those directions in which errors are contracted become expanding under time reversal. Thus a version of Eq. (4.9), solved for $\hat{x}_{n-m} = s_{n-m}$, will fail again since the expansion is the dominant effect.

What finally works is to solve the implicit relation

$$x_n - f(x_{n-m}, \ldots, x_{n-1}) = 0 \qquad (4.10)$$

for one of the coordinates in the middle, say $x_{n-m/2}$, for even m. This stabilises the replacement in both the stable and the unstable directions. This is illustrated in Fig. 4.4. Of course we do not know f. Even if we did, we could not in general solve it for one of its arguments. Therefore, we choose a very crude approximation for now and replace the function f (solved for one argument) by a locally constant function. This is exactly what we did for the simple nonlinear prediction method given above. In other words, in order to obtain an estimate $\hat{x}_{n_0-m/2}$ for the value of $x_{n_0-m/2}$, we form delay vectors $\mathbf{s}_n = (s_{n-m+1}, \ldots, s_n)$ and determine those

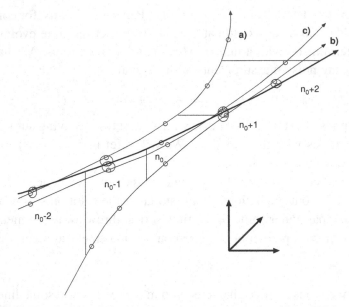

Figure 4.4 Illustration of the simple noise reduction algorithm. With respect to the measurement at time n_0, trajectory **a)** has been close to the reference trajectory (thick line) for two time steps in the past, whereas **b)** is close for two time steps into the future. Both are not close at time n_0 due to the chaotic dependence on initial conditions. Trajectory **c)** is close in the past *and* in the future and thus also at time n_0. It is trajectories of the last type which are by definition contained in the neighbourhoods used in Eq. (4.11).

which are close to s_{n_0}. The average value of $s_{n-m/2}$ is then used as a cleaned value $\hat{x}_{n_0-m/2}$:

$$\hat{x}_{n_0-m/2} = \frac{1}{|\mathcal{U}_\epsilon(s_{n_0})|} \sum_{s_n \in \mathcal{U}_\epsilon(s_{n_0})} s_{n-m/2} \,. \tag{4.11}$$

Again, for numerical ease we use the maximum norm. Only when the embedding dimension is really large (say, $m > 20$), and we want to treat data with a high noise amplitude, then a Euclidean norm is more useful for the identification of those neighbours in the noisy data set that would also be good neighbours for the noise free data. Due to the assumed independence of signal and noise,

$$(s_n - s_k)^2 = \sum_{l=0}^{m-1}(s_{n+l} - s_{k+l})^2 = \sum_{l=0}^{m-1}(x_{n+l} + \eta_{n+l} - x_{k+l} - \eta_{k+l})^2$$
$$\approx (x_n - x_k)^2 + 2m\sigma^2 \,,$$

where the approximation becomes the better the larger m. Hence, the squared Euclidean distance of two noisy delay vectors is identical to the squared Euclidean distance of the noise free vectors plus a constant. The noisy vectors closest to $\mathbf{s}_n$ are therefore in good approximation exactly those whose noise free counterparts would also be the closest vectors to the noise free version of $\mathbf{s}_n$, and these are the ones we need for successful noise reduction.

Not surprisingly, Eq. (4.11) is very similar to Eq. (4.7). As opposed to the prediction case, here $\mathcal{U}_\epsilon(\mathbf{s}_{n_0})$ is never empty, no matter how small a value of ϵ we choose: it always contains at least $\mathbf{s}_{n_0}$. This is good to know since we have to make some choice of ϵ when we use this algorithm. It is guaranteed that if we choose ϵ too small, the worst thing that can happen is that the only neighbour found is $\mathbf{s}_{n_0}$ itself. This, however, yields the estimate $\hat{x}_{n_0-m/2} = s_{n_0-m/2}$, which just means that no correction is made at all.

We have just embarked on an important discussion. How can we make sure that all we do is to reduce noise without distorting the signal? Since we do not know the clean signal, all we can ask for in practice is that the effect of any possible distortion is more acceptable than the effect of the noise before. In most cases, this will mean that the error due to distortion must be considerably smaller than the noise amplitude.

Obviously, it will be necessary to gain experience with nonlinear noise reduction using examples where the true signal is known. Since this is usually the case only with numerical data generated on the computer, such data are predominantly used in the original literature. These studies have demonstrated that the simple nonlinear scheme, as well as more advanced schemes, which we will describe in Section 10.3, are in fact able to reduce noise in deterministic data without causing undue distortions. Before cleaning experimental data, the more sceptical reader may wish to try the algorithm on artificial data of similar properties in order to gain confidence. Moreover, we recommend as a minimal test for the reliability of the result to study the statistical properties of what the algorithm has eliminated as noise, i.e., of the time series of differences between noisy signal and signal after noise reduction, $\hat{\eta}_n = s_n - \hat{x}_n$. Its autocorrelation function should decay fast, there should be no significant cross-correlations between $\hat{\eta}_n$ and the signal $\hat{x}_n$, and its marginal distribution should satisfy one's expectations of what the measurement noise looks like (in particular, its standard deviation should not exceed the assumed noise amplitude).

Studies with known true signals suggest that a good choice for the size of the neighbourhoods is given by 2–3 times the noise amplitude. While in some cases it is easy to guess the approximate noise level from a plot of the data, we can also estimate it using the correlation sum, as will be discussed in Section 6.7. If we are in doubt about the noise level, we should underestimate it. As we said previously, this will at worst weaken the performance but will not introduce artefacts of the filtering.

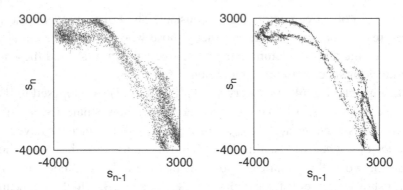

Figure 4.5 Simple nonlinear noise reduction. 5000 values of the NMR laser data (Example 1.1) were contaminated with 10% artificial noise with the same power spectrum, as shown in the left hand panel. On the right the simple noise reduction algorithm described above has been used to clean the data.

For the embedding dimension m it seems reasonable to use a value such that the dynamics are deterministic both in the future and in the past coordinates. Here, the conservative choice is a larger value of m by which we get rid of some of the more dubious neighbours. Within the TISEAN package, the scheme described here is implemented as the program `lazy`, which is discussed in Appendix A.8.

Example 4.4 (Simple nonlinear noise reduction on NMR laser data). As a tribute to those very reasonable experimental scientists who refuse to be convinced just because a method works on the Hénon map, the data set we choose as our first example is at least half real. The "clean" signal is taken to be 5000 values of the NMR laser data we showed before (see Appendix B.2 for details). They contain about 1.1% measurement noise. Now we artificially add "measurement noise" consisting of random numbers which have been filtered to have the same power spectrum as the data themselves. The rms amplitude of the noise is chosen to be 10% of the amplitude of the data. This kind of noise is called *in-band* noise and is completely indistinguishable from the signal for any spectral filter. Now we use the simple nonlinear noise reduction scheme to remove the measurement noise again. The result is shown in Fig. 4.5. Of course, our algorithm will not be able to distinguish the real and the additional noise which limits the evaluation of its performance. Still we can reduce the amplitude of the added random component by a factor of about 1.7, using values of ϵ corresponding to about 2–3 times the noise amplitude.   □

Example 4.5 (Simple nonlinear noise reduction on far infrared laser data). Let us now proceed to a *really* real example. In Fig. 4.6 (left hand panel) we show an enlargement by a linear factor of two of a phase portrait of data set A of the Santa

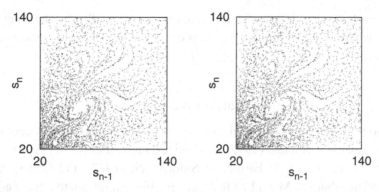

Figure 4.6 Data set A of the Santa Fe Institute time series competition, Weigend & Gershenfeld (1993), before and after noise reduction with the simple algorithm.

Fe Institute time series competition; see Weigend & Gershenfeld (1993). The data (Appendix B.1) represent the fluctuating intensity of a far infrared laser and can be approximately described by three coupled ordinary differential equations [Hübner *et al.* (1993)]. Since the raw data are 8-bit integer numbers, the measurement error is at least the discretisation error, which can be assumed to be uniformly distributed in the interval $[-0.5, 0.5]$. Noise reduction would be worth while even if we could just remove this error and thus increase the dynamical range of the data. We apply the noise reduction algorithm assuming an (absolute) noise amplitude of one unit and choose $\epsilon = 2$, i.e. define as neighbours the points which are not further apart than two units in each of the seven directions of embedding space. The resulting data set is shown in Fig. 4.6 (right hand panel) with the same magnification as the original data. All we can say at this stage is that the structure in the data appears enhanced after noise reduction. This is of course an unsatisfactory statement, but we will need more powerful tools in order to verify and quantify noise reduction on real data. In particular, we will need an estimate of the noise level before and after noise reduction. Such an estimate is possible on the basis of the correlation integral (Chapter 6, and Section 10.2). □

Let us make a remark about the practical implementation of this simple non-linear noise reduction method. Straightforward encoding of Eq. (4.11) would result in a time consuming algorithm since we want to clean *all* points and not just a few. The main computational burden is to form ϵ-neighbourhoods for each of the points we want to correct. We can save orders of magnitude of computer time if we apply a fast neighbour search algorithm instead of trying all possible pairs and rejecting those which are not close enough. In fact, we reduce the asymptotic operation count from $O(N^2)$ to $O(N \ln N)$ if we use a binary tree, or even to $O(N)$ if we use a box-assisted method. Schreiber (1995) describes in detail how this is done. The

TISEAN package uses box-assisted neighbour search methods whenever it is appropriate, where you find implementations of this and more refined noise reduction schemes.

Further reading

General sources for the problem of time series prediction are the proceedings volumes by Casdagli & Eubank (1992) and Weigend & Gershenfeld (1993). Classical articles in the field include Farmer & Sidorowich (1987) and (1988), Casdagli (1989), and Sugihara & May (1990). More specific references to advanced prediction methods will be given in Chapter 12. The simple noise reduction method of Section 4.5 has been proposed by Schreiber (1993).

Excercises

4.1   Try zeroth (see Section A.4.1) on some artificial data sets. Start out with the two linearly uncorrelated data sets created in Exercise 2.2. How do the prediction errors depend on the prediction time in the two cases?

4.2   Try the above program on 5000 iterates of the Hénon map (Example 3.2), adding different amounts of noise. Compare the prediction error to the (artificial) case where you use the Hénon map itself for predictions. For reassurance in case you find the result puzzling: we obtain a mean prediction error of about 0.1 for Gaussian noise of amplitude 0.05 if we choose neighbourhoods of radius between 0.05 and 0.15.

Chapter 5

Instability: Lyapunov exponents

5.1 Sensitive dependence on initial conditions

The most striking feature of chaos is the unpredictability of the future despite a deterministic time evolution. This has already been made evident in Fig. 1.2: the average error made when forecasting the outcome of a future measurement increases very rapidly with time, and in this system predictability is almost lost after only 20 time steps. Nevertheless we claim that these experimental data are very well described as a low dimensional deterministic system. How can we explain this apparent contradiction?

Example 5.1 (Divergence of NMR laser trajectories). In Fig. 5.1 we show several segments of the NMR laser time series (the same data underlying Fig. 1.2; see Appendix B.2) which are initially very close. Over the course of time they separate and finally become uncorrelated. Thus it is impossible to predict the position of the trajectory more than, say, ten time steps ahead, knowing the position of another trajectory at this time which was very close initially. (This is very much in the spirit of the prediction scheme of Section 4.2.) □

The above example illustrates that our every day experience, "similar causes have similar effects", is invalid in chaotic systems except for short periods, and only a mathematically exact reproduction of some event would yield the same result due to determinism. Note that this has nothing to do with any unobserved influence on the system from outside (although in experimental data it is always present) and can be found in every mathematical model of a chaotic system.

This unpredictability is a consequence of the inherent instability of the solutions, reflected by what is called *sensitive dependence on initial conditions*. The tiny deviations between the "initial conditions" of all the trajectories shown in Fig. 5.1 are blown up after a few time steps, and every unobserved detail of the state at time n_0 is important for the precise path of the trajectory in state space.

65

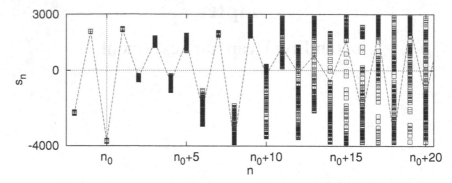

Figure 5.1 Broken line: a reference part of the time series of the NMR laser data. The symbols represent the elements of 160 other segments of the same series, which during three time steps ($n \leq n_0$) coincide with the reference trajectory within the resolution of the measurement.

A more careful investigation of this instability leads to two different, although related, concepts. One aspect, which we do not want to elaborate on in the first part of the book, is the loss of information related to unpredictability. This is quantified by the *Kolmogorov–Sinai entropy* and will be introduced in Chapter 11. The other aspect is a simple geometric one, namely, that nearby trajectories separate very fast, or more precisely, exponentially fast over time.

5.2 Exponential divergence

The fact that trajectories diverge over the course of time would not in itself be very dramatic if it were only very slow, as is typical of predominantly periodic systems. Thus we speak of chaos only if this separation is exponentially fast. The properly averaged exponent of this increase is characteristic for the system underlying the data and quantifies the strength of chaos. It is called the *Lyapunov exponent.*[1]

Example 5.2 (Exponential separation of trajectories, NMR laser data). Fig. 5.2 is another view of what we demonstrated in Fig. 5.1: the (scalar) distances between the elements of initially close segments of the NMR laser time series (Appendix B.2) are shown on a logarithmic scale as a function of time. First of all, almost all the curves are more or less parallel. This confirms that these experimental data really obey a deterministic time evolution. Apart from a lot of unexplained structure, one clearly observes an overall linear increase in this semi-logarithmic plot. □

---

[1] If the divergence rates are averaged over short times only, one can define *local Lyapunov exponents* which allow for a time or phase space resolved analysis. However, these local exponents are no simple invariants and results are often difficult to interpret. In any case, one should keep in mind the fact that the (global) Lyapunov exponent is an average over a quantity which may show strong fluctuations.

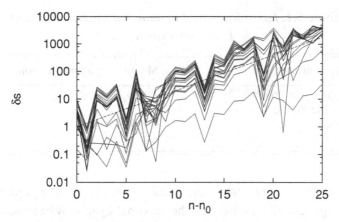

Figure 5.2 Absolute values of the differences between the $n - n_0$th element of a reference trajectory and the corresponding elements of neighbours closer than $\epsilon = 8$. These differences show something like an exponential increase over three orders of magnitude, as suggested by the dashed line, $e^{0.3n}$. The fact that all differences behave similarly reflects the underlying determinism. However, in the presence of strong intermittency the equivalent plots for different reference points could look very different, see Section 8.4. The attractor extension is about 7000 in these units (NMR laser data).

In Chapter 11 we will see that one can define as many different Lyapunov exponents for a dynamical system as there are phase space dimensions. Here we want to restrict ourselves to the most important one, the *maximal Lyapunov exponent* λ. Let $\mathbf{s}_{n_1}$ and $\mathbf{s}_{n_2}$ be two points in state space with distance $\|\mathbf{s}_{n_1} - \mathbf{s}_{n_2}\| = \delta_0 \ll 1$. Denote by $\delta_{\Delta n}$ the distance some time Δn ahead between the two trajectories emerging from these points, $\delta_{\Delta n} = \|\mathbf{s}_{n_1+\Delta n} - \mathbf{s}_{n_2+\Delta n}\|$. Then λ is determined by

$$\delta_{\Delta n} \simeq \delta_0 e^{\lambda \Delta n}, \qquad \delta_{\Delta n} \ll 1, \qquad \Delta n \gg 1. \tag{5.1}$$

If λ is positive, this means an exponential divergence of nearby trajectories, i.e. chaos. Naturally, two trajectories cannot separate farther than the size of the attractor, such that the law of Eq. (5.1) is only valid during times Δn for which $\delta_{\Delta n}$ remains small. Otherwise, we observe a violation of Eq. (5.1) in the form of a saturation of the distance. Due to this fact, a mathematically more rigorous definition will have to involve a first limit $\delta_0 \to 0$ such that a second limit $\Delta n \to \infty$ can be performed without involving saturation effects. Only in the second limit does the exponent λ become a well-defined and invariant quantity. We will elaborate on this later, in particular in Section 11.2.

In dissipative systems one can also find a negative maximal Lyapunov exponent which reflects the existence of a stable fixed point. Two trajectories which approach the fixed point also approach each other exponentially fast. If the motion settles

Table 5.1. *Possible types of motion and the corresponding Lyapunov exponents.*

Type of motion	Maximal Lyapunov exponent
stable fixed point	$\lambda < 0$
stable limit cycle	$\lambda = 0$
chaos	$0 < \lambda < \infty$
noise	$\lambda = \infty$

down onto a limit cycle, two trajectories can only separate or approach each other slower than exponentially. In this case the maximal Lyapunov exponent is zero and the motion is called *marginally stable*. If a predominantly deterministic system is perturbed by random noise, on the small scales it can be characterised by a diffusion process, with $\delta_{\Delta n}$ growing as $\sqrt{\Delta n}$. Thus the maximal Lyapunov exponent is infinite.[2] According to the mathematical definition, this is true no matter how small the noise component is. If we assume that there is only additive measurement error we can, however, try to estimate the Lyapunov exponent of an assumed underlying deterministic system (see Table 5.1 for a summary).

The Lyapunov exponents carry the units of an inverse time and give a typical time scale for the divergence or convergence of nearby trajectories. When computing numerical values one has to take into account the proper normalisation: do we want to compute this time scale in units of the time index of the measurements or in units of real time, say in seconds? When the data are turned into map-like data by the method of the Poincaré surface of section, one has to know the average time between two subsequent points of the Poincaré map, τ_P. Then the Lyapunov exponents computed for the map data are related to those of the flow data by $\lambda_{\text{map}} = \lambda_{\text{flow}} \tau_P$.

In the preceding paragraphs we were speaking about *the* Lyapunov exponent of the system as a characteristic quantity. But the system is represented only by the measurements, and we could perform the measurements on the same system in another way. Much easier, we could rescale, shift or otherwise process the data, and we have much freedom of choice of the precise method of state space reconstruction. How would the Lyapunov exponent suffer from all these modifications? The answer is simply that Lyapunov exponents do not depend on them at all. Lyapunov exponents are *invariant* under all these transformations, as long as they are smooth. The basic reason is that they describe the *long-term* behaviour. Any

---

[2] Sometimes, the exponent of diffusive motion is said to be zero, because distances between two paths starting at the same point grow like $\sqrt{t}$. However, the affine nature of Brownian paths suggests to look at the behaviour in the limit of short times, where the derivative of $\sqrt{t}$ with respect to t diverges. Also, $\lambda = \infty$ for Brownian motion well agrees with the fact that also the entropy of a stochastic process is infinite.

smooth invertible re-parametrisation of phase space can modify distance ratios by finite factors only. Such factors drop out in the limit as $\Delta n \to \infty$. This justifies our interest in this number, since everybody else will find exactly the same results with his/her own observation, as long as the underlying physical system really is the same.

Now, you may not be convinced by Fig. 5.2 and may wonder whether there is really an exponential growth. In fact there is, but do not forget that Eq. (5.1) is valid in the true state space of the system. The measurement process is some projection, and the projection of the distances may very well shrink simply due to the angle of observation. Already in a delay embedding the analogous figure looks much nicer. Additionally, the local rate of divergence varies throughout the attractor and the true exponent is an appropriate average over what we see in Fig. 5.2.

5.3 Measuring the maximal exponent from data

Since a positive maximal Lyapunov exponent is a strong signature of chaos, it is of considerable interest to determine its value for a given time series. The first algorithm for this purpose was suggested by Wolf *et al.* (1985). Unfortunately, it requires much care when using it and one can easily obtain the wrong results. This algorithm does not allow one to test for the presence of exponential divergence, but just assumes its existence and thus yields a finite exponent for stochastic data also, where the true exponent is infinite. While Wolf's algorithm does not use more than a delay reconstruction of phase space, there is another class of algorithms which also involves the approximation of the underlying deterministic dynamics. References are Sano & Sawada (1985) and Eckmann *et al.* (1986). This method is very efficient if the data allow for a good approximation of the dynamics. It will be explained in Chapter 11.

Here we want to describe an algorithm introduced recently by Rosenstein *et al.* (1993) and by Kantz (1994) independently. It tests directly for the exponential divergence of nearby trajectories and thus allows us to decide whether it really makes sense to compute a Lyapunov exponent for a given data set.

In Fig. 5.2 we observe an average exponential growth superimposed by strong fluctuations. These fluctuations have their origin in different aspects of the deterministic dynamics. As we have mentioned already, the projection involved in the measurement may make distances shrink apparently for short times, although they grow in the true state space. Moreover, in the true state space distances do not grow everywhere on the attractor with the same rate, and locally they may even shrink. The Lyapunov exponent is therefore an average of these *local divergence rates* over the whole data. Finally, experimental data are generally contaminated by noise. Its influence can be minimised by using an appropriate averaging statistics when computing the exponent. In order to obtain the average exponential growth of distances, one has to do the following.

Choose a point s_{n_0} of the time series in embedding space and select all neighbours with distance smaller than ϵ. Compute the average over the distances of all neighbours to the reference part of the trajectory as a function of the relative time, i.e. the (arithmetic) average over the curves shown in Fig. 5.2. The logarithm of the average distance at time Δn is some effective expansion rate over the time span Δn (plus the logarithm of the initial distance) containing all the deterministic fluctuations due to projection and dynamics. Repeating this for very many values of n_0, the fluctuations of the effective expansion rates will average out. Thus one has to compute:

$$S(\Delta n) = \frac{1}{N} \sum_{n_0=1}^{N} \ln \left(\frac{1}{|\mathcal{U}(s_{n_0})|} \sum_{s_n \in \mathcal{U}(s_{n_0})} |s_{n_0+\Delta n} - s_{n+\Delta n}| \right). \tag{5.2}$$

The reference points s_{n_0} are embedding vectors as usual; $\mathcal{U}(s_{n_0})$ is the neighbourhood of s_{n_0} with diameter ϵ. Note that the last element of s_{n_0} is s_{n_0}, whence $s_{n_0+\Delta n}$ is outside the time span covered by the delay vector s_{n_0}. Since *a priori* one might neither know the minimal embedding dimension m nor the optimal distance ϵ, one should compute $S(\Delta n)$ for a variety of both values. The size of the neighbourhood should be as small as possible, but large enough such that on average each reference point has at least a few neighbours. Otherwise one might systematically ignore certain parts of the attractor and thus compute a wrong value (see Example 5.4).

The program lyap_k in the TISEAN package (see Section A.5) performs the necessary calculations to create the curves $S(\Delta n)$. If for some range of Δn the function $S(\Delta n)$ exhibits a *robust* linear increase, its slope is an estimate of the maximal Lyapunov exponent λ per time step, which has to be converted to the desired units. We place stress on the word *robust*, since finding a linear increase for isolated optimised values of m and ϵ only cannot be considered as a positive signature of exponential divergence. Below we shall present some typical results, which will illuminate this point. Sometimes one finds linear increases for some choices of ϵ and m but no segments of straight lines for other choices. Before quoting a numerical value for λ, one should try to understand the deviations of the curves from linear behaviour (noise, too small embedding dimension, too few neighbours, saturation due to too large ϵ, power law separation, etc.).

Example 5.3 (Lyapunov exponent of NMR laser data). Our results for the NMR laser data (Appendix B.2) are shown in Fig. 5.3. Since this is a very typical outcome and, according to our taste, a reasonable way of plotting S, let us spend some more time on it. Curves are shown for embedding dimensions $m = 2$ (upper panel) and

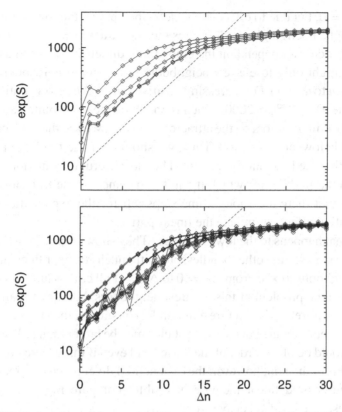

Figure 5.3 Estimation of the maximal Lyapunov exponent of the NMR laser data in $m = 2$ dimensions (upper panel, insufficient reconstruction) and $m = 3, 4, 5$ dimensions (lower panel, useful results). The linear parts of the curves are well described by an exponential $\propto e^{0.3\Delta n}$ and we estimate the maximal Lyapunov exponent to be $\lambda = 0.3 \pm 0.01$. For further explanations see text.

$m = 3, 4, 5$ (lower panel). We used $\epsilon = 15, 30, 60, 120$, and 240. We plotted $e^{S(\Delta n)}$ on a logarithmic scale (which is the same as plotting $S(\Delta n)$ on a linear scale) in order to be able to use physical units on the axis. The data base consisted of 38 000 scalar measurements but for each curve we extended the first sum of Eq. (5.2) over just enough reference points such that at least 500 reference points had more than 10 neighbours. The curves with $m = 2$ (upper panel) had a rather steep jump from $\Delta n = 0$ to $\Delta n = 1$. This means that $m = 2$ does not yield a proper embedding. This is confirmed by glancing back at the attractor in Fig. 3.6. In a two dimensional representation, different branches of the attractor intersect. Naturally, the images of nearby points in the vicinity of these intersections may be very far apart. This is a consequence of an unresolved projection.

Let us now discuss the lower panel. When you have a close look you will see five bundles of curves, each one corresponding to one of the five values of ϵ which can be seen for $\Delta n = 0$, but due to effects to be described below the lowest three bundles overlap strongly. Each single bundle represents the values $m = 3, 4, 5$, which shows that the result does not depend on the embedding dimension as soon as m is large enough (we might only lose a few neighbours in the initial neighbourhood when passing from m to $m + 1$). For increasing positive Δn, $S(\Delta n)$ grows until it saturates around a value $e^{S(\Delta n \to \infty)} \approx 2000$, which is about the mean absolute distance of two arbitrary embedding vectors on the attractor. Actually, we see that the scaling range already ends below, at $e^S \approx 1000$. Thus, we should not use too large a value of ϵ when estimating the Lyapunov exponent. The most useful information comes from the curves in the middle, for which the initial distance of the trajectories is small enough such that there are enough time steps left for the exponential divergence before saturation effects terminate the linear part.

Finally, what happens to the lowest curves? They show some almost linear range with the same slope as the other bundles, but with much stronger fluctuations due to the sparse neighbourhoods. From $\Delta n = 0$ to $\Delta n = 1$ all curves jump upward. Thus, again, we have the problem of false nearest neighbours, but now for all m, and the explanation is different. The data are noisy and if ϵ is smaller than the noise level, one still finds some neighbours but the true distance may be larger than ϵ. The distance of the images would be of the order of the noise level even if there were no exponential divergence. From then on, however, the exponential divergence dominates over the noise and the desired linear increase is visible (compare Fig. 10.5: after noise reduction these curves also behave well).   □

From computing $S(\Delta n)$ we have learned several things: we have a rough idea about the noise level, we know that $m = 2$ is too small to form a good embedding, and we have observed a range where distances increase exponentially. Using linear regression for the linear parts of the curves we can determine the numerical value of the Lyapunov exponent together with the error of the fit, $\lambda = 0.3 \pm 0.01$. A more rigorous estimate of the statistical error and possible systematic errors would require a much more thorough analysis.

Example 5.4 (Lyapunov exponent of far infrared laser data). From flow data from the Lorenz-like laser experiment of the PTB Braunschweig (see Appendix B.1), the time series is a record of the output power of a chaotic laser. The data present short-term oscillations with a slowly increasing amplitude. When the amplitude reaches a critical value, it breaks down and the oscillations start anew. At this disruption the phase information is lost, but the new amplitude is also difficult to predict. These are instances with the largest instability, whereas the smooth increase of the amplitude is well predictable. Due to this strong inhomogeneity of the local divergence rates we expect difficulties when estimating the Lyapunov exponent. In

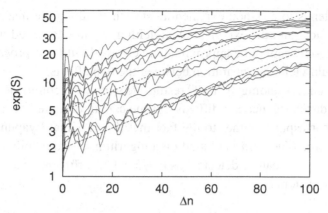

Figure 5.4 The curves $e^{S(\Delta n)}$ for $m = 2$–5 and $\epsilon = 2, 4, 8$, and 16 for Lorenz-like laser data (after simple nonlinear noise reduction). The dashed lines are curves $\propto e^{0.025\Delta n}$. For discussion see text.

order to avoid additional difficulties due to the rather coarse discretisation of the data we use the noise reduced data we produced in Example 4.5.

The curves $e^{S(\Delta n)}$ are shown in Fig. 5.4. At first sight the result is disappointing. The curves show strong oscillations and even when interpolating between these, linear parts are less obvious. Their slopes are slightly different for different values of ϵ. Our explanation is as follows: first, the oscillations clearly have their origin in the periodicity of the signal. Second, as already stated, the expansion rate is largest close to the disruption and very small elsewhere. Unfortunately, the smaller is ϵ, the fewer the unstable parts are sampled since no neighbours are found. (We checked this explicitly by plotting all points with/without neighbours in three dimensions.) Thus for small ϵ one computes the average of λ not on the invariant measure but on a less chaotic component of it. With this understanding of our result we have enough evidence for deterministic chaos, but without it we should be more careful with our statements. The numerical value of λ is much less precise than in the last example, and as an estimate we propose $0.025 < \lambda < 0.03$. The problem is that the data set contains only about 20 of the disruption events, which is not enough for a good estimate of their contribution. We did the same analysis after performing the Poincaré section to obtain map-like data, after applying an advanced nonlinear noise reduction method and after removing the skewness of the data. This confirms the above-quoted value, but still the $S(\Delta n)$ curves remain somewhat unsatisfactory. □

The problem we encountered in the last example is common to systems where the instability is strongly localised in the state space. Later we shall see that this is also a difficulty in more refined methods, and not only when computing Lyapunov exponents. Beyond that, the algorithm proposed here has some deficiencies for flow data, since it tends to mix non-exponential stretching along the marginally stable direction of a flow and the exponential stretching we are interested in. Evidently, the limit $\Delta n \to \infty$ cannot be approximated very well with this procedure, which otherwise would yield the separation between the two effects.

For a better understanding of "null" results we strongly recommend Exercise 5.3 below. When distances increase diffusively, one cannot observe linear increases of $S(\Delta n)$. This corresponds either to the fact that the maximal Lyapunov exponent is infinite or zero. When Δn is plotted on a logarithmic scale, both an increase in distances $\delta_n \propto \Delta n^{1/2}$ can be detected, as is typical of diffusion, and $\delta_n \propto \Delta n$ as in stable periodic systems.

Further reading

The algorithm for the largest Lyapunov exponent is derived in Kantz (1994) and in Rosenstein *et al.* (1993). A related method was introduced earlier by Kurths & Herzel (1987). The algorithm from Wolf *et al.* (1986) is quite prominent and has been widely used, but due to its instability and the impossibility of distinguishing exponential divergence from divergence due to noise it cannot be recommended. For more advanced references on how to compute more than the maximal Lyapunov exponents see Chapter 11.

Exercises

5.1 Create a synthetic scalar time series of map data by iterating the following three dimensional map [Baier & Klein (1990)]:

$$x_{n+1} = 1.76 - y_n^2 - 0.1 z_n \,,$$
$$y_{n+1} = x_n \,,$$
$$z_{n+1} = y_n \,.$$

Choose as an observable the square of the x-coordinate. Compute the maximal Lyapunov exponent using the method described in this chapter (program `lyap_k` of the TISEAN package) for a series of lengths 500 and 5000, respectively, and compare the results with the accurate value $\lambda \approx 0.225$.

5.2 Use a random number generator (or the TISEAN utility `addnoise`) to add uniformly distributed noise to the above times series (measurement noise). Repeat the analysis.

5.3 Create the path of a quasi-Brownian motion on your computer. Let η_n be the output of a random number generator, and let $s_n = s_{n-1} + \eta_n$ (to obtain a true Brownian motion the random numbers should be Gaussian). Try to compute the "maximal Lyapunov exponent" for this time series of length $> 10\,000$ (if you do not introduce an additional delay, Δn should range up to about 1000). How do you interpret the result? For a better understanding, plot $S(\Delta n)$ versus the logarithm of Δn.

Chapter 6

Self-similarity: dimensions

6.1 Attractor geometry and fractals

In the preceding chapter we discussed the dynamical side of chaos which manifests itself in the sensitive dependence of the evolution of a system on its initial conditions. This strange behaviour in time of a deterministically chaotic system has its counterpart in the geometry of the set in phase space formed by the (non-transient) trajectories of the system, the *attractor*.

Attractors of dissipative chaotic systems (the kind of systems we are interested in) generally have a very complicated geometry, which led people to call them *strange*. However, strange sets can also occur without dissipation in more general settings. As we have pointed out already in Chapter 3, a system described by autonomous differential equations (a *flow*) cannot be chaotic in less than three dimensions. With the same argument that trajectories are not allowed to intersect in a deterministic system we can conclude that not only the phase space but also the attractor of a chaotic flow must be more than two dimensional. However, slightly more than two dimensions is sufficient and the motion on a $2 + \epsilon$ dimensional fractal can indeed be chaotic. As we will see, *strange attractors* with fractional dimensions are typical of chaotic systems. Map-like systems can of course show chaos with attractor dimensions less than two. Noninteger dimensions are assigned to geometrical objects which exhibit an unusual kind of self-similarity and which show structure on all length scales.

Example 6.1 (Self-similarity of the NMR laser attractor). Such self-similarity is demonstrated in Fig. 6.1 for an attractor reconstructed from the NMR laser time series, Appendix B.2. Unfortunately, the omnipresent noise hides the small-scale structures, such that below some length scale non-trivial self-similarity is replaced by the trivial self-similarity of a smooth distribution in two dimensions. For this reason we show the attractor after a nonlinear noise reduction algorithm has been applied, which allows for several magnification steps before the

75

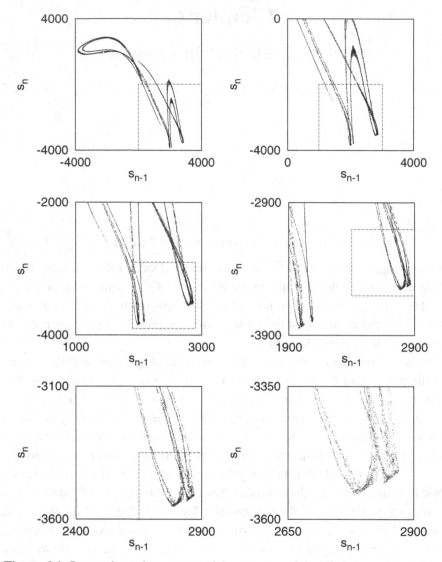

Figure 6.1 Successive enlargements of the attractor of the NMR laser data (after nonlinear noise reduction). The linear scale of the figures is increased by a factor of two between successive panels.

resolution of the data is exhausted. If you are as impressed by the figures as we are ourselves, you will probably want to learn more about nonlinear noise reduction in Section 10.3. □

The standard mathematical example for a self-similar set is a *Cantor set*. It is exactly self-similar when rescaling it by a given factor. Strange attractors behave differently, as we see in Fig. 6.1. We find exact self-similarity only locally with position dependent scaling factors. But for practical purposes this is even more useful since there is scale invariance in a statistical sense for arbitrary scaling factors.

6.2 Correlation dimension

There are several ways to quantify the self-similarity of a geometrical object by a dimension. Of course we require the definition to coincide with the usual notion of dimension when applied to nonfractal objects: a finite collection of points is zero dimensional, lines have dimension one, surfaces two, etc. But let us for the moment cut a long story short and propose a definition which is of particular interest in practical applications where the geometrical object has to be reconstructed from a finite sample of data points which are most likely to contain some errors as well. This notion, called the *correlation dimension*, was introduced by Grassberger & Procaccia (1983) and (1983a).

Let us first define the correlation sum for a collection of points $\mathbf{x}_n$ in some vector space to be the fraction of all possible pairs of points which are closer than a given distance ϵ in a particular norm. The basic formula (to be modified for practical applications) is

$$C(\epsilon) = \frac{2}{N(N-1)} \sum_{i=1}^{N} \sum_{j=i+1}^{N} \Theta(\epsilon - \|\mathbf{x}_i - \mathbf{x}_j\|), \qquad (6.1)$$

where Θ is the Heaviside step function, $\Theta(x) = 0$ if $x \leq 0$ and $\Theta(x) = 1$ for $x > 0$. The sum just counts the pairs $(\mathbf{x}_i, \mathbf{x}_j)$ whose distance is smaller then ϵ. In the limit of an infinite amount of data ($N \to \infty$) and for small ϵ, we expect C to scale like a power law, $C(\epsilon) \propto \epsilon^D$, and we can define the correlation dimension D by

$$d(N, \epsilon) = \frac{\partial \ln C(\epsilon, N)}{\partial \ln \epsilon},$$

$$D = \lim_{\epsilon \to 0} \lim_{N \to \infty} d(N, \epsilon). \qquad (6.2)$$

We can easily verify that this definition yields the correct integral dimensions for regular geometrical objects. (See also Exercises 6.1 and 6.2.) It is equally obvious that the two limits we have to take will get us into trouble whenever we have a finite sample instead of a full distribution: N is limited by the sample size, and the range of meaningful choices for ϵ is limited from below by the finite accuracy of the data and by the inevitable lack of near neighbours at small length scales.

Before we give details of how to estimate the correlation sum and the dimension from time series data, let us mention some general considerations that we feel are important. More specific questions, in particular about data requirements, will be given below.

Dimensions, Lyapunov exponents, etc., are ways of quantifying the properties of a signal. Each of these concepts is one of very many ways of turning a sequence of data into a single number (or a few of them). What is so special about Lyapunov exponents or dimensions that we choose these and not any other concept? This choice is guided by the hope that the result enhances our knowledge about the *underlying system* rather than simply compressing many measurements into a number. This cannot be achieved with most statistics that we can define *ad hoc*.

A quantity which can serve this purpose must not depend significantly on the measurement procedure, coordinates chosen, etc. According to their definition, dimensions and a number of other quantities fulfil this requirement of *invariance*. If we want to communicate whether a geometrical object is a surface or a volume by giving its dimension, this number must not depend on the details of the measurement, the resolution of the data, etc. We cannot relax this fundamental requirement for fractal objects just because they are less intuitive. It is now important to remark that while the correlation *dimension* is invariant, the correlation *sum* at a given scale ϵ_0 is not. Thus we cannot simply take $d(N, \epsilon_0)$ as an estimate for D. If we go as far as to establish a value D (with some error bars) as the correlation dimension of a system, we must make sure that this value is robust under reasonable variations of the measurement technique and data processing.

As a conceptual remark, let us stress that in the first place, the correlation dimension is a tool to *quantify* self-similarity when it is *known* to be present. When applied for this purpose it provides a safe and stable algorithm. The concept is much less suited and has to be used with much more caution if self-similarity still has to be *established*, that is, if it is uncertain if the data are low dimensional deterministic data at all.

6.3   Correlation sum from a time series

The above definitions of the correlation sum and dimension involve phase space vectors as the locations of points on an attractor. Thus, given a scalar time series, we first have to reconstruct an auxiliary phase space by an embedding procedure. Most popular and fully satisfactory for this purpose is the delay embedding technique which we used on several occasions before (it was introduced in Section 3.2). Technically, the reconstruction procedure involves the choice of a delay time τ. While absurd choices of τ render the correlation algorithm useless we have to insist that the results are invariant under *reasonable* changes to the embedding procedure. How to make such reasonable choices is discussed in Section 3.3. A value for the delay time which yields a convincing phase portrait should do for the correlation sum as well.

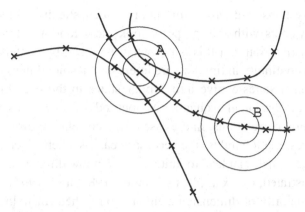

Figure 6.2 A typical situation for the computation of C for a flow. While for point A there are still some neighbouring points lying on dynamically uncorrelated parts of the data (in this enlargement shown as different trajectories), all neighbours of point B are direct images and pre images of B and thus simulate a dimension of 1. To avoid the resulting incorrect estimate of C one has to ignore all neighbours over time (i.e. $|i - j| < n_{min}$ in Eq. (6.1), where $t_{min} = n_{min}\Delta t$ is some correlation time).

Once the embedding vectors $\mathbf{s}_n$ are reconstructed, the estimation of the correlation dimension is done in two steps. First one has to determine the correlation sum $C(m, \epsilon)$, Eq. (6.1), for the range of ϵ available and for several embedding dimensions m. Then we inspect $C(m, \epsilon)$ for the signatures of self-similarity. If these signatures are convincing enough, we can compute a value for the dimension. Both steps require some care in order to avoid wrong or misleading results.

We want to interpret the data (in the form of embedding vectors) as a random sample drawn from an underlying probability distribution, the *natural measure* on the attractor in question. It is the self-similar geometry of this distribution that we want to quantify. In order to estimate the correlation sum of the underlying distribution we have to extrapolate information obtained from the finite sample. The straightforward estimator, Eq. (6.1), turns out to be biased towards too small dimensions when the pairs entering the sum are not statistically independent. For time series data with nonzero autocorrelations, independence cannot be assumed. For instance, the embedding vectors at successive times are often also close in phase space due to the continuous time evolution. We call correlations of this kind *temporal*. See Fig. 6.2 for an illustration of this problem in a deterministic flow.

Let us dwell a little more on the distinction between temporal and geometrical correlations. The most important temporal correlations are caused by the fact that data close in time are also close in space, a fact which is not only true for purely deterministic systems but also for many stochastically driven processes. Superficially, systems with this property look "less random" than independent draws from a distribution, but this is not what we mean by "deterministic" here. Rather than this continuity in time, we are interested in smoothness in phase space, that similar present states evolve into similar states in the near future. Unfortunately, not only the eye but also quantitative methods of time series analysis can be biased by temporal correlations. In the case of the correlation dimension they lead to serious underestimation unless the necessary care is taken. Even infinite dimensional, stochastic signals can lead to finite- and even low dimensional estimates, as has been demonstrated, for example, by Theiler (1986) and (1991). It is more than likely that the majority of dimension estimates published for field measurements are seriously too low because they mistake temporal coherence for geometrical structure. We will give some examples for spurious correlation sums later.

As serious as the problem of temporal correlations is for the estimation of the correlation sum, the remedy is simple [Theiler (1986)]. In Eq. (6.1) we have to exclude those pairs of points which are close, not because of the attractor geometry but just because they are correlated. Usually, this is the case when they are close in time. Thus the second sum is started only after a typical correlation time $t_{\min} = n_{\min}\Delta t$ has elapsed:

$$C(m, \epsilon) = \frac{2}{(N - n_{\min})(N - n_{\min} - 1)} \sum_{i=1}^{N} \sum_{j=i+1+n_{\min}}^{N} \Theta(\epsilon - \|\mathbf{s}_i - \mathbf{s}_j\|). \quad (6.3)$$

We have here replaced the state space vectors by delay embedding vectors $\mathbf{s}_n$, where their dimensionality m enters as a parameter. Note that $t_{\min}$ is not necessarily equal to the *average* correlation time, defined, for example, by the time at which the autocorrelation function has decayed to $1/e$. Since looking for close pairs means looking for rare events, correlations between these need not make themselves felt strongly in averaged quantities. Thus we should be generous with the choice of $t_{\min}$. Of course, we also lose a few "real" neighbours, but the loss of statistics is not dramatic; it is about a fraction $2n_{\min}/N$ of all pairs. We will study the artefacts which occur when temporal correlations are not taken care of in more detail in Section 6.5, where also a formal procedure for choosing $t_{\min}$ is described. Before we learn how to evaluate correlation sums correctly let us show a typical example of a data set probably representing a fractal attractor.

Example 6.2 (Correlation sum of NMR laser data). Let us consider the NMR laser data that we have met on several occasions before. (See Appendix B.2 for a summary.) After inspecting the data (in the stroboscopic sampling) visually for obvious non-stationarities, looking at phase portraits with different delays and at different portions of the data available, we venture to compute the correlation sum,

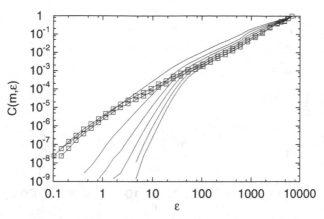

Figure 6.3 Correlation integral for NMR laser data. $C(m, \epsilon)$ is plotted versus ϵ. A double logarithmic plot was chosen since we are looking for power law behaviour. The different curves were obtained with embedding dimensions $m = 2$ (uppermost curve) to $m = 7$ (lowest curve). In the range $100 < \epsilon < 1000$ we can indeed find reasonably straight lines as an indicator of self-similar geometry. We also plot the corresponding curves (squares) for $m = 6, 7$ after the data has been cleaned by the local projective noise reduction scheme, Section 10.3.2. See Example 10.3 for details of the cleaning. Now the scaling range is extended at least down to $\epsilon = 1$.

Eq. (6.3), for this data set. We will only interpret the result subject to further consistency checks described later in this chapter. For now we have to choose a delay time for the embedding, a range of interesting embedding dimensions and a correlation time t_{min} in order to discard temporal neighbours which could affect the result adversely. For this map-like data set, there is no reason to choose a delay time different from 1. Let us compute $C(m, \epsilon)$ in 2–7 dimensional embeddings. If seven turns out to be insufficient we can repeat the computation for higher values. Most likely we are on the safe side when we discard all pairs closer than 500 steps in time (see Example 6.6 below). The correlation times of the data are much shorter but we can easily afford the resulting loss of 2.5% of the pairs for statistical purposes.

Fig. 6.3 shows the correlation sums $C(m, \epsilon)$ obtained with these choices. As is typical for low dimensional deterministic experimental data, we find something like a power law for $C(m, \epsilon)$, only within a small range of length scales ϵ, here in the region $100 < \epsilon < 1000$. The power law behaviour of $C(m, \epsilon)$ as the signature of self-similarity can best be found by plotting the slope $D(m, \epsilon)$ of a double logarithmic plot of $C(m, \epsilon)$ versus ϵ, the latter still on a logarithmic scale. This has been done in Fig. 6.4. This representation shows much more clearly the *plateau* of $D(m, \epsilon) = \partial \ln C(m, \epsilon) / \partial \ln \epsilon$ which corresponds to the desired power law for C. We can easily find that the plateau value for D does not change much with the embedding dimension m as soon as $m > 2$, but there are still some fluctuations

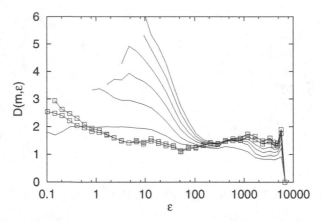

Figure 6.4 Local slopes of the correlation integral (NMR laser data) shown in Fig. 6.3. In this representation the scaling behaviour and also deviations from it are more clearly visible. Again, the different curves represent $m = 2, \ldots, 7$, now counted from below. Using the curves after nonlinear noise reduction (squares) we would read off an estimate of $D = 1.5 \pm 0.1$.

present. However, the estimated curve $D(m, \epsilon)$ is characteristic enough to suggest that the data are a sample taken from a strange attractor of dimension $D < 2$. More convincing evidence will be presented once we have applied nonlinear noise reduction to the data, see Example 10.3. □

6.4 Interpretation and pitfalls

You may have noted that we have largely avoided talking about the correlation *dimension* of a data set so far. The reason is that from our point of view the quantity which can be computed with a more or less automated scheme is the correlation *sum* $C(m, \epsilon)$, whereas a *dimension* may be assigned only as the result of a careful *interpretation* of these curves and any other information available. While the correlation sum is well defined for every finite set of points, the dimension of this finite set is zero and can have a nontrivial value only for an assumed underlying measure. Thus a dimension estimate involves a nontrivial extrapolation from a finite data set to a supposedly existing attractor.

In this section we will explain the main features that we can expect to find in a typical correlation sum for a sample drawn from a strange attractor. This knowledge will help us to find a region of length scales characterised by self-similarity – if such a region is present.

Let us have a look at a typical output of the correlation algorithm for an experimental sample of a low dimensional dynamical system. In particular, we want to study the local slopes $D(m, \epsilon)$ which are plotted for the NMR laser data,

Example 6.2, in Fig. 6.4. Remember that the different curves represent different embedding dimensions m. One can clearly distinguish four different regions that we now want to discuss in order ranging from large to small scales.

In the *macroscopic regime*, scaling or self-similarity is obviously destroyed by the cutoff introduced by the finite size of the attractor. For large ϵ (here about $\epsilon > 1000$), the correlation sum does not exhibit scaling since the macroscopic structures of the attractor determine its value, and its local scaling exponent is both m and ϵ dependent.

On smaller length scales, a true *scaling range* may be found. For a self-similar object the local scaling exponent is constant and the same is true for all embedding dimensions larger than $m_{min} > D$.[1] If this plateau is convincing enough, this scaling exponent can be used as an estimate of the correlation dimension of the fractal set. In Fig. 6.4, the curves are relatively flat for $100 < \epsilon < 3000$ but the value of D increases with m for $\epsilon > 1000$. Thus we can speak of a *plateau* at most for $100 < \epsilon < 1000$.

At smaller scales we can usually discern a *noise regime*. If the data are noisy, then below a length scale of a few multiples of the noise level the method detects the fact that the data points are not confined to the fractal structure but are smeared out over the whole available state space. Thus the local scaling exponents increase and at the noise level they reach the value $D(m, \epsilon) = m$, the embedding dimension. The behaviour is different for pure discretisation error. Then the reconstructed phase space vectors are confined to a regular grid in m dimensional space and thus appear to be zero dimensional when regarded on length scales smaller than the resolution of the data. Above these length scales, very awkward fluctuations will occur. We recommend the addition of artificial white noise to coarsely discretised data before passing it to the correlation algorithm [Möller *et al.* (1989), Theiler (1990a)]. We recommend even more strongly that part of the information lost in the discretisation be restored by processing the data with a nonlinear noise reduction algorithm. See Example 6.4 below.

Further down in scale, we will face a *lack of neighbours* within the ϵ-neighbourhoods. Statistical errors become the dominant effect and the curves start to fluctuate tremendously. This scale is smaller than the average inter-point distance in embedding space (which is $\approx N^{-1/m}$ in the noise regime) and is characterised by the fact that, say, less than 50 pairs contribute to $C(m, \epsilon)$.

Since this is a textbook, the data shown in Fig. 6.4 have to be regarded as comprising a textbook example. With different data you will encounter deviations from the scheme outlined above. Let us first discuss those deviations which have to be

---

[1] Instead of $m > 2D_F$ as in the embedding theorem, $m > D_F$ is sufficient for correlation dimension estimates, since self-intersection of the reconstructed set with itself have measure zero and hence do not distort this particular statistics [Sauer & Yorke (1993)].

expected *even if* the signal underlying the data set is low dimensional deterministic. The extremely confusing thing about dimension estimates (not just of the
correlation algorithm alone) is that there are data sets which are *not* deterministic
but which roughly resemble low dimensional data, since the data violate certain
requirements. Fortunately, with sufficient care one can fully avoid to be misled by
such problems, as we will show at some other point.

A low dimensional deterministic signal, measured with a relative error of more
than about 2–3% will not show a significant scaling region of $C(m, \epsilon)$ (nor a plateau
of $D(m, \epsilon)$) since the artefacts of the noise meet the artefacts of the overall shape
of the attractor. As we will see in Section 6.7, the correlation sum is affected by
the noise up to length scales which are at least three times the noise level. Even in
optimistic cases, scaling cannot be expected above about one-fifth of the attractor
extent, and one octave is not sufficient for a useful scaling region.

Example 6.3 (Correlation sum of Taylor–Couette flow). When observing some
time varying system, the minimal achievable noise level depends on several factors.
Obviously, an electronic circuit or a laser system is easier to control than, for
instance, turbulent flow in a tank. Thus the data in this example, though more noisy

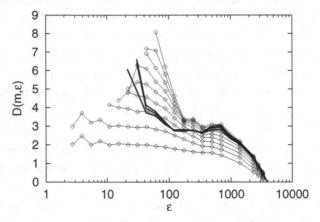

Figure 6.5 Local slopes $D(m, \epsilon)$ of the correlation sum for Taylor–Couette flow
data. Embedding is done with a delay of five time steps in $m = 2, \ldots, 10$ dimensions (diamonds, from bottom to top). From the experimental set-up, this data
set is expected to be low dimensional. With this prejudice one may establish a
small scaling region around $\epsilon \approx 300$. The higher noise level (about 5% of the total
amplitude) affects the correlation sum up to $\epsilon \approx 200$. The attractor dimension is
probably about three, which makes an estimate harder than for the NMR laser
data, for example. We also show the results for $m = 8, 9, 10$ after nonlinear noise
reduction (thick lines without symbols). The noise reduction method will be described in Section 10.3.2. The plateau which is visible when the effects of noise are
pushed towards smaller length scales suggests a dimension slightly below three,
but three is not completely ruled out.

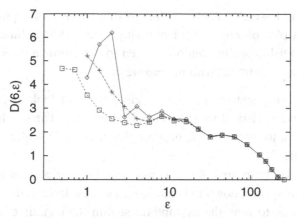

Figure 6.6 Local slopes of the correlation sums for the far infrared laser data, Appendix B.1. We show curves for embedding dimension 6 with delay 5. Diamonds: unprocessed 8-bit data. Crosses: off-grid data with uniform noise of the size of the resolution added. Squares: data after simple nonlinear noise reduction.

than, for example, the NMR laser data we used previously, must be regarded as a data set of exceptionally high quality. The physical phenomenon observed here is the motion of a viscous fluid between coaxial cylinders, called *Taylor–Couette flow*. The data was provided by Thorsten Buzug and Gerd Pfister; see Buzug *et al.* (1990a) and Pfister *et al.* (1992), Appendix B.4. Due to the higher noise level and the apparent higher dimension of the data set, only a small scaling region may be discerned in Fig. 6.5. □

Example 6.4 (Correlation sum of far infrared laser data). In Example 4.5 we performed nonlinear noise reduction on the far infrared laser data, Appendix B.1. The resolution of this data set is predominantly limited by the 8-bit discretisation, while other measurement errors seem to be rather small. This is a good example for studying the effect of coarse-grained measurements on the correlation sum, which we do in Fig. 6.6, where we show $D(m, \epsilon)$ for a six dimensional embedding. The discretisation makes the effective scaling exponent $D(m, \epsilon)$ go to zero at length scales smaller than unity (diamonds). Obviously, a grid renders phase space discrete and thus zero dimensional. But we observe artefacts up to much higher length scales, say for $\epsilon < 10$. A quick fix is to shake the points off the grid by adding uniform uncorrelated random numbers in the interval $[-0.5, 0.5]$ (crosses). Still the errors are of the same order in size but have much nicer properties and lead to a smoother curve for $D(m, \epsilon)$. A better remedy is to pass the data through a nonlinear noise reduction algorithm. This makes sense since the raw correlation sum suggests that the data might at least be effectively described by a low dimension. The squares in Fig. 6.6 indicate the correlation sum obtained after the simple noise reduction method of Section 4.5 has been applied (see Example 4.5, in particular Fig. 4.6).

The noise level is lowered visibly. Although this curve looks quite good, please remember that a plot of $D(m, \epsilon)$ for one single embedding dimension m is not sufficient to establish a scaling region or even to determine a dimension. We also have to verify that d saturates with increasing m.   □

If the data are very sparse, the noise regime will not look as characteristic as in Fig. 6.4 because statistical fluctuations will dominate. The resulting correlation sums are very hard to interpret. We propose that a credible scaling region must be clearly discernible.

Even very good deterministic data can lack a scaling region because macroscopic structures prevent scaling down to small length scales. In fact, the theory does not yield any notion as to how the asymptotic scaling behaviour $C(m, \epsilon) \propto \epsilon^D$, cf. Eq. (6.2), has to be approximated at finite length scales. A prominent example is biperiodic data which forms a *torus* in phase space. A torus is a piece of a tube which is closed to form a ring. No matter how large the ring is, the proper scaling with dimension two (of the surface of the tube) cannot be found above about half the diameter of the tube. A thin tube will even appear one dimensional on intermediate length scales. (See also Exercise 6.5.)[2]

From the preceding discussion it should be immediately obvious that a scaling range will always have to be established by inspection. Even data sets from the same experiment can vary so strongly in their geometrical structure that we cannot recommend the use of any automated procedure to determine the scaling exponent D as an estimate for the correlation dimension.

So far we have considered only the case where the data are indeed a sample from a low dimensional attractor and have asked under what circumstances we can hope to estimate its dimension from the correlation sum. Of course, we usually do not know beforehand if deterministic chaos is at work in the system. Strictly speaking, we are then using the algorithm, in fact even the delay embedding, outside its natural habitat. The hope is, however, to find signatures we recognise as typical of deterministic systems and thus justify the approach *a posteriori* – a hazardous concept. Unfortunately, there are examples of data which are *not* deterministic but which can nevertheless lead to something which resembles a *plateau* in the scaling curve $D(m, \epsilon)$ – if one is not careful enough. The next section will teach you how to be careful enough. Unconcerned people find straight lines in virtually every log-log plot. Some authors do not even look for scaling but *define* a certain range of length scales to be the "scaling region". With this method you always get a number you

---

[2] The following example is inspired by a remark which, according to Theiler (private communication), goes back to Grassberger (private communication): How many dimensional is a plate of spaghetti? Zero when seen from a long distance, two on the scale of the plate, one on the scale of the individual noodles and three inside a noodle. Maccaroni is even worse.

can call the *dimension* but this number is pretty meaningless. In particular, a small number does not imply determinism or chaos. Of course, you should not follow this bad practice, but we would go one step further and give the advice not to believe *any* dimension estimate you find published without a convincing scaling plot being shown!

6.5 Temporal correlations, non-stationarity, and space time separation plots

Soon after its publication in 1983 the correlation dimension became quite popular and since then many papers have reported finite or even low dimensions for a vast number of systems. [Examples using field data include Nicolis & Nicolis (1984), Fraedrich (1986), Essex *et al.* (1987), Kurths & Herzel (1987), Tsonis & Elsner (1988), and Babloyantz (1989).] Usually, these claims were made whenever the authors succeeded in fitting a straight line to a portion of the log-log plot of the correlation sum $C(m, \epsilon)$. Some authors failed to observe that the curves they were fitting with straight lines were actually not quite straight but rather more or less S-shaped. Only later did several authors find examples of nondeterministic data sets which would yield correlation sums with almost straight portions which would allow for the interpretation that $C(m, \epsilon)$ follows a power law; see Theiler (1986) and (1991), Osborne *et al.* (1986) and Osborne & Provenzale (1989). This seemed to be in contradiction with the fact that stochastic data are of infinite dimension.

Closer examination shows that all of these examples, and in fact many of the dimension estimates published for time series data, suffer from the same problem: *temporal correlations*. The correlation sum is a particular estimate of an abstract quantity, the *correlation integral*, which is a functional on the underlying invariant measure. The estimate is unbiased only if the finite sample which we use is an uncorrelated sample of points in the m dimensional space according to the invariant measure in this space. Data from a time series are dynamically correlated and clearly violate this requirement. However, if the violation is very weak, such as for data from a strongly chaotic map, its effect will be almost invisible. If the correlations are strong, they can spoil everything. As we pointed out previously, for many processes such as highly sampled chaotic flows, pairs of points which were measured within a short time span tend to be close in phase space as well and thus introduce a bias when we estimate the correlation sum from its definition, Eq. (6.1). Once we know the typical time over which data items are correlated, the problem is easily solvable by rejecting the pairs that are close in time, as described in Section 6.3 above; in particular compare Eq. (6.1) with Eq. (6.3). When you find a paper where the

6. Self-similarity: dimensions

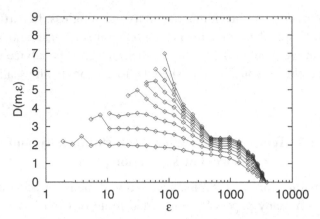

Figure 6.7 Spuriously low "plateau" in $D(m, \epsilon)$ for Taylor–Couette flow data when temporal correlations are not accounted for. Compare the apparent scaling region with the correct curve shown in Fig. 6.5. The present figure suggests a dimension of about 2.4, which is much too low for this data set. Discarding points which are less then 20 time steps apart would have given the correct result (see Example 6.7 below).

fractal dimension of chaotic flow data is close to but smaller than two (which is a contradiction), then most surely points with temporal proximity have not been excluded from the correlation sum.

Example 6.5 (Improper calculation of correlation sum of Taylor–Couette flow data). Temporal correlations are present in almost every data set. Although their effect on the correlation sum is most severe for short and densely sampled data, we should never fail to take the precaution of discarding temporal neighbours. As a warning let us repeat the calculation of the correlation sum for the Taylor–Couette flow data (Example 6.3; see also Appendix B.4). We use the first 10 000 data points and the original formula, Eq. (6.1), which only excludes the self-pairs from the sum (Fig. 6.7).

The question we want to address in this section is how to detect these temporal correlations and how to estimate a safe value of the correlation time t_{min}. This problem has been solved by Provenzale *et al.* (1992) by introducing the *space time separation plot*. The idea is that in the presence of temporal correlations the probability that a given pair of points has a distance smaller than ϵ does not only depend on ϵ but also on the time that has elapsed between the two measurements. This dependence can be detected by plotting the number of pairs as a function of two variables, the time separation Δt and the spatial distance ϵ. You can either make a scatter plot or draw a surface or contour lines with your favourite plotting program. Technically, you create for each time separation Δt an accumulated histogram of

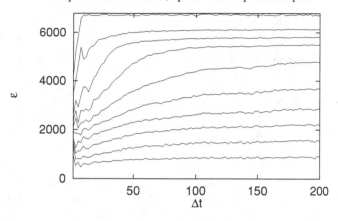

Figure 6.8 Space time separation plot for NMR laser data. Contour lines are shown at the spatial distance ϵ where for a given temporal separation Δt a fraction of $1/10, 2/10, \ldots$ (lines from below) of pairs are found. Observe the saturation above $\Delta t \approx 100$.

spatial distances ϵ. Remember to normalise properly. You can use the program `stp` from the TISEAN package, see Section A.3.

Example 6.6 (Space time separation plot for NMR laser data). In Fig. 6.8 we show a space time separation plot for the NMR laser data (Appendix B.2). You can read the plot as follows: the contour lines shown indicate the distance you have to go to find a given fraction of pairs, depending on their temporal separation Δt. Only for values of Δt where the contour lines are flat (at least on average, they may oscillate), does temporal correlation not cause artefacts in the correlation sum. For the NMR laser data this is the case for about $\Delta t > 100$. Thus we were indeed safe in choosing $t_{\min} = 500$ when computing the correlation sum in Example 6.2. □

Example 6.7 (Space time separation plot for Taylor–Couette flow). The space time separation plot for the Taylor–Couette flow data (Appendix B.4) looks slightly more complicated. The dominant frequency in the system is also reflected in the level curves shown in Fig. 6.9. While the oscillation is harmless as long as the observation period is much longer than a cycle length, we must be concerned about the first 20–40 time steps, where the level lines increase consistently. If we do not discard at least those pairs which are less than 20 time steps apart, then we obtain the spurious correlation sum shown in Fig. 6.7. To be safe we usually choose $t_{\min} = 500$ or so – the loss in statistics is marginal. □

All the examples of data sets which exhibit spurious scaling of $C(m, \epsilon)$ (see references above) could not have been mistaken for low dimensional chaos if the temporal correlations had been handled properly by excluding pairs that are close in time from the correlation sum. The space time separation plot tells us how many

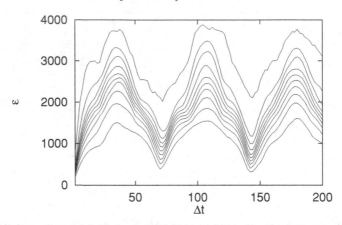

Figure 6.9 Space time separation plot for Taylor–Couette data. Contour lines are again shown at 1/10, 2/10, ... Initially ($\Delta t < 20$) the curves increase fast, later we find stable oscillations with the dominant period of the system.

pairs have to be excluded. Temporal correlations are present as long as the contour curves do not saturate. But what if the curves never saturate at all? This is indeed what happens with data with significant power in the low frequencies, such as $1/f$ noise or Brownian motion. In this case, *all* points in the data set are temporally correlated and there is no way of determining an *attractor* dimension from the sample. A similar situation arises if the data set is too short. Then there are almost no pairs left after removing temporally correlated pairs.

If we regard the problem from a different point of view, correlation times of the order of the length of the sample (nonsaturating contour curves) mean that the data does not sample the observed phenomenon sufficiently. If you want to find out which TV programme is broadcast every day it is not enough to watch TV for one day. This deficiency of the data can also be seen as a kind of non-stationarity of the measurement.

It is very common in science that methods are only applicable if some requirements are fulfilled. The correlation dimension is no exception. If done properly it never gives *wrong* results (such as finite dimensions) for $1/f$ noise, etc. It does not give any result at all! When we use the space time separation plot we get a warning when we cannot apply the algorithm.

Nevertheless, you may think that a spurious finite dimension of a fully correlated data set might be characteristic for the process underlying the data. In fact, this is correct for certain *non-recurrent processes* such as free diffusion: the cluster in space which is formed by a typical Brownian path in two or more dimensions, has a Hausdorff dimension of two [Mandelbrot (1985)]. This value is in fact obtained in a numerical calculation of the correlation dimension on suitable length scales (depending on the length of the time series), if one uses $n_{min} = 1$ in Eq. (6.3). A

Brownian path is a rather strange fractal, since once embedded in an m-dimensional space, it is not restricted to any lower dimensional manifold (in particular, not to a two-dimensional one, as one might expect from its dimensionality). Generalisations of diffusion processes (*anomalous diffusion*) yield geometries with similar properties but different dimension values. The lack of recurrence expresses itself also in the power spectrum, namely by a diverging power in the low frequencies. For certain coloured noise processes with an $f^{-\alpha}$-spectrum with $1 < \alpha < 3$, the (correlation) dimension will assume the finite value $D_2 = 2/(\alpha - 1)$ [Osborne & Provenzale (1989), Theiler (1991)], if it is used to characterise a single path rather than the underlying smooth probability distribution. A more suitable characterisation of such processes is the analysis in terms of *Hurst exponents*, as it is discussed in Section 6.8.

6.6 Practical considerations

Obviously, we will have to use a computer to obtain the correlation sum in practice. (See, however, Exercise 6.2.) As usual, there are smart ways and not so smart ways to implement Eq. (6.3). Only when you have access to a really fast computer can you afford to use the not so smart solution and implement the formula as two nested loops without thinking further. It should be clear by now that you need quite a number of points in order to estimate $C(m, \epsilon)$ over a large enough range of length scales. A few hundred are definitely not enough to yield a statistically significant result at the small length scales. For N data points at hand, the double sum in the definition of $C(m, \epsilon)$ contains about $N^2/2$ terms, which makes it very CPU time consuming to compute $C(m, \epsilon)$ for all data sets except those which are too small anyway.

Since the number of close pairs determines the smallest accessible length scale, we cannot usually waste part of the precious data by limiting the range of one or both of the sums (which is, however, what some people do). It is more suitable to observe that it is *only* on the smallest length scales that we need the maximal possible statistics. For larger distances we can easily find enough pairs. This observation has led several authors to the idea of using a fast neighbour search method in order to find *all* close pairs within the data set and to apply the naive algorithm for fewer data for the larger length scales. This is not the right place to discuss the different possibilities for finding close neighbours efficiently, a problem which occurs quite frequently in nonlinear time series analysis. We will give some pointers to the literature, and describe the method used in the TISEAN package in Section A.1.1.

Figs. 6.3 and 6.4 were obtained using this algorithm, in fact using the routine d2 of the TISEAN package, see Section A.6. We computed the correlation sum $C(m, \epsilon)$ for two values of ϵ in each octave, that is, adjacent values differed by a factor of $\sqrt{2}$. While it should be immediately clear how a diagram such as Fig. 6.3 is obtained, we also have to take numerical derivatives for Fig. 6.4. We suppress statistical fluctuations by not taking finite differences between adjacent values of ϵ but by estimating instead

$$D(m, \epsilon) \approx \frac{\ln C(m, 2\epsilon) - \ln C(m, \epsilon/2)}{\ln 2\epsilon - \ln \epsilon/2} = \frac{1}{2} \log_2 \frac{C(m, 2\epsilon)}{C(m, \epsilon/2)}. \qquad (6.4)$$

Let us remark that the algorithm behaves somewhat awkwardly when the data set contains many *identical* pairs, a situation which can only occur if the data are discretised very coarsely. The program does not provide a check for this since raw data of this kind are not suitable for dimension calculations anyway. Of course, formally, points on a discrete lattice in space form again a zero dimensional set, even if they were infinitely many. Hence we have again the problem of the proper extrapolation of the correct dimension from the larger scales. In order to be able to do so, we have to minimise the effects of the artefacts on the correlation sum on larger length scales. We could, e.g., process such data by a nonlinear noise reduction algorithm before the correlation sum is estimated. If this is not possible, add white noise of a magnitude at least comparable to the resolution to the data.

6.7 A useful application: determination of the noise level using the correlation integral

The correlation algorithm is a tool for analysing the scaling properties of point sets in phase space. Attractors of deterministic systems show power law scaling of $C(m, \epsilon)$. It is characteristic for such systems that the exponent in the power law is invariant under coordinate transformations and in particular does not depend on the embedding dimension m once m is large enough to ensure a proper reconstruction. For stochastic systems the situation is different. Delay vectors formed from a random signal are not restricted to a low dimensional manifold but fill all available directions in phase space. Apart from edge effects, the scaling exponent is always equal to m.[3]

Now consider the typical case of a deterministic signal which is measured only with some finite resolution due to measurement errors and digitisation of the output. Way above the noise level the errors can be neglected and the proper scaling exponent $D(m, \epsilon) \approx d$ may be observed within some scaling range. At small length scales comparable to the amplitude of the random errors, this scaling is broken and

---

[3] If temporal correlations decay sufficiently fast, see last section.

the typical behaviour for stochastic data, $D(m, \epsilon) = m$, dominates instead. This qualitative difference led us to the distinction of a *scaling range* and a *noise range* in Section 6.4 above. (See also Example 6.2 and the diagrams there.)

In this section we will study the consequences for the curves $D(m, \epsilon)$ in more detail which will enable us to measure the strength of the noise (the average measurement error) present in an otherwise deterministic data set. Such an estimate will be very useful in the interpretation of results of nonlinear methods of analysis. We will be able to measure the success of nonlinear noise reduction, we will be able to understand better the limits of predictability of the data set and we can cross-check with what we learned about the noise in the data set when we studied the divergence of trajectories in Section 5.3.

Let us compare the correlation sums $C(m, \epsilon)$ and $C(m + 1, \epsilon)$ obtained with embeddings in m and in $m + 1$ dimensions respectively. Let m dimensions be enough for a faithful reconstruction. If the data were completely deterministic without noise, the two curves would differ only by a factor which is constant throughout the scaling range. The constant is determined by a certain average of the local expansion rate, related to the correlation entropy (Section 11.4). Points in m dimensional space are also close in $m + 1$ dimensional space – in fact we could predict the $m + 1$st coordinate from the m dimensional embedding vectors. In reality, all coordinates, and in particular the $m + 1$st one, are contaminated with some noise. Therefore, below the noise level, the correlation sum in $m + 1$ dimension feels the presence of one space dimension more. The following intuitive result for the ϵ dependence of the correlation sum can be derived by a less handwaving argument for which we refer the interested reader to Schreiber (1993a). It reads:

$$C(m + 1, \epsilon) \propto C(m, \epsilon)\, C_{\text{noise}}(\epsilon), \qquad (6.5)$$

where $C_{\text{noise}}(\epsilon)$ is the correlation sum of the one dimensional distribution of the noise, provided the autocorrelation time of the noise is shorter than the embedding delay time τ. If the noise is drawn from a continuous distribution $\mu(\eta)$ then its correlation sum becomes an integral:

$$C_{\text{noise}}(\epsilon) = \int_{-\infty}^{\infty} d\eta\, \mu(\eta) \int_{\eta-\epsilon}^{\eta+\epsilon} d\eta'\, \mu(\eta'), \qquad (6.6)$$

which can be evaluated when the noise distribution $\mu(\eta)$ is known. Most commonly we find Gaussian noise with $\mu(\eta) \propto \exp(-\eta^2/2\sigma^2)$, where σ is amplitude or absolute noise level. For such noise Eq. (6.6) yields $C_{\text{noise}}(\epsilon) \propto \text{erf}(\epsilon/2\sigma)$.[4] We can also calculate $D(m + 1, \epsilon) = \partial \ln C(m + 1, \epsilon)/\partial \ln \epsilon$ and find that $D(m + 1, \epsilon) =$

---

[4] $\text{erf}(x) = \int_{-\infty}^{x} e^{-y^2/2} dy$ is the *error function*. Its values are listed in tables and in software libraries.

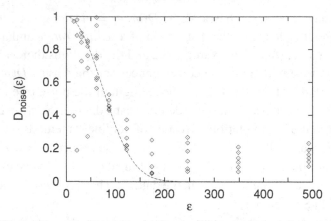

Figure 6.10 Increment in $D(m, \epsilon)$ with increasing m for Taylor–Couette flow data. The values $D(m + 1, \epsilon) - D(m, \epsilon)$ are shown for $m = 5, \dots, 10$, together with the best fitting curve $D_{\text{Gauss}}(\epsilon)$ with $\sigma = 41$. Although there are deviations from the theoretical shape we can estimate the noise level to be about 40–45 experimental units, which is about 5% of the total amplitude of the time series.

$D(m, \epsilon) + D_{\text{noise}}(\epsilon)$. For Gaussian noise of amplitude σ we obtain

$$D_{\text{Gauss}}(\epsilon) = \frac{\epsilon \exp(-\epsilon^2/4\sigma^2)}{\sigma \sqrt{\pi} \, \text{erf}(\epsilon/2\sigma)} ; \tag{6.7}$$

for uniform noise in the range $-\sigma/2 < \eta < \sigma/2$ the result is

$$D_{\text{uniform}}(\epsilon) = \frac{2 - 2\epsilon/\sigma}{2 - \epsilon/\sigma} \Theta(\sigma - \epsilon) . \tag{6.8}$$

However, any filtering and any embedding procedure except ordinary delay coordinates will turn most noises closer to Gaussian distributions via the central limit theorem. Indeed, we usually find that when m delay coordinates yield a proper embedding then the difference of the subsequent curves $D(m + 1, \epsilon)$, $D(m + 2, \epsilon)$, $\dots$ is well described by Eq. (6.7) when we choose σ appropriately.

Let us just remark again that discretisation error is *not* equivalent to uniform white noise. It induces spurious geometric correlations and lets the signal appear to be zero dimensional at small length scales.

Example 6.8 (Noise level of Taylor–Couette flow data). As an example let us plot the data in Fig. 6.5, the local slopes $D(m, \epsilon)$ of the correlation sum for Taylor–Couette flow data, Appendix B.4, in a different way. Instead of $D(m, \epsilon)$, $m = 2, \dots, 10$ we show in shown Fig. 6.10 $D(m + 1, \epsilon) - D(m, \epsilon)$, $m = 5, \dots, 10$ (for smaller values of m we do not have a good embedding). Further we concentrate on the crossover region from the noise regime to the scaling range using a linear scale in ϵ. Then we try different guesses of the noise level σ and compare the data to the corresponding curves $D_{\text{Gauss}}(\epsilon)$ given by Eq. (6.7). □

In the above example we used a visual fit to extract an estimate for the noise level σ. We can of course replace this visual fit by a least squares scheme. We find the visual method good enough since with real data we have to expect considerable deviations from the expected shape anyway due to non-Gaussian parts of the noise. But even when such deviations are present we can often obtain useful information about the noise in the data set.

Above, we studied the effect of noise on the correlation sum quantitatively with the result that the deviation of the observed scaling function $D(m, \epsilon)$ from the true one is given by the formula $D(m + 1, \epsilon) = D(m, \epsilon) + D_{\text{noise}}(\epsilon)$, or

$$D(m, \epsilon) = D(m_0, \epsilon) + (m - m_0)D_{\text{noise}}(\epsilon), \qquad (6.9)$$

provided m_0 is sufficient for a proper embedding of the data. For most kinds of noise, the shape of $D_{\text{noise}}(\epsilon)$ is a function similar to the one given by Eq. (6.7). Other references on the effect of noise on the correlation integral are Diks (1996) and Oltmans & Verheijen (1997).

An important consequence of this analytical result for the shape of the correlation integral is that a small amount of noise already conceals possible scaling behaviour: even at $\epsilon = 3\sigma$ the effective dimension increases visibly with m beyond m_0, namely, by an amount of 0.2 per additional dimension. This means that, since even in the best case scaling can be expected only up to about one-fourth of the attractor extent (compare the examples shown in this book) and down to three times the noise level, a data set with 2% noise can give at most a tiny scaling region of two octaves. Note that this is a very optimistic view: usually we have to use at least $D(m_0 + 2, \epsilon)$ in order to verify that $D(m, \epsilon)$ saturates with m within the scaling region. Now you are confused by the last example, where we found a noise level of 5%, but estimated a finite dimension before? Well, if you go back to Fig. 6.5 you will see that the plateau is really tiny, and that our interpretation was assisted by the results from the same data after noise reduction.

6.8 Multi-scale or self-similar signals

In this section we want to discuss some scaling phenomena and fractality where the signal itself forms a complicated geometrical object while the concept of an attractor in phase space cannot be applied successfully. Dynamical systems governed by ordinary differential equations such as Eq. (3.2) produce continuous and differentiable signals. Several stochastic and deterministic processes can create signals that are much less well behaved: although continuous, they are not differentiable. Such signals are either fractals or are otherwise scale invariant and require special treatment, in particular since they are often non-recurrent and must therefore be considered to be effectively non-stationary: many of the processes for which the following analysis is suited do not create an invariant measure when regarded in higher embedding spaces (compare Section 6.5).

A scalar signal can be *fractal* in the sense that its graph, as a function of time, has a nontrivial Hausdorff dimension. This means that the length of the graph seen at a finite resolution increases forever when the resolution on the time axis is increased, since more and more fluctuations are resolved. Its dimension can therefore lie anywhere between one and two. From a mathematical point of view, this behaviour is quite exotic and it is not trivial to create such a graph. The *Weierstrass functions* are of this type.

We should stress that such signals usually cannot result from low dimensional chaotic systems except for the case of deterministic enhanced diffusion in periodic potentials, see Zaslavsky (1993) and Klafter & Zumofen (1993), where, however, the phase space is unbounded.

Example 6.9 (Hydrodynamical turbulence). Fully developed turbulence is characterised by scale invariance (with an upper and lower cutoff). Energy is fed into the system on the largest length scales where they create large eddies. Dissipation (and thus transformation of kinetic energy into heat) takes place on the small length scales (viscous scale). The scales in between are called the *inertial range*. A hierarchy of eddies on all scales is the image in real space of the energy cascade linking large and small scales. All together this gives rise to scaling laws for all observables, like for the correlations of the velocity field or the energy dissipation field. An experimental time series reflects this complicated structure. Think of the measurement of the local velocity at a fixed position in the fluid. Due to the steady mixing of the system, the probe will record all the different scales. Therefore *the signal itself* forms a fractal but there is no low dimensional attractor.   □

6.8.1 Scaling laws

Turbulence is an infinite dimensional deterministic process described by a partial differential equation. Stochastic processes can produce signals of similar kind. Let us recall the properties of the standard diffusion (*Wiener*) process in order to introduce a new concept related to self-similarity, *self-affinity*. In the Wiener process, a particle at position $x(t)$ has a probability $P(\Delta x, \Delta t) = \exp(-\Delta x^2/2D\Delta t)/\sqrt{2\pi D\Delta t}$ to move to position $x + \Delta x$ within the time step Δt. The variance of the increments scales with the time intervals as $\langle \Delta x^2 \rangle \propto \Delta t^{1/2}$. Assume that we have created such a graph $x(t)$ with infinite precision. Now, if we look at it very closely, i.e. with a good resolution in both time and space,

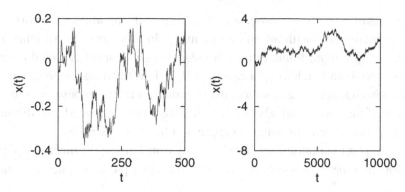

Figure 6.11 A Brownian path (Wiener process) $x(t)$ in high and low resolution (reduction by a factor of 20 in both axes).

we will observe diverging fluctuations, since for the "slopes" one finds easily: $\langle \Delta x^2 \rangle / \Delta t \propto \Delta t^{-1/2}$. On the other hand, if we look from far away, thus reducing the scales in time and space by the same factor, we observe an almost smooth curve (compare Fig. 6.11).

In the limit of high resolution, the graph fills the plane and has dimension two, whereas in the opposite limit, it degenerates towards a horizontal line and thus has dimension one. So in each of the two limits we find self-similarity. Finally, we find a scale independent view, if we scale space and time with different scaling factors, more precisely, if we rescale space by factors which are the square root of the scaling factors in time. Such objects are called *self-affine* and are self-similar only in certain limits.

Now let us proceed to *anomalous* diffusion laws, i.e. laws, where the square root relation between increments and time intervals is substituted by nontrivial power laws,

$$\langle \Delta x^2 \rangle \propto \Delta t^{2H} \tag{6.10}$$

with a *Hurst exponent H* different from 1/2. Brownian motion (or the Wiener process), has the property that the distribution of the increments Δx possesses a finite variance for every fixed Δt, and the increments are uncorrelated at successive time steps. One can easily show that a generalisation of Eq. (6.10) towards $H \neq 1/2$ requires to relax at least one of these two properties.

If we insist on uncorrelated increments and therefore on the Markov property, we have to admit increments with unbounded variance, although we require still that every single increment remains finite with probability one. This yields a probability distribution whose large Δx behaviour is $p(\Delta x) \propto \Delta x^{-1/H}$ with $1/2 \leq H \leq 1$, i.e., a probability distribution with strong tails. Such a process, which is called *Levy flight*, creates enhanced diffusion.

A more natural stochastic process can be found, if one allows for correlations among the increments with an infinite memory. In this case, the variance of the increments may remain finite. Such a *non-Markovian process* is called *fractional Brownian motion* and yields exponents $0 < H < 1$. We can find an intuitive explanation for what happens: the memory effects induce correlations between successive increments. If they are positively correlated, diffusion is enhanced (*persistence*), if they are anti-correlated, diffusion is suppressed (*anti-persistence*).

The scaling on the microscopic level directly translates to the macroscopic level. Hence, for increments $x_t - x_{t_0}$ observed on macroscopic time intervals one also finds

$$\langle (x_t - x_{t_0})^2 \rangle \propto (t - t_0)^{2H} , \tag{6.11}$$

where $0 < H < 1$ in the fractional Brownian case and $1/2 < H < 1$ for the Levy flights. Ordinary diffusion fulfils $H = 1/2$, and ballistic motion (i.e., $x(t) \approx at$ with nonzero a) has $H = 1$.

Example 6.10 (Fluctuations of an upright still standing human). Every upright standing normal person performs unwillingly some small amplitude oscillations, which induce a complicated motion of the centre of pressure below the soles of the feet in the x-y-plane (it moves from the toes to the heel, from the left to the right). The position of this centre of pressure as a function of time can be measured by a device called a force plate. It turns out that the most suitable analysis of this signal is the computation of the fluctuations Eq. (6.11), which yields nice power law behaviour and thus supports the idea that the oscillations are of a random nature (Fig. 6.12). Attempts to find low dimensional deterministic motion failed, a significant disagreement with the null hypothesis of a Gaussian linear stochastic process has not been found. The very interesting observation is the cross-over from accelerated anomalous diffusion with an exponent close to 3/4 to diffusion with an exponent close to or less than 1/2 at a time lag of about 2 seconds. It is plausible that this is a signature of the posture control which tries to suppress large fluctuations, needs some time for the correct response, and naturally is anti-correlated to the actual position (compare this to the fractional Brownian motion). We obtained the time series from Steve Boker [Bertenthal *et al.* (1996), Appendix B.9]. See also Collins & De Luca (1995) for a study of this phenomenon. □

The self-similarity of the signal reflects long range temporal correlations in the time series. When studying the cluster in the delay embedding space formed by these data, it has, for every finite time series, also a statistically self-similar structure. We mentioned already the result obtained by Mandelbrot, that a diffusion cluster has dimension two in any embedding dimension equal to or larger than two. The dimension reflecting the temporal correlations here, it is not surprising to learn

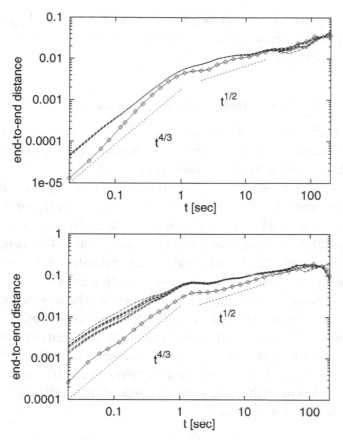

Figure 6.12 The growth of the end-to-end distance, Eq. (6.11), for the motion of the centre of pressure of an upright standing person. [Bertenthal *et al.* (1996), see also Appendix B.9.] Upper panel: the motion in forward–backward direction, lower panel: left–right direction (continuous lines with diamonds). The bundles of dashed lines are the results for 10 different realisations of surrogates each.

that the dimensionality is related to the Hurst exponent by $D_H = 1/H$. However, in numerical estimates of this dimension it is crucial *not to discard points close in time*, since strictly speaking the whole cluster is temporally correlated. In this sense it is not a contradiction to our standard reasoning that a stationary stochastic process creates a set whose dimension is identical to the embedding dimension. The intricate aspect of these fractal objects here is also that they cannot be embedded in any lower dimensional manifold of the space in which one is considering them. Whereas the fractal attractor of a dynamical system typically can (at least locally) be embedded in a manifold whose dimension is the next integer larger than the fractal dimension, this is not true for the processes discussed here.

Since the exponent H reflects temporal correlations, you might wonder whether it can be found in the power spectrum. Indeed also the power spectrum of such signals shows a power law decay towards large frequencies, given by $\beta = 1 + 2H$. Hence, the power spectrum of a Brownian path looks like $1/k^2$ when k is the wave number.

6.8.2 Detrended fluctuation analysis

It is often not advisable to estimate the Hurst exponent of measured data by Eq. (6.11). Real world data from processes with nontrivial Hurst exponents often carry trends, which result either in a trivial value of the exponent, namely $H = 1$, reflecting the "ballistic" increase of the quantity due to the trend, or, more often, do not exhibit a clear scaling. It has been shown that in this case a particular, additive way of detrending is reasonable. The method was developed by Peng *et al.* (1994) for the analysis of DNA and soon after applied to many dynamical phenomena including physiological data, climate data, and economical data.[5] Before we can think about detrending, we often need to pass over to a different signal. Consider, e.g., velocity measurements from a turbulent wind stream. The recorded wind speeds are definitely finite (a speed of more than 30 m/s on the earth's surface is really exceptional), so that scaling such as Eq. (6.11) would be restricted to rather short time intervals, beyond which this quantity would saturate at twice the variance of the data. In order to extend scaling, one should therefore interpret such measurements with finite variation to be the increments of a second, then diffusion like process. One thus performs the transformation

$$y(t) = \int_0^t (x(t') - \bar{x})dt' , \tag{6.12}$$

where $\bar{x} = 1/T \int_0^T x(t')dt'$ is the mean of $x(t)$ computed on the whole time series (subtracting the mean is implicitly contained in linear detrending as described below, but linear detrending is more than subtracting the global mean value). The result of the fluctuation analysis will be an exponent α which either can be taken as the Hurst exponent for $y(t)$ or for unity plus the Hurst exponent of the underlying process $x(t)$.

Instead of directly performing a fluctuation analysis for $y(t)$, one introduces different levels of subtractive detrending. As we will point out in Chapter 13 subtractive detrending is not very reasonable in many cases, but the fluctuation analysis profits from it. The trends will be removed by subtracting a suitable smoothed version of $y(t)$. We discuss here the smoothing by polynomials of kth order *on the same time interval on which the fluctuations are considered.* More precisely,

---

[5] In November 2002 when the final revision of this book is made, the ISI database contains more than 150 papers making use of DFA in time series analysis.

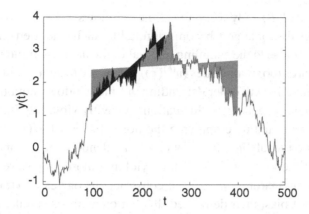

Figure 6.13 Sketch of the linear detrending for two different time windows, both starting at $t = 100$. The trend to be removed depends not only on the starting point t but also on the window length τ. The fluctuation integrals $v^{(1)}(100, \tau)$ are the integrals over the squared distances inside the grey-shaded regions in between the signal $y(t)$ and the linear fits.

let $g^{(k)}(t, \tau, t')$ be a polynomial of order k with the property that g minimises $\int_t^{t+\tau} dt' (g^{(k)}(t, \tau, t') - y(t'))^2$. For $k = 0$, the constant $g^{(0)}$ satisfying the minimisation problem is evidently the mean value of $y(t)$ in the considered time interval. For $k = 1$, $g^{(1)}(t')$ is a linear regression of $y(t')$ for $t' \in [t, t + \tau]$ (see Fig. 6.13), and so on. Now compute

$$\frac{1}{\tau} \int_t^{t+\tau} (y(t') - g^{(k)}(t, \tau, t'))^2 dt' =: v^{(k)}(\tau) . \tag{6.13}$$

Since in practice we work with a discrete time variable, all integrals should be converted into the corresponding sums. Moreover, for more robust results, one considers averages $\langle v^{(k)}(\tau) \rangle$ averaged over all (maybe half-overlapping) time windows of length τ in the data set of length T. It can be shown that if $x(t)$ fulfils Eq. (6.11), then

$$\langle v^{(k)}(\tau) \rangle \propto \tau^{2\alpha} = \tau^{2+2H} \tag{6.14}$$

If the data $x(t)$ are non-stationary so that y contains trends and the scaling is destroyed for both x and y, detrending with a sufficiently high-order polynomial may re-establish the scaling for $\langle v^{(k)}(\tau) \rangle$. If the signal itself is already diffusion like, the initial integration step can be abandoned (hence, take $y(t) = x(t)$), and the resulting DFA-exponent α itself is identical to H associated with x.[6]

---

[6] If x is white noise, the integrated $y(t)$ forms a diffusion process with $\alpha = 1/2$, and hence one could associate a Hurst exponent of $H = \alpha - 1 = -1/2$ to white noise. Since the smallest outcome of the fluctuation analysis is $\alpha = 0$, negative exponents cannot be determined. The integration Eq. (6.12) increases the exponent but unity.

Since for $\tau = O(T)$ only very few time windows constitute the average in Eq. (6.14), the results for large τ become unreliable, such that often only $\tau \leq T/4$ is used. For small τ close to the sampling interval of the data, the trend removal introduces a bias towards too small values $\langle v^{(k)}(\tau) \rangle$ and hence to another violation of scaling. As an extreme, for kth order detrending, a time window containing $l = k + 1$ sample points yields exactly zero fluctuation, evidently violating Eq. (6.14). Partly, this can be compensated by employing the normalisation $1/(l(\tau) - k - 1)$ in the time discrete version of Eq. (6.13), where $l(\tau)$ denotes the number of samples covered by the time interval τ. A strategy yielding more precise results has been suggested by Kantelhardt et al. (2001): construct surrogates of $x(t)$ by a random shuffling. Without bias, their detrended fluctuation analysis should yield $\alpha = 1/2$ on the whole range of scales. The violation of scaling in the numerical estimates reflects the statistical and systematic errors of the numerical estimation and defines the correction function $c(\tau) := \tau^{1/2}/\langle v^{(k)}_{\mathrm{surr}}(\tau) \rangle$. Then the outcome of the analysis of the original data $x(t)$ can be corrected by multiplying $\langle v^{(k)}_{\mathrm{data}}(\tau) \rangle$ by $c(\tau)$. If no violation of scaling were present, this correction would simply imply the proper normalisation so that the fluctuation integrals are independent of the order k of detrending.

At first sight you may be surprised that the detrending does not modify the scaling exponent, since the signal might be nonlinear and hence a subtractive trend removal might be inappropriate. In particular, you might wonder whether a Brownian path whose both endpoints are at almost the same value (which is the case for linear and higher order detrending) is not a very unlikely and hence a peculiar path. In fact you are right with both concerns. The method, however, lives from the fact that the scaling behaviour of the fluctuations around the smoothed curve is not affected by these modifications. The absolute values of the fluctuation integrals $\langle v^{(k)}(\tau) \rangle$ are in fact decreasing in k, but their scaling in τ is unaffected for all $k > k_{\min}$, for which the correct $\alpha = 1 + H$ becomes visible. There may be, however, some strange effects for $k < k_{\min}$.

As mentioned before, the Hurst exponent can also be found in the power spectrum. However, with the detrending done in the time domain on time scales matching exactly the time scale of the analysis, the scaling laws are typically much better expressed by the detrended fluctuation analysis than by the power spectrum.

Further reading

A general introduction to fractal geometry can be found in Falconer (1985) or in Mandelbrot (1985). An early attempt at dimension estimation is reported in Froehling et al. (1981). Classics are Farmer et al. (1983) and Takens (1983). A number of articles about dimensions and related topics are included in

Mayer-Kress (1986). Additional material on estimating the correlation dimension and problems encountered can be found in Layne *et al.* (1986), Theiler (1986), (1987), (1990), (1990a), and (1991), Möller *et al.* (1989), in Grassberger *et al.* (1991) and in Gershenfeld (1992). References for alternative methods of dimension estimation will be given in Chapter 11.

Exercises

6.1 Modify the definition of the correlation sum, Eq. (6.3), for the case of a continuum of points by replacing the sums over points by integrals over the distribution.

6.2 Calculate this *correlation integral* in one dimension for the following point distributions, taking edge effects into account. Also compute its logarithmic derivative $D(m, \epsilon) = d \ln C(m, \epsilon)/d \ln \epsilon$.

- For points uniformly distributed on the interval [0,1] of the real line, show that $D(m, \epsilon) = (2 - 2\epsilon)/(2 - \epsilon)$ for $\epsilon < 1$.
- For a normal distribution $\exp(-\eta^2/2)/\sqrt{2\pi}$ show that

$$D(m, \epsilon) = \frac{\epsilon \exp(-\epsilon^2/4)}{\sqrt{\pi}\,\mathrm{erf}(\epsilon/2)}.$$

- The natural measure of the logistic equation, $x_{n+1} = 4x_n(1 - x_n)$, see also Exercise 2.2, can be computed analytically and has singularities of the square root type: $\mu(x) \propto 1/\sqrt{x}, x \ll 1$. Concentrate on the case where ϵ is small.

How well does $D(m, \epsilon)$ converge to the expected limit ($D = 1$) for $\epsilon \to 0$?

6.3 Create an artificial time series by iterating the Ikeda laser map [Ikeda (1979)]:

$$z_{n+1} = 1 + 0.9z_n \exp\left(0.4i - \frac{6i}{1 + |z_n|^2}\right),$$

where the z_n are complex variables. Take the real part of z_n as the actual time series. Use the program listed in A.6 to compute the correlation sum. Since scaling is violated above $\frac{1}{10}$ of the attractor extent you should not use less than 5000 data points. In fact, even on a small computer you will be able to go far beyond this. Now add a small amount of Gaussian white noise (say 1–2%) to the data and repeat the analysis.

6.4 Repeat Exercise 6.3 but use $x_n = \mathrm{Re}\,(z_n)$ and $y_n = \mathrm{Im}\,(z_n)$ instead of delay coordinates.

6.5 Compute the correlation integral of the surface of a *torus* (looks like a doughnut). You can create the data either by forming random points on the torus or by creating a biperiodic time series with incommensurate frequencies. In the latter case you can use the program d2 without changing anything but you have to take care to avoid artefacts (formally, the correlation time of periodic data diverges). We suggest the use of random multivariate data vectors instead of delay coordinates.

6.6 Induce artificial temporal correlations in the Ikeda data of Exercise 6.3 by applying a *moving average* filter: $s_n = \frac{1}{10} \sum_{n'=n-9}^{n} x_n$. Repeat the calculation of the correlation sum with different settings of $n_{min} = 1, 2, \ldots$.

6.7 Compare the space time separation plots of Ikeda data sets with and without temporal correlations.

6.8 Do a space time separation plot for the Brownian motion data created in Exercise 5.3. We usually choose t_{min} larger than the time Δt where the contour lines saturate. What does this imply for this data set?

6.9 Use the program d2 to compute $D_{noise}(\epsilon)$. Do you find the right noise level for the Ikeda map data created in Exercise 6.3?

Chapter 7
Using nonlinear methods when determinism is weak

In the preceding two chapters we established algorithms to estimate the Lyapunov exponent and the correlation dimension from a time series. We tried to be very strict about the conditions which must be met in order to justify such estimates. The data quality and quantity had to be sufficient to observe clear scaling regions. The implied requirement that the data must be deterministic to a good approximation is also valid for successful nonlinear predictions (Chapter 4). If this were the whole story, the scope of these methods would be quite limited. In the main, well-controlled laboratory data from experiments which have been designed to show deterministic chaos would qualify. Although these include some very interesting signals, many other data sets for which classical, linear time series methods seem inappropriate do not fall into this class.

Indeed, there is a continuous stream of publications reporting more or less successful attempts to apply nonlinear algorithms, in particular the correlation dimension, to field data. Examples range from population dynamics in biology, stock exchange rates in economy, and time dependent hormone secretion or ECG and EEG signals in medicine to geophysical records of the earth's magnetic field or the variable luminosity of astronomical objects. In particular the interpretation of the results as measures of the "complexity"[1] of the underlying systems has met with increasing criticism. It is now quite generally agreed that, in the absence of

---

[1] You may notice that we largely avoid using the term "complex" in this book, although the quantities we discuss are sometimes referred to as "complexity measures". There seems to be no unique way of defining complexity in a rigorous sense. Nor is it even used in a consistent way in the literature. While some authors use it for any signal which is not constant or periodic, most people agree that neither periodic nor completely random sequences are to be called complex. Complexity should thus quantify the difficulties we have in describing or understanding a signal. The problematic point here is that such a definition does not yield an independent description of a signal but of *our relationship to the signal*. Are Chinese letters more or less complex when you know Chinese? A review which discusses these issues and a number of definitions which have been proposed is Grassberger (1986); see also Crutchfield & Young (1989). Badii & Politi (1999) give an excellent account of approaches to defining complexity.

clear scaling behaviour, quantities derived from dimension or Lyapunov estimators can be at most *relative* measures of system properties. But even then it is not clear which properties are really measured.

Let us consider the correlation integral as an example. When there is a clear scaling region, the dimension can easily be read off a plot of the local slopes $d(\epsilon)$, Eq. (6.2). We can think of ways to turn this into an automated procedure, resulting in a smart program which reads a time series, asks for the embedding parameters and prints out a number, the correlation *dimension*. Playing with this program, we will find, for example, that this number has the tendency to be larger for ECGs during exercise than during rest periods. Does this mean that something in the heart is more "complex" during exercise? If at all, in which sense? Let us give some arguments against such an interpretation of the observation. First, while the correlation dimension in theory is invariant under smooth coordinate changes, this is not true for finite data estimates. Moreover, when a clear scaling region is absent, attempts to determine a dimension from the correlation integral are affected by different noise levels, different degrees of non-stationarity, etc. But even if such caveats are properly taken care of, there is no theorem that guarantees that a dimension estimator yields numbers which increase monotonically with the dimension of the underlying attractor. When a dynamical system undergoes a parameter change which leads to a substantially different attractor dimension, then usually the overall geometrical shape of the attractor changes also (*crisis*, see Section 8.6), which affects finite resolution dimension estimates in an unpredictable way.

This said, let us observe that, usually, estimated dimensions *do* change with the attractor dimension. They just might do so in a very complicated way. If two modes of operation of a system differ in their estimated dimensionality but are hard to distinguish otherwise, a dimension estimator can be a powerful way of discriminating between the two. But rather than some dubious interpretation, what we need is a proper statistical test which enables us to distinguish the modes with a specified level of significance. Most of this chapter is devoted to such tests, concentrating on the two important cases that we either want to distinguish non-linearity in general from Gaussian linear stochastic processes (Section 7.1) or that we want to discriminate between two different dynamical states of a (nonlinear) system (Section 7.2).

Finally, in Section 7.3 we will discuss some ways of extracting qualitative information from a time series. This can be useful for different reasons. First, qualitative information may have a more direct interpretation, e.g. in a clinical situation. Second, if the data set is short, or nonstationary, or otherwise unsuitable for a thorough statistical treatment, qualitative information might be all that is available.

7.1 Testing for nonlinearity with surrogate data

This book is about methods of nonlinear time series analysis, which are much less understood than classical methods for linear stochastic processes. Thus, before we attempt any nonlinear analysis on a data set, it is a good idea to check if there actually is any nonlinearity which makes such an enterprise necessary. The main lesson to learn is that linear stochastic processes can create very complicated looking signals and that not all structures that we find in a data set are likely to be due to nonlinear dynamics going on within the system. This chapter will thus be devoted to such questions as the following. Is the apparent structure in the data most likely due to nonlinearity or rather due to linear correlations? Or, is the *irregularity* (nonperiodicity) of the data most likely due to nonlinear determinism or rather due to random inputs to the system or fluctuations in the parameters? Before we study such questions in more detail, let us give some examples of what kind of complicated, irregular signals we can get from linear stochastic processes.

Example 7.1 (Distorted autoregressive process). A very simple linear stochastic process is given by the following prescription:

$$x_n = 0.99x_{n-1} + \eta_n, \tag{7.1}$$

where the η_n are independent Gaussian random numbers.[2] Figure. 7.1 shows a sample output of this process. Successive values are strongly correlated, that is, they tend to be similar. Already, this altogether linear process yields some structures, but look what happens if we observe the same values through a nonlinear measurement function:

$$s_n = x_n^3, \qquad x_n = 0.99x_{n-1} + \eta_n. \tag{7.2}$$

The sample output in the lower panel of Fig. 7.1 looks pretty spiky and quite complicated. But, still, there are no nonlinear dynamics in this process; all the dynamics are contained in the linear AR(1) part and the nonlinearity is purely static. If the lower panel of Fig. 7.1 were the output of an experiment we wanted to study, the best approach would be to invert (at least approximately) the nonlinear distortion and then use linear methods such as the power spectrum and autocorrelations (see Chapter 2) to study the properties of the signal. □

---

[2] Such a process is often called an *autoregressive* process of order 1, or an AR(1) process. Refer to Section 12.1 for more details. It can be seen as the discrete time counterpart of the *Ornstein–Uhlenbeck* process; see any textbook on stochastic processes, e.g. van Kampen (1992).

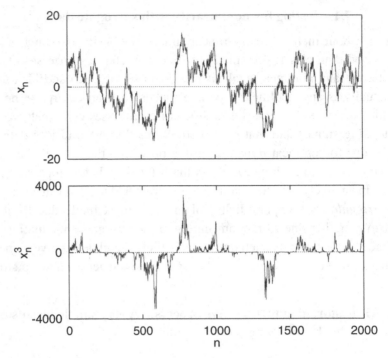

Figure 7.1 Output of an AR(1) process. Upper panel: x_n from Eq. (7.1). Lower panel: the same series measured by the function $s_n = x_n^3$.

The linear process in the example given above was a very simple one. More generally, we can think of *autoregressive moving average* (ARMA) processes of the form

$$x_n = a_0 + \sum_{i=1}^{M_{AR}} a_i x_{n-i} + \sum_{j=0}^{M_{MA}} b_j \eta_{n-j} , \qquad (7.3)$$

where, again, the η_n are independent Gaussian random numbers with zero mean and unit variance. The first sum is referred to as the *autoregressive* (AR) part and describes the internal dynamics of the process. The second sum is called a *moving average* (MA) over the random input values. Often, a_0 is dropped. The particular class of processes is specified by giving M_{AR} and M_{MA}; the example above would be called an AR(1) or ARMA(1,0) model. What we need to know for this chapter is that for any ARMA process with Gaussian inputs, the two-point autocorrelation function and thus the power spectrum is completely determined by the coefficients a_i and b_i and, even more important, all higher moments and correlations are determined as well. In this sense, the power spectrum can be regarded as a complete statistical description of the process. We will use this fact in the following. A further discussion of these processes can be found in Section 12.1.1.

Only nonlinear deterministic signals are characterised by finite dimensional attractors, finite, positive Lyapunov exponents and good nonlinear-short term predictability. Thus, in the past many people have used, for example, their dimension algorithm in order to look for nonlinear determinism. Whenever the algorithm yielded a finite number this was taken as a signature of chaos. But, as we pointed out earlier (see e.g. the discussion on temporal correlations in the correlation sum in Chapter 6), linear structures can be mistaken for determinism. Estimated finite dimensions and Lyapunov exponents on their own are not suitable for proving strict nonlinear determinism unless very clear scaling regions can be established. This is not a flaw of these algorithms; a time series can be anything between purely random (i.i.d.: independent, identically distributed) and strictly deterministic, and asking just a yes/no question is simply not meaningful. We should put the analysis on more solid ground by first testing the hypothesis that the data could be explained by a linear model. For this purpose we will now develop a statistical significance test.

The main idea is the following. Say we compute some nonlinear observable λ from the data. Then λ may be a dimension, a prediction error or whatever. Does the value found for λ suggest that the data are indeed nonlinear? To answer this, we first have to ask another question. What distribution of values for λ could we get from a comparable linear model? Is the value we found perhaps consistent with a linear description? If not, the data might be nonlinear.

7.1.1 The null hypothesis

Let us turn this idea into a test. Since we do not usually have any theory for the distribution of the values of λ for linear stochastic processes, we will estimate this distribution using the method of *surrogate data*. The need for accurate statistical tests in nonlinear time series analysis and the usefulness of the Monte Carlo approach has been particularly stressed by Theiler *et al.* (1992) and (1992a). First we will have to specify a *null hypothesis* that we want to test against. A typical one would be that the data results from a Gaussian linear stochastic process, see below. Next we should specify the *level of significance* that we are aiming for. If we allow for a 5% chance ($\alpha = 0.05$) that we reject the null hypothesis although it is in fact true, then the test is said to be valid at a 95% significance level. More than one wrong result out of 20 is usually not considered acceptable by the scientific community. In designing the statistical test we have to take care that the actual rejection probability on data from the null hypothesis (the *size* of the test) does not exceed the value specified by the level of significance we are aiming for.

Now we have to create a number of *surrogate* data sets which are comparable to the measured data in certain respects to be specified, but which are also consistent with the null hypothesis we are testing for. Then we compute the statistics λ_0 for the data and λ_i, $i = 1, \ldots, K$ for all the K surrogate data sets and test if the value computed for the data is likely to be drawn from the distribution of λ obtained from the surrogates. If we have good reasons to believe that λ follows a Gaussian distribution, we can compute the mean $\bar{\lambda} = K^{-1} \sum_{i=1}^{K} \lambda_i$ and the variance $\sigma_\lambda^2 = (K - 1)^{-1} \sum_{i=1}^{K} (\lambda_i - \bar{\lambda})^2$. Then, for example, for a two-sided test[3] at the 95% level we must require that $|\lambda_0 - \bar{\lambda}| < 2\sigma$. However, since we do not usually know that the values of the statistics λ actually follow a Gaussian distribution, we suggest not making this assumption but using a rank based test instead.

Suppose the data follows the null hypothesis. When there are $1/\alpha - 1$ surrogates and one data set, we have a total of $1/\alpha$ sets following the null hypothesis. Consequently, each of the sets with probability α by chance yields the smallest measured value of λ. Assume that we want to test whether the measured λ_0 is smaller than the expected value for data respecting the null hypothesis. A natural strategy will then be to reject the null hypothesis whenever the data gives a value λ_0 which is smaller than all the λ_i obtained from the surrogates. This will give a false rejection when λ_0 happens to be smaller than all λ_i by chance, which occurs with probability α. This is exactly what we allowed for when we specified the significance level. For a two-sided test we have to make $2/\alpha - 1$ surrogates and reject the null hypothesis whenever the data yields the smallest *or* the largest value of λ.

7.1.2 How to make surrogate data sets

We have not yet specified the properties that the surrogate data sets must have, because they depend on the actual null hypothesis that we want to test. We will use composite null hypotheses which contain some free parameters. Rather than asking, for example, if the data follows an ARMA(1,1) process with coefficients $a_1 = 0.3$, $b_0 = 0.5$, $b_1 = 0.7$ it is more natural to ask if the data might come from *any* ARMA(M_{AR}, M_{MA}) process. One way to do such a test is by requiring that the statistic λ does not depend on these parameters at all, which is hard to achieve for all but the simplest statistics. A more convenient approach is to allow for free parameters but to impose the constraint that the surrogates must have the same values of these parameters as the data.

Consider the null hypothesis that the data are independent random numbers drawn from some fixed but unknown distribution (which is the free parameter). Then any random shuffle (without repetition) of the data yields exactly the same empirical distribution and is, by construction, independent.

---

[3] A one-sided test is suitable if we ask if a quantity is significantly *larger* than a certain value. If we just ask if it is *different*, i.e. larger *or* smaller, the test will have to be two-sided.

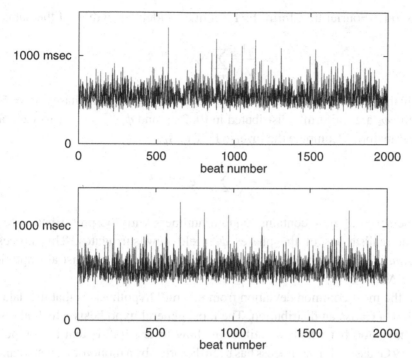

Figure 7.2 Heart beat intervals during atrial fibrillation. Upper panel: original time series. Lower panel: random shuffle of the same values.

Example 7.2 (Heart rate during atrial fibrillation). We concatenated disjoint sections of interbeat interval data of a human during atrial fibrillation [MIT–BIH Arrhythmia Database; Moody & Mark (1991), see Appendix B.6] to yield a reasonably stationary signal. The upper panel in Fig. 7.2 shows the resulting data set and the lower panel shows a surrogate data set obtained by a random permutation of these values. By measuring the autocorrelation at lag 1, $\langle s_n s_{n-1} \rangle$ for the data and 19 surrogates we could reject the null hypothesis that the data consists of independent random numbers at the 95% significance level. $\square$

A more interesting null hypothesis is that the data comes from a stationary linear stochastic process with Gaussian inputs. The free parameters are the mean, the variance, and the autocorrelation function, or, equivalently, the mean and the power spectrum.[4] Adequate surrogate data sets must contain correlated random numbers which have the same power spectrum as the data. This is the case if the Fourier transforms of the data and the surrogates differ only in their phases but have the same amplitudes which go into the power spectrum. Here is the recipe. Take the

---

[4] We will in the following assume that the mean has already been subtracted.

fast (discrete) Fourier transform (FFT, routines widely available) of the data:

$$\tilde{s}_k = \frac{1}{\sqrt{N}} \sum_{n=1}^{N} s_n e^{i2\pi nk/N} . \tag{7.4}$$

Multiply the complex components $\tilde{s}_k$, $\quad 0 < k < N$ by random phases, $\tilde{s}'_k = \tilde{s}_k e^{i\phi_k}$, where the ϕ_k are uniformly distributed in $[0, 2\pi[$, and $\phi_{N-k} = -\phi_k$ to yield a real inverse transform. Compute the inverse FFT of $\tilde{s}'_k$:

$$s'_n = \frac{1}{\sqrt{N}} \sum_{k=1}^{N} \tilde{s}'_k e^{-i2\pi nk/N} . \tag{7.5}$$

The time series s'_n now contains random numbers with the prescribed spectrum. Different realisations of the phases ϕ_k yield new surrogates. This process of *phase randomisation* preserves the Gaussian distribution (at least asymptotically for large N).

Now, the most common deviation from this null hypothesis is that the data does not follow a Gaussian distribution. The most general hypothesis which allows for such a deviation but which we still know how to test for[5] is that the output of a stationary Gaussian linear process has been distorted by a monotonic, instantaneous, time independent measurement function s:

$$s_n = s(x_n), \qquad x_n = \sum_{i=1}^{M_{AR}} a_i x_{n-i} + \sum_{j=0}^{M_{MA}} b_j \eta_{n-j} . \tag{7.6}$$

Free parameters are now the spectrum and the empirical distribution of the values in the series due the unknown function s. The usual approach is to invert s empirically by rescaling the values to follow a Gaussian distribution: $\hat{x}_n = \hat{s}^{-1}(s_n)$. Now the phase randomisation is done on the $\hat{x}_n$, yielding surrogates $\hat{x}'_n$. These are now rescaled back to the original distribution of the s_n. These surrogates s'_n agree with the data in their distribution but we remark that the power spectrum is slightly changed since we have to estimate s and the $\hat{x}_n$ will differ from the original x_n. These *amplitude adjusted* FFT surrogates are often good enough, but they can give unacceptably many false rejections whenever strong correlations are paired with a nonlinear distortion.

There is a way to polish surrogates so that they follow the desired spectrum more closely. A simple implementation similar to a *Wiener filter* can be used to enforce the correct spectrum [Schreiber & Schmitz (1996)]. This makes a new rescaling necessary to enforce the right distribution. The two steps can be iterated several

---

[5] Well, almost.

times.[6] The filter is very simple indeed. Let $|\tilde{s}_k|$ be the desired amplitudes of the FFT as obtained from the data. To filter a surrogate series, compute its FFT $\tilde{s}'_k$ and replace it by $\tilde{s}''_k = \tilde{s}'_k |\tilde{s}_k| / |\tilde{s}'_k|$. Then transform back.[7]

An implementation of this algorithm is provided by TISEAN as the routine surrogates. In order to make polished surrogates, one first needs to store a sorted version s_n^{sort} of the data and the Fourier amplitudes $a_0^{\text{data}}, \ldots, a_{N/2}^{\text{data}}$. Now one creates a random permutation of the data as a starting point for the iterative scheme. The filter is applied by calling the FFT routine and then the inverse FFT, with the amplitudes replaced by a^{data}. Rescaling is done simply by using the ranks of the resulting series and the sorted copy of the data.

7.1.3 Which statistics to use

Once we have a number of surrogate data sets we have to compute a statistic λ on each of them as well as on the data. In principle any nonlinear statistic which assigns a real number to a time series can be used. Higher-order autocorrelations such as $\langle s_n s_{n-1} s_{n-2} \rangle$ are cheap in computation but quantities inspired by nonlinear science seem to be more popular because they are particularly powerful if the data set has a nonlinear deterministic structure. We have some experience with zeroth-order nonlinear prediction error and quantities which can be computed automatically from the correlation sum. Let us mention a particular higher-order autocorrelation, $\langle s_n s_{n+1}^2 - s_n^2 s_{n+1} \rangle$, which measures time asymmetry, a strong signature of nonlinearity. A numerical comparison of several nonlinear statistics has been performed in Schreiber & Schmitz (1997).

If the data deviates from the null hypothesis, the probability that the surrogate data test will actually reject this hypothesis is called the *power* of the test against this particular alternative. It depends on the statistics used in the test. It is desirable to choose a statistic which gives maximal power, which sometimes means that the statistic itself is not a good estimator for anything any more. Let us take nonlinear prediction errors as an example. As we argued earlier, we only trust out-of-sample errors. These can either be obtained at very high cost using take-one-out statistics or by splitting the data into a training set and a test set. The latter leads to stronger statistical fluctuations for short data sets which limit the power of the statistical test. On the other hand, if we only want to use the error as a discriminating statistic in a statistical test, an in-sample prediction error is good enough and, in fact, yields a more powerful test. See also Theiler & Prichard (1996).

---

[6] We conjecture that perhaps with some restrictions on s the scheme converges, at least in the limit of large N. Since we do not have a proof we leave it as an exercise.

[7] The Wiener filter is actually the *square* of this filter, see Section 2.4.

Let us mention that quite a number of different statistics have been proposed as tests for determinism; see "Further reading" for references. Almost all of these quantify the predictability of the signal in one way or the other. While all authors are able to give examples of where their method works, a systematic comparison is difficult. See Schreiber & Schmitz (1997) for an attempt. Eventually, it is up to you to choose the statistic which promises to be most appropriate for your data.

Apart from prediction based quantities (in a general sense), we can use any estimator of the correlation dimension which yields stable numbers as long as we do not quote these numbers as actual dimension estimates. The fact that surrogate random data consistently yield higher dimension estimates than a data set under study does not imply that the estimated dimension for the data is right. A dimension estimate can only be made on the basis of a clear scaling region. Thus, although we never use it to estimate the correlation dimension, the maximum likelihood estimator by Takens (1985) yields a good example of a nonlinear statistic in surrogate data tests. If we choose an embedding dimension m and a delay time $\tau = \tau \Delta t$ we can form embedding vectors as usual. Further, fixing a minimal time separation $t_{min} = n_{min} \Delta t$ we define the correlation sum as in Eq. (6.3):

$$C(\epsilon) = \frac{2}{(N - n_{min})(N - n_{min} - 1)} \sum_{i=1}^{N} \sum_{j=i+1+n_{min}}^{N} \Theta(\epsilon - \|\mathbf{x}_i - \mathbf{x}_j\|) . \quad (7.7)$$

Then, according to Theiler (1988), Takens's estimator is given by

$$D_{\text{Takens}} = \frac{C(\epsilon_0)}{\int_0^{\epsilon_0} C(\epsilon)/\epsilon d\epsilon} . \quad (7.8)$$

The upper cutoff length scale ϵ_0 has to be fixed as well. We found half of the rms amplitude of the data to be a reasonable compromise between speed (smaller ϵ_0 gives faster estimates) and statistical fluctuations (which are smaller for larger ϵ_0).

Unless the minimal time separation t_{min} in the correlation sum is chosen large enough, statistics derived from it can be very sensitive to deviations of the surrogates from the correct spectrum given by the data. This can cause spurious rejections, in particular when ordinary amplitude adjusted surrogates are used. Since, furthermore, in our (limited) experience the zeroth-order prediction error seems to have more power against even weak nonlinearities we prefer that statistic. For this purpose we used the TISEAN routine `zeroth`, see Section A.4.1.

Example 7.3 (Heart rate during atrial fibrillation). For the data of Example 7.2 we were unable to reject the null hypothesis of a distorted correlated Gaussian linear process at the 95% significance level. This can mean that there is no nonlinearity present or that the statistical test was not powerful enough to detect it.   □

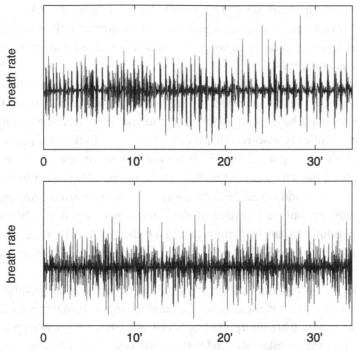

Figure 7.3 Human breath rate data and corresponding polished surrogate.

Example 7.4 (Human breath rate). The longer segment (4096 points) of human breath rate from the Santa Fe Institute time series competition data set is shown as the upper panel in Fig. 7.3. A typical polished surrogate set is shown as the lower panel. Note that this set has exactly the same distribution and almost exactly the same spectrum as the data themselves. A qualitative difference is already apparent in the diagram. A test with 99 surrogates leads to the rejection of the null hypothesis that the data set is distorted Gaussian coloured noise at the 99% significance level. Here we have used Takens's estimator, Eq. (7.8), with embedding dimension 3 and delay 3. In this particular case we find that the distribution of the statistics for the surrogates is indeed close to Gaussian and the data is off the mean by more than 7σ. If the result were less clear, we would have worried about the slight nonstationarity we can see in the data. □

7.1.4 What can go wrong

Although the surrogate data test is relatively easy to use we want to mention some pitfalls. The first class of problems comes from imperfections of the method itself. The usual amplitude adjusted surrogates can yield false rejections with a probability which is much larger than that specified by α. In our studies we have found 2–3 times more rejections with strongly correlated data such as the AR(1) process in Example 7.1, even when the distortion is much less dramatic than in the example. The polished surrogates seem to solve this problem but we cannot prove that the scheme always converges.

There is another conceptual flaw in FFT based surrogates. It is true that Gaussian linear processes are fully specified by either the autocorrelations or the power spectrum. But all we have is an estimate of the spectrum based on sampled data. This estimate implicitly assumes that the data set is periodic with period N when we take the FFT of N points. This is, however, not really the case and when the two endpoints of the series do not match, the corresponding jump in the periodic continuation causes additional high frequency in the spectrum. An approximate cure is to consider only a segment of the data which has been chosen to have matching end points. Apart from introducing a bias, this usually requires the use of an implementation of the FFT which can perform transforms of sequences with N not a power of two.

More generally, non-stationarity of any kind is a potential cause of false positives. If the data set is not a sufficient sample of an underlying stationary process, singular events can occur which are interpreted as deviations from the linear hypothesis. The occurrence of a single spike falls under this category. If the process itself changes over time, the data does not satisfy the null hypothesis and we expect a rejection. This leads to the large class of formally correct rejections which are, however, interpreted in the wrong way. This issue is discussed below.

It is not permitted to incorporate specific knowledge about the data into the design of the test. One could, for example, be tempted to determine a "scaling region" of the correlation sum by picking the length scale where the local slopes are smallest for the data. This procedure probably only picks a fluctuation and the surrogates are unlikely to have exactly the same fluctuation. Thus one will find a "significant" deviation. Optimising the parameters of a prediction algorithm on the data and using the same parameters for the surrogates has the same effect. Either optimise for each set, data and surrogates, separately, or do not optimise at all!

Finally remember that the significance has been specified under the assumption that we do a single test only. If we try more than one statistic or change the way we construct the surrogates for the same data set, the actual significance level is determined by the probability that at least *one* of all these tests gives a false rejection. For example, five independent tests at the 95% level each yield a total probability of a false rejection of 23%, which is not usually considered to be acceptable for a significant result. Since on the one hand the significance level has to be specified before each test and on the other hand we sometimes want to leave room for further tests until we are able to reject the null hypothesis, a possible strategy would be to use up half of the remaining false rejection probability with each further test. A (conservative) approximation is to use a modified level of significance, $\alpha_i = \alpha/2^i$, in the ith test when α is the total acceptable false rejection probability.

7.1.5 What we have learned

Now if we did everything right, what does a rejection or a failed rejection tell us? As pointed out already, if we cannot reject the null hypothesis this can either mean that the hypothesis actually describes the data properly or that the test we have done has failed due to its finite power. (We can expect that the null will be rejected only with probability $\beta < 1$, even when the alternative is true. The value of β is called the *power* of the test against a certain alternative hypothesis.) If the test has been done at a high significance level, there may be room for another test; see the discussion above.

Suppose we were able to reject the null hypothesis we tested for, say, that the data set was a representative sample from a stationary correlated Gaussian linear process, distorted by a time independent, instantaneous, monotonic measurement function. If we used correct surrogates, the rejection could, with probability α, have happened by chance. Maybe the data was nonstationary or had non-Gaussian inputs. The measurement function could depend on more than one measurement, could be nonmonotonic or would fluctuate. Or finally, the process could involve nonlinear dynamics. In these cases the rejection is fully justified but in all but the last one, no nonlinear dynamics are present! Here is a simple example of such a process.

Example 7.5 (Noninstantaneously distorted autoregressive process). Consider again the AR(1) process studied in Example 7.1, but let us change the measurement function slightly:

$$s_n = x_n x_{n-1}^2, \qquad x_n = 0.99 x_{n-1} + \eta_n \, . \tag{7.9}$$

The output (Fig. 7.4, upper panel) looks quite similar to the previous process and the polished surrogate (lower panel) seems to be appropriate. However, the null hypothesis of an instantaneously distorted Gaussian linear random process was rejected in 50% of the trials when aiming for 95% significance and using nonlinear prediction error as a statistic. Let us stress that the dynamics of this process were completely linear, only the measurements were distorted in a nonlinear way. □

There are many other examples and it is an interesting question as to how to draw the border between static nonlinearity and nonlinear dynamics. For example, integrate-and-fire models driven by AR or ARMA processes can show very complicated spike trains. Does the integrating neuron just perform a static nonlinear

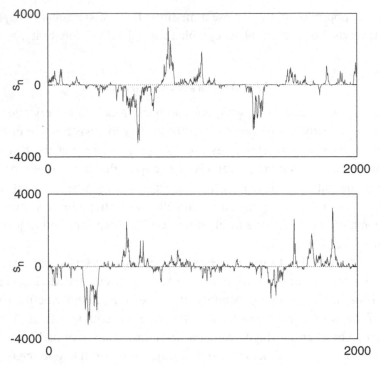

Figure 7.4 Output of a noninstantaneously distorted AR(1) process together with a typical surrogate data set.

measurement of the AR signal or does its internal degree of freedom take part in something we may call nonlinear dynamics? We do not want to get philosophical here but we find it important to stress that there is so much room for different kinds of nonlinearity that the rejection of a linear null hypotheses does not bring us much closer to a proof of the presence of chaos. The fact that we used statistics derived from chaos theory should not mislead us in this respect!

7.2 Nonlinear statistics for system discrimination

In the previous section we discussed how to discriminate between nonlinear behaviour and linear stochasticity on the basis of a time series. Similarly, we can use nonlinear statistics to discriminate between different, possibly nonlinear, states of a system. Typical questions concern, for example, differences in the electroencephalographic (EEG) pattern between different brain activities, sleep/rest, eyes open/closed, etc. Or one could hope to distinguish patients with a high risk of sudden cardiac death from normal subjects on the basis of ECG recordings or series of interbeat intervals. The basic idea is to compute some property, say a dimension estimator or a prediction error, or any other nonlinear statistics for data sets from both stages, and to look for significant nontrivial differences. If we employ nonlinear theory, "nontrivial" means that we are looking for differences which are not easily detected by power spectral methods or even by looking at a plot of the time series. Many characteristic changes of, for instance, the EEG during different brain states are described in the clinical literature in detail. Some of them are rather obvious. To be called nontrivial, a result should not fall short of the textbook knowledge. However, some "obvious" (to the clinician) features are quite hard to detect automatically, and automatic detection may be essential for long-term monitoring.

As any statistical test, nonlinear discrimination of time series data must yield significant results. What exactly that means depends on the aim of the study. Suppose we have K time series from patients suffering from a certain disease and also K time series from healthy controls. Now we can compute some nonlinear statistic λ, say an estimator of the correlation dimension, for each of the total of $2K$ sets. One possible question would be if the results are compatible with the null hypothesis that the dimension is the same for both groups. We could test this by computing the mean $\langle \lambda \rangle = K^{-1} \sum_{i=1}^{K} \lambda_i$ and the standard error of the mean $\delta \langle \lambda \rangle = [K^{-1}(K-1)^{-1} \sum_{i=1}^{K} (\lambda_i - \langle \lambda \rangle)^2]^{1/2}$ for both classes. With this knowledge and assuming normal errors we can compute the probability of a false rejection and, if it is small, reject the null hypothesis at the corresponding level of significance.[8]

No matter how small the actual difference between the two groups is, we will be able to detect it as statistically significant if we are able to take K large enough. This is due to the fact that the standard error of the mean decreases as $1/\sqrt{K}$ with the number of samples. Usually, the clinically interesting question is a different one. Can we use the difference to decide *for an individual patient* to which class she/he belongs and, if so, with what probability of being wrong? This question is not determined by the error of the mean but by the variance $\hat{\sigma}^2 = (K-1)^{-1} \sum_{i=1}^{K} (\lambda_i - \langle \lambda \rangle)^2$, which does not decrease with K. In this case the only way to increase the power of a test is by choosing the discriminating statistic λ. Provided K is not too small, a decision will be wrong on average in 5% of the cases when the mean values of λ for the two groups are about 4σ apart. A separation of 2σ yields about 30% false classifications. Thus we want to have the mean values as different as possible, together with small variances. Note that for most clinical applications, 30% would be an unacceptably high error probability.

So far we have not said anything about the statistical errors we expect for the various quantities which are candidates as nonlinear statistics λ. This is for two reasons. The first is that we find it misleading to quote a statistical error of, say,

---

[8] The relevant statistical test, Student's t-test, is described in any textbook on statistics. As a rule of thumb, for groups of equal and not too small size which have comparable variances, the mean values for both groups should be at least three standard errors of the mean apart at the 95% significance level.

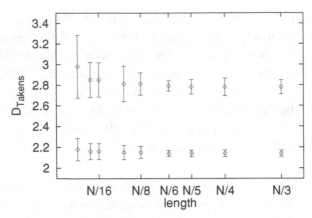

Figure 7.5 Mean and standard deviation of Takens's estimator D_{Takens} computed for 32, 20, 16, 10, 8, 6, 5, 4, 3 segments of the Taylor–Couette flow data. The upper row of points was obtained with embedding dimension $m = 6$, the lower with $m = 3$. The estimator for the whole data set of 32 768 points yields $D_{\text{Takens}} = 2.75$ for $m = 6$ and we extrapolate the statistical error to be ± 0.07. The estimates with $m = 3$ are much too small but, due to the smaller variance, are more useful for statistical discrimination.

the correlation dimension while so many systematic uncertainties are present. The second reason is that, honestly, we do not know. The actual variance of an estimator depends on the estimation algorithm (how we pick the scaling regions, etc.), the embedding parameters used, the length, sampling rate, and noise level of the time series, the number of degrees of freedom of the system and its macroscopic geometry. All dependencies are nontrivial and, on the whole, P2C2E,[9] as Iff puts it in Rushdie (1990). If a statistical error is really needed, for example in order to give an error probability for discriminating statistics, the Monte Carlo method is usually the most useful. Let us give an example.

Example 7.6 (Statistical error of the dimension of Taylor–Couette flow data).
In Example 6.3 we computed the correlation integral of the Taylor–Couette flow data, Appendix B.4. At that time we were rather vague about the actual dimension we would estimate on the basis of these results. We feel that the self-similarity of the set is much better represented by Fig. 6.5 than by any number we could give. Since, in the present context, having a number is more important than if that number is actually close to the *true* dimension, let us compute Takens's estimator D_{Takens} of the correlation dimension, Eq. (7.8). We use delay 5, embedding dimensions 3 and 6, and compute the integral with $\epsilon_0 = \langle (s - \langle s \rangle)^2 \rangle / 2$, that is, half of the variance of the time series. Now we split the series of $N = 32\,768$ points into various numbers of

---

[9] Process Too Complicated To Explain.

shorter segments. For each segment we compute D_{Takens} and determine the variance for each segment length. Finally, we can try to extrapolate to the variance of the estimator for the whole series. See Fig. 7.5.

Now imagine there were two distinct states of the system, one with dimension 2.9, the other with dimension 3.6. Let us make the (optimistic) assumption that the estimator D_{Takens} is indeed roughly proportional to the true attractor dimension and the (even less realistic) assumption that the standard deviation does not increase with the dimension. Then the data with $m = 6$ shown in Fig. 7.5 suggests that given a time series of 2048 points ($D_{\text{Takens}} = 2.85 \pm 0.17$) we can determine to which of the two states it belongs with a 5% chance of being wrong. For $m = 3$ the attractor is insufficiently unfolded and the numbers thus computed are far from the true dimension of the attractor. Nevertheless it could be a good quantity for discrimination, since one can even expect an improved discrimination power. $\square$

We find that the discrimination power of Takens's estimator of the correlation dimension in Example 7.6 is not very impressive. However, numerical experiments with various data of known dimensionality show that the figures are quite typical. Further, contrary to what some authors claim, the power does not seem to change dramatically when other dimension estimators are used (mean or median of pointwise dimension, etc.). Much more important is a good choice of embedding parameters. Remember that for statistical discrimination it is not required that the statistic λ be a reasonable estimator of anything meaningful. Thus we can, for example, choose too small an embedding dimension, which leads to a severe underestimate of the dimension. Nevertheless, due to the improved statistics, the variance is smaller, leading to increased discrimination power. Since there seems to be no simple rule as to how to design good nonlinear discriminating statistics, we propose to perform a Monte Carlo investigation on artificial signals which are comparable in data quality and dimensionality.[10] Exercise 7.2 gives an example of a system with tunable attractor dimension.

7.3 Extracting qualitative information from a time series

In this section we will discuss some methods of visualising information contained in a time series. We call it qualitative information because we do not require the methods to yield quantitative results. Of course, we can turn almost anything into a number if we wish to, but here we will relax the usual criteria of reproducibility and

---

[10] It is very bad practice to use the *data* to design the discriminating statistics. The resulting parameter λ cannot be used in a test of significance on the *same* data. Claims based on such *in-sample* statistics, such as Morfill & Schmidt (1994), must be taken with great suspicion.

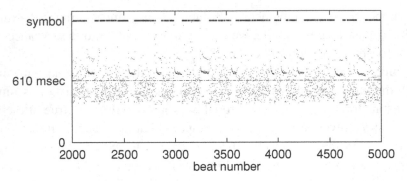

Figure 7.6 Interbeat intervals of a patient with recurrent episodes of atrial fibrillation, together with a symbolic encoding. See the text of Example 7.7.

invariance. We will only mention a few exemplary methods because we are in the domain of *ad hoc* algorithms here: possible time series situations and the questions one wants to answer vary vastly, as do the possible means of data visualisation.

One general class of methods relies on a symbolic representation of an otherwise continuous (high-resolution) time series. This severe reduction of information can be used to enhance relevant features. It is particularly useful to track qualitative changes in the dynamics, either due to a parameter change or occurring spontaneously. The idea is to label specific patterns that you can find in the time series by symbols and to study the sequence of symbols instead of the time series. The simplest "pattern" is that the value of the observable exceeds some given threshold. Let us give a simple example.

Example 7.7 (Atrial fibrillation data). In Example 7.2 we tested concatenated episodes of atrial fibrillation for nonlinearity. Let us have a look at the full time series (see Appendix B.6) with alternating fibrillation and normal episodes. In Fig. 7.6 we show part of the time series of interbeat intervals, plotted against the beat number (which increases monotonically with, but is not proportional to, time). We perform the following symbolic encoding:

$$\alpha_n = \begin{cases} 1 & s_n < a \quad \text{and} \quad s_{n-1} < a \\ 0 & \text{otherwise,} \end{cases} \tag{7.10}$$

that is, we plot a little symbol on the horizontal if two successive heartbeats are shorter than $a = 610$ ms. Roughly speaking, whenever the symbol is $\alpha_n = 0$ for a stretch of time, we find the patient in the state of fibrillation.[11] This encoding was

---

[11] Of course there are much smarter ways to detect fibrillation. In fact, the data base that this set was taken from, Moody & Mark (1992), has been set up for the development of such methods.

chosen in a more or less *ad hoc*, trial-and-error way, which is not untypical for symbolic methods. □

Symbol sequences resulting from *coarse-graining* of continuous time series have been used by several authors for quantitative study. The concept is particularly attractive if the symbols can be given some reasonable interpretation. Also, "complexity" or "information" measures for symbol sequences involve fewer technical problems than the equivalent for continuous signals. However, this means that we shift all the problems over to the choice of symbolic encoding, which is largely arbitrary. Let us have a closer look at one of the more popular information measures, the Shannon entropy of the symbol sequence (derivates come under such names as "pattern entropy", "approximate entropy", "coarse-grained entropy", etc.). From an *alphabet* of k symbols (or *letters*) one can form k^m different *words* $w_n = (\alpha_{n-m+1}, \ldots, \alpha_n)$ of length m. Let p_j be the probability of finding the word w_j when picking an arbitrary word in a symbol string. Then the Shannon entropy of block length m of the string is given by [Shannon & Weaver (1949); cf. Eq. (11.21)]:

$$\tilde{H}(m) = -\sum_{j=1}^{k^m} p_j \lg p_j. \tag{7.11}$$

It is measured in bits when we take the base two logarithm (lg). The entropy of the source is then $\tilde{h} = \lim_{m \to \infty} \tilde{H}(m)/m$. Although the Shannon entropy is well defined for a source of symbol strings, it is not a unique characteristic for the underlying continuous time series. The reason is that its value depends on the choice of symbolic encoding, for instance the value of the thresholds chosen. This implies that it is *not invariant* under smooth coordinate changes, not even under a simple change of experimental units. Some limited invariance can be recovered when $\tilde{h}$ is maximised over all possible thresholds. But we have to face the fact that proper invariance under smooth coordinate changes has its price; the $\lim_{\epsilon \to 0}$ we got rid of by coarse-graining. In the theory of symbolic dynamics for dynamical systems a well-defined procedure exists for converting a string of real data into symbols. The partition in phase space which is used to assign the symbols has to possess the property of being *generating*. The problem with such partitions is that they are difficult to construct, even in mathematical models, and that the border lines between different partition elements in phase space are highly nontrivial and have to be transformed if one transforms system variables. Thus although in principle this procedure allows us to compute an invariant quantity, the KS-entropy (which will be introduced in Section 11.4), it is highly impracticable for time series analysis. In particular for data with weak determinism, on the basis of symbolic dynamics one cannot do better than to compute noninvariant quantities.

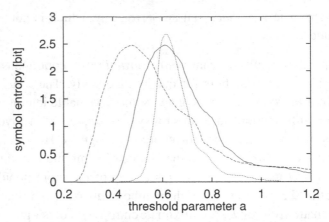

Figure 7.7 Pattern entropy of fibrillation data as a function of the threshold in the binary encoding. Solid line: unprocessed data. Dashed line: monotonically rescaled data. Dotted line: first differences. See the text of Example 7.8.

Example 7.8 (Pattern entropy of fibrillation data). For symbolic encodings of the form of Eq. (7.10) of the atrial fibrillation data (Example 7.7), we compute the Shannon entropy $\tilde{H}(3)$ of block length 3, Eq. (7.11), for a range of thresholds a. The solid line in Fig. 7.7 shows $\tilde{H}(3)$ as a function of a. A random string would yield $\tilde{H}(3) = 3$ bit. When by some change in the measurement procedure (different R-wave detector, etc.) we rescale the signal prior to the encoding, say $s'_n = s_n \sqrt{s_n}$, for any fixed a the entropy is changed (dashed line). However, the maximum value $\tilde{H}_{\max}$ remains the same as long as the change is a monotonic function. This is not the case for more complicated coordinate changes, such as taking first differences, $s''_n = s_n - s_{n-1}$. Now the value at the maximum is also altered (dotted line). □

Further reading

Apart from the classical works on the method of surrogate data, Theiler *et al.* (1992) and (1992a), interesting discussions of a more fundamental nature are included in Theiler & Prichard (1995). More material, also on possible pitfalls, can be found in Theiler *et al.* (1993) and in Schreiber & Schmitz (2000). Other tests for determinism are described in Brock *et al.* (1988), Sugihara & May (1990), Casdagli (1991), Kaplan & Glass (1992), Kennel & Isabelle (1992), Wayland *et al.* (1993), Salvino & Cawley (1994), Paluš (1995) and Barahona & Poon (1996). The construction of a generating partition is discussed in Grassberger & Kantz (1985) and in Giovanini & Politi (1991). A recent review on Surrogate data tests is contained in Schreiber (1999).

Exercises

7.1 Use a data set consisting of 2048 iterates of the Ikeda map (Exercise 6.3, just the real part), contaminated with 20% Gaussian white noise. Is it possible to reject the linear null hypothesis at the 95% significance level? If your computer is fast enough, make several tests in order to make a significant statement.

7.2 Consider the Mackey–Glass delay differential equation [Mackey & Glass (1977)]:

$$\dot{x}(t) = \frac{ax(t - t_f)}{1 + x(t - t_f)^{10}} - bx(t).$$

Introduce a discrete time step $\Delta t = t_f/m$ and show that the above can be written as an approximate $m + 1$ dimensional map in delay coordinates:

$$x_{n+1} = \frac{1}{2m + bt_f} \left((2m - bt_f)x_n + t_f a \left(\frac{x_{n-m}}{1 + x_{n-m}^{10}} + \frac{x_{n-m+1}}{1 + x_{n-m+1}^{10}} \right) \right).$$

This map can be used to create time series with chaotic attractors of different dimension. The delayed feedback time t_f can be used to vary the dimension, even to high values.

7.3 Determine the mean and variance of Takens's estimator (using the routines in the TISEAN package) for collections of short Mackey–Glass time series with different dimensions. Obtain a more precise estimate from a single long trajectory.

Chapter 8

Selected nonlinear phenomena

In the preceding sections we have already discussed the properties of chaotic dynamics, namely instability and self-similarity. But nonlinear systems possess a much richer phenomenology than just plain chaos. In fact, non-chaotic motion on stable limit cycles is very typical in real world systems. Hence we devote the first section to them. Synchronisation, which is also a nonlinear phenomenon and clearly related to ordered motion, will be discussed in Section 14.3. The aspects we describe in the following sections can be roughly divided into two classes. The first consists of problems we may have with the reproducibility or stationarity of the signal even though all system parameters are kept fixed. Examples are transient behaviour and intermittency. The other group comprises what can happen under changes of the parameters. We will discuss the various types of attractor and how transitions between them occur, the *bifurcations*.

8.1 Robustness and limit cycles

When speaking of nonlinear dynamics, the immediate associations are bifurcation scenarios and chaos, i.e., the sensitivity of solutions either to small changes of the system parameters or to small perturbations of the system's state vector. What is often overlooked is that a much more typical feature of nonlinear systems is the existence of stable limit cycles, which are often very robust against all kinds of perturbations.

When observing a periodic signal, a Fourier transform may be a very good representation of its features. As is well known, the power spectrum (or the autocorrelation function, respectively), uniquely determines the corresponding *linear model*. However, an autoregressive model[1] AR(m), $s_{n+1} = \sum_{i=1}^{m} a_i s_{n-i+1} + \xi_n$ is

---

[1] Those who are unfamiliar with AR-models and wish to fully understand the remainder of this paragraph are advised to first read Section 12.1.

clearly a bad model for purely periodic motion. In order for a linear model to produce periodic oscillations with fixed amplitudes, its damping and the noise amplitude must be exactly zero. The corresponding AR-model hence must be marginally stable, or otherwise said, the eigenvalues of its Jacobian must be exactly on the unit circle in the complex plane. This is not only a very non-generic situation that requires unphysical fine tuning, but it is also highly unstable as soon as we couple noise to the dynamics. The amplitude of this harmonic signal will start to exhibit a diffusion process, i.e., it will show non-stationary fluctuations. Hence, such a model cannot produce periodic oscillations of almost constant nonzero amplitude.

The situation is very different for limit cycles of nonlinear systems.[2] The prototypic situation (which appears generically after a Hopf bifurcation, see below) is the over-damped motion inside a Mexican hat potential, described by

$$\dot{\phi} = \omega, \qquad \dot{r} = r - r^3. \tag{8.1}$$

Due to the nonlinearity, the amplitude of this oscillation is very robust. Such a mechanism is active in the well known van der Pol oscillator,

$$\dot{v} = \rho(1 - x^2)v - x, \qquad \dot{x} = v \tag{8.2}$$

whose mechanical equivalence was making use of the centrifugal forces to control the rotational speed of steam engines.

Hence, when we encounter a periodic time series the amplitude of which does not vary much, one has to suspect a limit cycle of a nonlinear system rather than a linear oscillation. A typical signature of a limit cycle in delay representation is a hole, or at least a low density region about the centre of the distribution of points.

Example 8.1 (Human breathing). The time series representing the air flow through the nose of a human (described in Appendix B.5) is of typical limit cycle nature. This is most clearly seen in Fig. 10.8 of Example 10.7 (page 192) and is discussed there. □

Example 8.2 (Articulated human voice). The units of articulated human voice are called *phonemes* (comparable to syllables in written language). The dynamics corresponding to an isolated phoneme is of limit cycle type, as we will discuss in Example 10.6 (page 190) and in Section 13.2.4. □

Data of limit cycle type lead to two different difficulties. First, if the data are restricted to a limit cycle, it is typically not possible to construct a unique dynamical nonlinear model for them just using the data, since they fill only a very small part of the phase space. Second, limit cycles may be of different complexity (see below

---

[2] In engineering or biological systems which are meant to show stable periodic oscillations, this is the most natural situation. A nonlinear mechanism is indispensable to stabilise the amplitude.

for quasi-periodic motion). Noise driven limit cycle motion can easily be mistaken for chaotic motion on a ribbon-like attractor, and in fact there are model systems (such as the Rössler system), which, for certain choices of parameters, possess chaotic attractors whose time series look very much like noise limit cycles: their amplitude fluctuates irregularly, but the period of each single oscillation is almost constant.

8.2 Coexistence of attractors

One puzzling feature which may occur in nonlinear systems is that the repetition of the experiment with the same parameters may yield a qualitatively different result. One possible explanation for such behaviour is that different attractors coexist. It depends on the initial condition, on which of them the trajectory will settle down. The region of initial conditions in phase space leading to a given attractor is called its *basin of attraction*. The basins of two coexisting attractors can be interwoven in a very intricate way, possessing fractal boundaries [Grebogi *et al.* (1983)]. The simplest coexisting attractors are two different stable fixed points or limit cycles which are very often related by some symmetry. For example, in many hydrodynamical experiments one observes rolls whose orientation may be exchanged after a restart. In the famous Lorenz system [Lorenz (1963); see also Exercise 3.2] there exist ranges of parameters where stable fixed points and a chaotic attractor coexist (see Fig. 15.1); the same goes for the Ikeda map at standard parameters [Ikeda (1979)], and in many mathematical models even multiple chaotic attractors can be observed.

Unless the system is reset externally during the observational period, a single time series can only represent one of the possible attractors. Without further knowledge about the system, in particular its symmetries, it is hard to infer the existence or nonexistence of multiple attractors without repeating the experiment using different initial conditions. However, the occurrence of an asymmetric object in a situation where the physics should induce symmetry suggests the coexistence of its mirror image as a separate attractor.

8.3 Transients

By definition, transient behaviour will disappear after some time. Therefore at first sight it might neither be relevant nor interesting for time series analysis. However, both impressions are wrong.

An attractor in a dissipative nonlinear system has Lebesgue measure zero in phase space. Therefore, the probability that an arbitrarily chosen initial condition already lies on the attractor is zero. Normally, one has to wait some time until the trajectory has settled down there. During this *transient time*, the motion may have completely different properties than on the attractor itself. For highly dissipative systems, the transient time can be extremely short. Think of an ordinary stable fixed point which is approached exponentially fast by a trajectory starting in its neighbourhood. However, in certain cases transients can last arbitrarily long. In particular, spatially extended systems can have transient times which grow exponentially or even faster with the size of the system, see e.g. Crutchfield & Kaneko (1988). Long transients may be a problem for time series work since the dynamics is non-stationary (one part is transient, one part is the motion on the attractor). In certain cases, a long transient can yield effectively stationary dynamics. This transient can be nontrivial even if the attractor itself is simple (a fixed point or a limit cycle).[3]

Quite common in many nonlinear model systems is the coexistence of a chaotic *repeller* or *chaotic saddle* (i.e. an unstable object) and a stable (attracting) fixed point. In such a situation a typical trajectory first shows transient chaos, and only after some time does it settle down on the fixed point. When restarting the system many times, one finds an exponential distribution for the transient times T_{trans}, i.e. $p(T_{\text{trans}}) = e^{-\eta T_{\text{trans}}}$, where η is called the *escape rate*. In a laboratory experiment it can be interesting to measure η. Moreover, it has been shown that the transient chaotic motion can again be characterised by a maximal Lyapunov exponent and a dimension; see Kantz & Grassberger (1985), Grebogi *et al.* (1986). However, such a characterisation relies on the existence of an invariant subset in the phase space (called repeller or chaotic saddle), whose properties are displayed by a typical transient, which means that again the details of the initial conditions are irrelevant. Of course there are many situations where the trajectories keep a kind of memory of the initial condition until they settle down on the attractor. In such a case, each single transient is special.

8.4 Intermittency

Another range of common nonlinear phenomena is collectively called *intermittency*, which means that the signal alternates between periodic (regular, laminar) and chaotic (irregular, turbulent) behaviour in an irregular fashion. The chaotic phases can be long or they can look like short bursts, depending on the properties of the system.[4]

---

[3] It is sometimes very difficult to prove mathematically that a given dynamical system has a chaotic attractor. The Hénon map [Hénon (1976); see also Exercise 3.1] displays irregular behaviour in computer simulations when the standard parameters are used. In theory this could be a very long chaotic transient before some periodic attractor is reached. It has been proved, though [see Benedicks (1994)], that there is a set of parameters of positive measure where the map has a chaotic attractor.

[4] In spatially extended systems, such as weakly turbulent flow, intermittency also means a simultaneous appearance of turbulent and laminar phases at different places in real space, called *spatio-temporal intermittency*.

Several different scenarios have been proposed to explain intermittency in purely deterministic systems. In some systems, intermittency occurs generically (Hamiltonian chaos, some hydrodynamical systems). In others, it occurs only for critical values of some control parameter. In the latter case, a periodic orbit becomes unstable under the change of the control parameter or it disappears without the creation of a new stable one. For most of the theoretical mechanisms one can compute the statistics of this type of dynamics, i.e. the distribution of the inter-burst times and the durations of the bursts, and, moreover, one can predict how the average length of the periodic phases is related to the closeness of the control parameter to its critical value. Thus for a detailed study of intermittency one has to be able to change the control parameters of the experiment. For more information see any textbook on chaos, in particular Bergé *et al.* (1986), or the original paper by Pomeau & Manneville (1980). A classical experimental observation of intermittency is reported in Bergé *et al.* (1980) for Rayleigh–Bénard convection.

The alternation between two different dynamical regimes involves new time scales related to the duration of each of the two phases. Since the distribution of these times contains tails, often of a power-law type, the autocorrelation function can decay quite slowly and there can be considerable power in the low frequencies. For certain types of intermittency the power spectra at low frequencies are of the $1/f^\alpha$-type, where $\alpha > 0$ can be close to unity [Ben-Mizrachi *et al.* (1985)].

In data analysis, the term intermittency is used in a broader sense. It does not require that the laminar phase be strictly periodic, but just that it be more regular than the bursts. Such a time series poses several problems. The chaotic bursts clearly belong to the invariant measure, but very often, due to their rare appearance, we have only insufficient information about them. Inside the bursts the dynamics may be much faster, such that the sampling rate of the measurement might be too low to resolve them as accurately as necessary for a successful analysis. But there may also be too few bursts in the time series. Furthermore, due to the inhomogeneity of the series one may have various additional difficulties with some methods of analysis. For example, the considerable amount of power in the low frequencies closely resembles properties of non-stationary data. Unbiased dimension estimates become difficult, since with a finite amount of data one predominantly observes the laminar phase when one computes the correlation sum Eq. (6.1) on the small length scales.

Example 8.3 (Vibrating string). The data taken from a vibrating string (Appendix B.3) consist of pairs $(x(t), y(t))$, taken with a high sampling rate. The phase space object generated by the data at first sight looks like a two dimensional ribbon with perhaps some fractal structure on it. (See Fig. 8.1; notice that we do not call it an

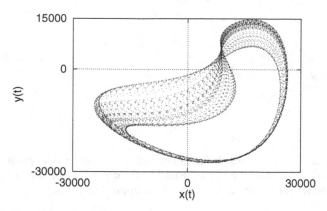

Figure 8.1 The object in phase space formed by the data of the vibrating string.

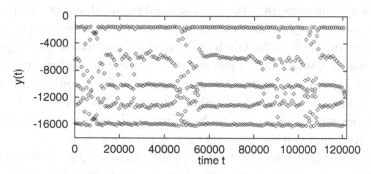

Figure 8.2 Poincaré surface of section of the string data. At each intersection with the y-axis from right to left the y-coordinate is plotted.

"attractor" at this stage of analysis.) In the Poincaré surface of section (Fig. 8.2) one observes relatively long phases of periodic motion interrupted by less regular parts (compare also Fig. 1.6). For example, the data segment from $t = 50\,000$ to $60\,000$ consists of about 21 oscillations of a very clean period five *limit cycle*. There are two possible explanations for this intermittency. The system could be close to a bifurcation, due to which the limit cycle became unstable and is embedded in a chaotic attractor. Alternatively, the system is close to a crisis point and a parameter of the experiment is slightly fluctuating, switching between two distinct attractors. On the basis of a single time series it can be very difficult to decide which explanation is correct. It is much easier to check these possibilities while the experiment is being run and parameters can be changed.

In this case, the data are good enough for a more profound analysis. Let us first explain what a backward tangent bifurcation is. Consider a one dimensional map. A graphical method of iterating a map is to follow a horizontal line at the current

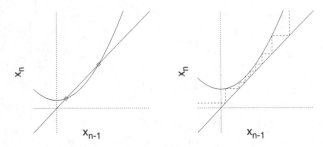

Figure 8.3 Backward tangent bifurcation. Left panel: subcritical parameter. The left fixed point is stable, the right unstable. Right panel: The supercritical map traps trajectories for many iterates around the approximate tangency.

value $x_{n+1} = f(x_n)$ to the diagonal and to use this position as input for the next iteration. Therefore, a trajectory x_0, $x_1 = f(x_0)$, ... is represented by a step-like sequence of straight lines. In particular, the fixed points of a map are given by the intersections of its graph with the diagonal. Let our map depend on a parameter, such that at a critical value the graph of the map is tangential to the diagonal. For a slightly smaller – sub-critical – parameter value there are no tangencies but true intersections, leading to one stable and one unstable fixed point (left hand panel in Fig. 8.3). Just above the critical point, the graph slightly misses the diagonal and trajectories which come close to the approximate tangency go through many iterations before they can escape (right hand panel in Fig. 8.3).

For the experimental data, the dominance of a period-five cycle in the Poincaré surface of section suggests plotting the fifth iterate of the Poincaré map, i.e. the points of Fig. 8.2 versus their fifth iterate. When represented in this way, a period five orbit would be represented by five isolated points on the diagonal. The result for the present data is shown in Fig. 8.4. The data lie on a one dimensional curve with five maxima and minima each, and there are five approximate tangencies with the diagonal. Thus we are slightly above a backward tangent bifurcation of a period five orbit.

Note that in order to study the statistics of the intermittent switching between regular and irregular behaviour, we need a record of many of the switching events. The present data set contains only a few of these although the data set contains many points. Nevertheless, Fig. 8.2 contains a nice additional confirmation of our interpretation. The few initial points of each laminar phase lie slightly off the locations of the "imaginary" period five orbit which is then slowly approached, and, towards the end of each laminar phase, left into the opposite direction. Compare this to the path of the trajectory in Fig. 8.3! □

The above example shows that we need the knowledge of nonlinear phenomena for a reasonable analysis of such data. If a typical cycle length is taken as the longest

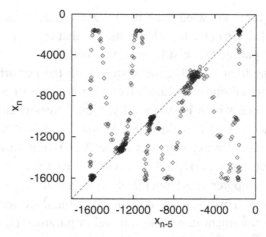

Figure 8.4 The fifth iterate of the Poincaré map of the vibrating string, obtained by plotting the points of Fig. 8.2 versus their fifth iterate. This graph of $f^{(5)}$ is typical of intermittency in the vicinity of a backward tangent bifurcation.

time scale in the signal, then treating the time series as representing an ordinary chaotic attractor, long time correlations due to intermittency will lead to wrong results. Indeed, there exists a chaotic attractor, but it is not sufficiently sampled by the data to compute any of its characteristics. If, instead, there is some drift of experimental parameters, the time series must be considered non-stationary and we would have to be extremely careful when applying methods described in this book.

In the particular example given above we were able to represent the dynamics by a one dimensional map, making a detailed analysis possible. In a general situation, intermittency is much harder to study. Typically, one would try to verify the expected scaling behaviour. With a single but long series of measurements, the statistics of the bursts may be studied. It is, however, highly desirable to have a sequence of data sets taken with different control parameters. Then one can study the scaling behaviour which occurs when the control parameter approaches a critical value.

8.5 Structural stability

Another typical feature of nonlinear systems is their lack of robustness against slight changes of parameters. In the literature one finds only very few nonlinear systems that are *structurally stable*, all of which are irrelevant for modelling purposes. The most important examples are the *hyperbolic* systems. We do not want to formally introduce the notion of hyperbolicity here; it is explained in Chapter 15 on control, and requires knowledge of the tangent space structures to be introduced in Section 11.2. *Structural stability* means that *every* tiny perturbation of any of the system's parameters can be compensated by a smooth transformation of the variables. One then says that the system with the perturbed parameters is *conjugate* to the original one. As stated, we are not aware of any chaotic model with physical relevance which is structurally stable everywhere in its parameter space, but experience shows that this kind of stability is not really required for practical purposes. The Hénon map (Example 3.2) is not formally structurally stable, but its attractor can be reproduced with all its features (up to numerical accuracy) on every computer. Experimental systems are always subject to tiny fluctuations of parameters. This means that in an experiment we see the behaviour of the system which is dominant in a small interval of parameters, and, in particular, no behaviour which is typical of values which have measure zero in parameter space.[5]

Nevertheless, this reasoning is important because of the possibility of bifurcations in structurally unstable systems. As we shall describe below in more detail, bifurcations are changes in the properties of an attractor when changing a control parameter.

Concerning data analysis, severe problems can arise when we try to set up model equations for structurally unstable systems. It is possible that details of the fit which one would expect to be irrelevant can influence the resulting model in such a way that it is unable to reproduce the observed data under integration or iteration, although the prediction error is very low.

Example 8.4 (Problems arising when modelling an electric resonance circuit). Time series from the nonlinear electric resonance circuit described in Appendix B.11 were recorded for many different temperatures in order to investigate how the properties of the nonlinearity depend on the temperature of the ferroelectric material inside the capacitance. For every data set, a second order differential equation with a bivariate polynomial nonlinearity was fitted, as will be described in Example 12.4. In order to verify the validity of the model, the differential equation was integrated. A model was considered good only if the time series generated by it formed an attractor that matches the experimental data. At certain temperatures, it was difficult to find a good model. Many failed by producing limit cycles instead of chaotic attractors, depending on the maximal order of the polynomial used for the fit. We conjecture that this is due to the lack of structural stability in this Duffing-like system.   □

---

[5] This is called *stochastic stability*, and the Hénon map can be shown to be stochastically stable [Benedicks & Viana, private communication].

8.6 Bifurcations

Bifurcations are abrupt changes of the attractor geometry or even topology at a critical value of the control parameter. This can happen due to the change of stability of existing objects but also through the birth and death of an invariant set. We start with bifurcations which are due to the change of a single parameter and which depend only on local properties of the dynamics. These are called *local codimension one bifurcations*. Locality implies that the invariant subsets are fixed points and periodic orbits in the vicinity of fixed points. There are exactly five different types, four of them are illustrated in Fig. 8.6. For simplicity, we will restrict the discussion to maps. Hence, everything applies to the Poincaré maps of flows as well, whose fixed points represent simple limit cycles with exactly one intersection with the Poincaré surface of section. Equivalent bifurcation types (except for the period doubling bifurcation) also exist for fixed points of flows, but the mathematical description is slightly different. Consult any textbook on nonlinear dynamics for details.

One of the most striking features of bifurcations is their *universality*: Locally, in the neighbourhood of a bifurcation of given type (closeness is required both in the parameter space and the state space), all systems are equivalent and can be mapped onto each other. This means in particular that additional dimensions of the phase space beyond those involved in the bifurcation are irrelevant for the bifurcation scenario. There are many more bifurcation types which either require more parameters to be adjusted or which are non-local. We will discuss briefly some of them at the end of this section. The universality of the local bifurcations makes it possible to give standardised functional forms of the dynamics in the vicinity. The prototypic maps we will give with every bifurcation are usually called *normal forms*: they are the most simple dynamical systems exhibiting the specified type of bifurcation at parameter value $a = 0$ and at the position in space $x = 0$. The normal forms for bifurcations of fixed points in flows are first order differential equations. For a bifurcation diagram of a realistic system (Lorenz-like laser system) see Fig. 15.1.

The *period doubling bifurcation* has been observed in almost every nonlinear mathematical model system and in very many experiments. At the critical parameter value a stable period p orbit becomes unstable, and at the same instant a period-$2p$ orbit emerges from it. The map $x_{n+1} = (-a - 1)x_n + x_n^3$ is a prototypic example for this behaviour, where for $a = 0$ the fixed point $x_0 = 0$ becomes unstable and a period two orbit with the two points $x_\pm = \pm\sqrt{a}$ emerge. In predominantly oscillatory systems, one often prefers the formulation that *subharmonics* are created. In a flow, a period doubling bifurcation of a fixed point cannot exist, since the trajectory after bifurcation has to be continuous. However, a limit cycle can bifurcate into

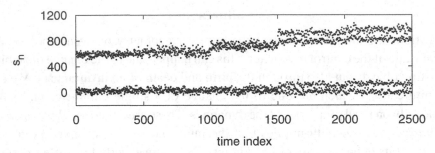

Figure 8.5 Five time series segments of length 500 each from an Nd:YAG multimode laser (see Example 8.5). From segment one to five, the pump power was increased by about 0.2% in each step. The period doubling bifurcation seems to take place in between the third and the fourth segment. It cannot be excluded that it happens already before but is hidden by noise.

a more complicated limit cycle (in the vicinity of the original cycle, which after bifurcation is unstable), which may correspond to a period doubling bifurcation in the corresponding Poincaré map.

Example 8.5 (Period doubling bifurcation in a laser experiment). From the frequency doubling Nd:YAG laser (see Appendix B.12), a nice sequence of data are recorded which represent a period doubling bifurcation from period two to period four. We converted the original highly sampled flow data of the laser intensity into map-like data on a Poincaré surface of section. Time series of one coordinate of these intersection points are shown in Fig. 8.5. Every data segment is recorded for fixed pump power of the laser, which was increased by 0.2% from segment to segment. □

Period doubling bifurcations often come in a cascade. In this so-called *Feigenbaum scenario*, further period doublings occur at higher values of the control parameter, until at some accumulation point infinite period is reached and the motion becomes chaotic. Many details and universal features of the *period doubling cascade* are known [references can be found e.g. in Ott (1993)]. For example, the parameter values r_p where the pth period doubling occurs scale with their distance towards the critical value r_c as $|r_p - r_c| \propto \delta^{-p}$. You can easily study this behaviour yourself using the logistic equation $x_{n+1} = 1 - ax_n^2$. In experiments, parameters cannot usually be tuned and kept constant enough to be inside the extremely tiny intervals of the higher periods, and most experiments show period doubling only up to period 8, some up to period 16.

The *pitchfork bifurcation* looks very similar in its bifurcation diagram. Here, a stable fixed point becomes unstable and two new stable fixed points are created. If you look at the second iterate F^2 of a map F, then a period doubling bifurcation

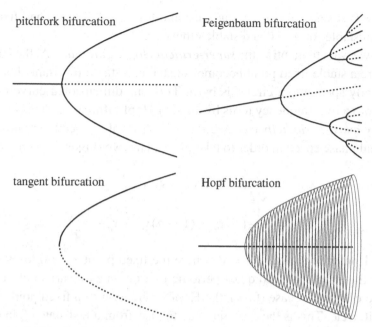

Figure 8.6 Sketch of different bifurcation scenarios. The abscissa represents the control parameter, the ordinate shows the location of the attractors. Continuous lines represent stable orbits, dotted lines unstable ones.

of F is a pitchfork bifurcation of F^2. This situation is realised in the laser system of Fig. 15.1. The map $x_{n+1} = (1 - a)x_n - x_n^3$ exhibits this behaviour for $a = 0$, where $x_0 = 0$ becomes unstable and $x_\pm = \pm\sqrt{a}$ are two stable fixed points for $a > 0$ (coexisting attractors).

The *tangent bifurcation* (*saddle-node bifurcation*) is the typical mechanism by which, inside a chaotic regime, a stable periodic solution can occur after a small change of the control parameter. It is accompanied by an unstable periodic orbit of the same period. Creation (or, when seen from the other end of the parameter scale, annihilation) of a pair consisting of a stable and an unstable periodic orbit is the mechanism for one type of intermittency (type I, see Section 8.4). Experimental data close to such a bifurcation are discussed in Example 8.3. On the periodic side of the bifurcation point, very often long chaotic transients occur, when the initial condition is not properly chosen on the periodic orbit. The prototypic map for this behaviour is $x_{n+1} = -a + x_n + x_n^2$, which has no fixed point for $a < 0$ and two fixed points for $a > 0$, only one of which is stable.

The *transcritical bifurcation* is another non-generic situation where a pair of one stable and one unstable fixed points collide at the bifurcation and they exchange their stability properties. The map $x_{n+1} = (1 + a)x_n - x_n^2$ shows this behaviour, where $x_0 = 0$ is stable for $a < 0$ and unstable for $a > 0$. The second fixed point $x_a = a$ is unstable for $a < 0$ and stable otherwise.

Finally, we want to mention the *super-critical Hopf bifurcation*. At the parameter value where a stable fixed point becomes unstable, a stable invariant closed curve (topologically speaking, a circle) is born. This one dimensional curve can itself bifurcate into a two-frequency torus by another Hopf bifurcation. This is the route to chaos by *quasi-periodicity* (see Section 8.7). A Hopf bifurcation requires a two dimensional phase space in order to take place. The two dimensional map

$$x_{n+1} = \frac{1}{\sqrt{2}}\big((1 + a)x_n + y_n + x_n^2 - 2y_n^2\big)$$

$$y_{n+1} = \frac{1}{\sqrt{2}}\big(-x_n + (1 + a)y_n + x_n^2 - x_n^3\big) \qquad (8.3)$$

exhibits a Hopf bifurcation at $a = 0$, where the fixed point $x_0 = 0$ loses stability. For $a > 0$ this map produced quasi-periodic motion on an invariant closed curve. In the time continuous case (flow), the Hopf bifurcation of a fixed point creates a stable limit cycle. This is the essential way to pass from a rest state of an ordinary differential equation to a truly dynamical state. A second Hopf bifurcation introduces a second frequency, so that generally one will find quasi-periodic continuous motion on a torus. In a Poincaré surface of section, this second Hopf bifurcation is a Hopf bifurcation of a fixed point into an invariant curve.

Many systems, in particular externally driven ones, exhibit *symmetry breaking bifurcations*. Such a bifurcation can be a pitchfork bifurcation. Imagine a symmetric potential of the type $V(x) = \alpha x^2 + x^4$, where α passes from positive to negative values and thus a single well turns into a double well. A particle subject to friction will settle down in either of the two wells and thus break the symmetry. When under further modification of parameters chaotic attractors develop from either of the fixed points, they are also asymmetric in an otherwise symmetric physical environment. The Duffing oscillator, which is described by the differential equation $\ddot{x} = -\rho\dot{x} + x - x^3 + A\cos\omega t$, shows such a kind of bifurcation [Guckenheimer & Holmes (1983)]. The nonlinear electric resonance circuit of Appendix B.11 is very similar to the Duffing oscillator and possesses such a symmetry breaking bifurcation and coexistence of mutually symmetric attractors.

The above bifurcations are all continuous in the sense that the response to an infinitesimal parameter change is infinitesimal in spite of the singularity. In contrast, the sudden global change of chaotic attractors is called *crisis*. It is a global bifurcation and thus difficult to study from the theoretical point of view. Several crisis scenarios can be distinguished. A crisis can for example consist of a sudden disappearance of the chaotic attractor. At a super-critical value of the control parameter the stable object is a periodic orbit, and there will typically be long chaotic transients. An attractor can also undergo a sudden change of its size, or two coexisting attractors can merge. The origin of such transitions is usually the collision of the attractor with the stable manifold of another invariant set at the critical parameter value. For values close to crisis, one very often observes a particular kind of intermittent behaviour, called *crisis induced intermittency* [Grebogi *et al.* (1987)].

Obviously, in order to study bifurcations with time series data, one has to be able to change the control parameters of the system more or less continuously.

8.7 Quasi-periodicity

We are familiar with quasi-periodicity[6] in harmonic motion. The time evolution can be decomposed into different parts which themselves are periodic, but the different periods do not match and thus there does not exist a period T such that a trajectory returns *exactly* to where it has been before. If the frequencies ω_i of the different components fulfil $\sum m_i \omega_i = 0$ only if all $m_i = 0$, the motion is non-periodic. The frequencies are said to be *incommensurate*, i.e. they are not related by rational numbers. In purely harmonic systems the standard Fourier analysis yields as conclusive results as in the periodic case; one observes peaks at $\omega = \omega_i$ and only a noise background elsewhere. Geometrically, the orbit in this case fills a possibly self-intersecting hyper-torus (with two periods, it looks like the surface of a doughnut).

In nonlinear systems we can find quasi-periodic solutions as well. Due to the nonlinearity we now find an interaction between the different oscillatory modes, and the Fourier transform possesses a peak whenever the condition $\omega = \sum m_i \omega_i$ is fulfilled for arbitrary integers m_i. In fact, due to the incommensurability of the ω_i, this means that there are peaks in any arbitrarily small frequency interval. In harmless cases the amplitudes of the peaks decay fast with a magnitude of m_i, so that few dominant peaks still stand out prominently. But this is not guaranteed, and it may be difficult to distinguish quasi-periodic motion from weak chaos just on the basis of a single power spectrum. Quasi-periodicity is quite a common route to chaos. A stable fixed point turns into a limit cycle via a Hopf bifurcation. The limit cycle goes through a second Hopf bifurcation into a two-frequency torus. Under further changes of the parameter the torus becomes unstable and may either develop into a three-torus via another Hopf bifurcation, or into a chaotic attractor with a dimension slightly larger than two, which still can have some overall structure resembling the torus. In particular, the two frequencies can dominate the chaotic

---

[6] Quasi-periodicity is a technical term reserved for the superposition of periodic behaviour of different period. Some people use the term incorrectly for almost periodic motion.

motion. This indicates that in time series analysis, quasi-periodicity with more than two basic frequencies, in particular if it is contaminated with noise, is difficult to identify.

Further reading

The phenomena described in this chapter are discussed in every textbook on non-linear dynamics, e.g. in Bergé *et al.* (1986), Schuster (1988), and Ott (1993), and very thoroughly in Guckenheimer & Holmes (1983). A more mathematical presentation can be found in Glendinning (1994). Experimental evidence for the quasi-periodicity route to chaos is reported in Gollub & Swinney (1975).

Part II

Advanced topics

Chapter 9

Advanced embedding methods

The reconstruction of a vector space which is equivalent to the original state space of a system from a scalar time series is the basis of almost all of the methods in this book. Obviously, such a reconstruction is required for all methods exploiting dynamical (such as determinism) or metric (such as dimensions) state space properties of the data. In the first part of the book we introduced the *time delay embedding* as *the* way to find such a space. Because of the outstanding importance of the state space reconstruction we want to devote the first section of this chapter to a deeper mathematical understanding of this aspect. In the following sections we want to discuss modifications known as *filtered embeddings*, the problem of unevenly sampled data, and the possibility of reconstructing state space equivalents from multichannel data.

9.1 Embedding theorems

A scalar measurement is a projection of the unobserved internal variables of a system onto an interval on the real axis. Apart from this reduction in dimensionality the projection process may be nonlinear and may mix different internal variables, giving rise to additional distortion of the output. It is obvious that even with a precise knowledge of the measurement process it may be impossible to reconstruct the state space of the original system from the data. Fortunately, a reconstruction of the original phase space is not really necessary for data analysis and sometimes not even desirable, namely, when the attractor dimension is much smaller than the dimension of this space. It is sufficient to construct a new space such that the attractor in this space is equivalent to the original one. We will have to specify what we mean by *equivalent*.

Glancing back at the first part of this book we see that, first of all, we want to exploit determinism in the data. Thus the time evolution of a trajectory in the reconstructed space should depend only on its current position in the new space and on nothing else. Uniqueness of the dynamics in the reconstructed space is not the only property we want. Dimensions, Lyapunov exponents and entropies are only invariant under smooth nonsingular transformations. Therefore, in order to guarantee that the quantities computed for the reconstructed attractor are identical to those in the original state space we have to require that the structure of the tangent space, i.e. the linearisation of the dynamics at any point in the state space, is preserved by the reconstruction process. Thus an *embedding* of a compact smooth manifold A into $\mathbb{R}^m$ is defined to be a map F which is a one-to-one *immersion* on A, i.e. a one-to-one C^1 map with a Jacobian $DF(x)$ which has full rank everywhere. The crucial point is to show under what conditions the projection due to the scalar measurements and the subsequent reconstruction by delay vectors form an embedding.

The problem with the embedding of scalar data in some $\mathbb{R}^m$ has two aspects. First, if the state of a system is uniquely characterised only if we specify simultaneously D variables, we have to construct D independent variables from the scalar time series for every time t that the signal is sampled. Once we know these variables, we also know that the dynamics live in some D dimensional manifold. Unfortunately, this manifold will most probably be curved. Thus the second problem is, since we would like to work with a global representation of the dynamics in some vector space, to find an embedding of a curved manifold in a Cartesian space.

9.1.1 Whitney's embedding theorem

This second part of the problem has already been solved by Whitney (1936). He proved that every D dimensional smooth manifold can be embedded in the space $\mathbb{R}^{2D+1}$, and that the set of maps forming an embedding is a dense and open set in the space of C^1 maps. Note that Whitney was neither thinking of fractals nor of scalar time series data. So he assumed that one had a description of the manifold in some local coordinates, and showed that one could embed it globally in an $\mathbb{R}^{2D+1}$.

Example 9.1 (Klein's bottle). The two dimensional object known as *Klein's bottle* cannot be embedded in three dimensions. □

Example 9.2 (Embedding of a limit circle). Let a one dimensional manifold be given by $\phi \in [0, 2\pi]$, with the neighbourhood relation $\text{dist}(\phi, \phi') = \min(|\phi - \phi'|, |\phi' + 2\pi - \phi|), \phi > \phi'$. The latter condition induces some curvature and makes the manifold topologically equivalent to a circle, which can be embedded only in $\mathbb{R}^m, m \geq 2$. Although such a representation appears natural to us, any generic deformation of a circle in two dimensions will cause self-intersections, such that, generically, one needs a space of $2 + 1 = 3$ dimensions for an embedding (see Fig. 9.1).

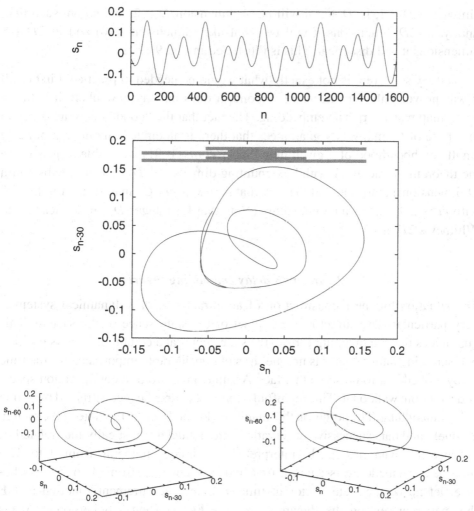

Figure 9.1 Embedding of a one dimensional manifold (limit cycle) in different dimensions. Upper panel: about 36 ms of a recording (after noise reduction) of one of the authors saying [u:], representing some nice limit cycle oscillation (see Example 10.6). Next panel: in a two dimensional embedding, several self-intersections violate uniqueness. Evidently, they cannot be removed by small deformations of the loop such as changing the time lag. In the upper part of the figure, the overlap structure in a one dimensional representation is shown schematically. Lower panel: two different views in three dimensions show that these self-intersections disappear generically when $m > 2D$.

A nice heuristic explanation of Whitney's theorem is presented in Sauer & Yorke (1993). A generic embedding of a D dimensional curved manifold in a D dimensional space will overlap itself on a set of nonzero measure. If the embedding is done in a space of dimension $D + 1$, the self-intersection will lie in a subspace of dimension $D - 1$, in $D + 2$ it will be of dimension $D - 2$, and so on, such that, finally, in $2D$ dimensions it will be point-like (dimension zero) and in $2D + 1$ dimensions it will disappear. This is illustrated in Fig. 9.1.   □

Whitney's theorem is not exactly what would be needed in practice. First of all it was proved only for integer D. Second, it does not say how likely it is that a given map really forms an embedding. The fact that the "good" maps are dense in the space of C^1 maps only guarantees that there is an embedding in an arbitrarily small neighbourhood of a given map. Sauer *et al.* (1991) were able to generalise the theorem to fractals A with box counting dimension[1] $D_F < d$ as a subset of a D dimensional manifold and to show that *almost every* C^1 map from A to the $\mathbb{R}^m$ with $m > 2D_F$ forms an embedding. (Note that for integer D_F, m is identical to Whitney's $2D + 1$.)

9.1.2 Takens's delay embedding theorem

The situation for the reconstruction of an attractor A of a dynamical system is very particular. The attractor is a sub-set of the state space of the system. This guarantees that the mapping **F** from the state vector at a certain time to its position one sampling interval later is unique. It is of considerable importance for the time delay embedding to preserve this fact. Assume a manifold in configuration space, and a random walk on it. This manifold *cannot* of course be reconstructed by a time delay embedding. The points on this manifold do not describe the state of the system uniquely and thus successive observations are not deterministically interrelated.

Since the dynamics can be interpreted as a unique mapping from A to A, a time delay embedding is again a time independent map from A to $\mathbb{R}^m$. As before, let $\mathbf{x}_n$ be the state vector at time n, s the measurement function and **F** the map representing the dynamics, $\mathbf{x}_{n+1} = \mathbf{F}(\mathbf{x}_n)$. Then a delay vector reads $(s(\mathbf{x}_n), s(\mathbf{F}(\mathbf{x}_n)), s(\mathbf{F} \circ \mathbf{F}(\mathbf{x}_n)), \ldots)$. Therefore, it is intuitively clear that knowing $s(\mathbf{x}_{n_i})$ at successive times n_i is equivalent to knowing a set of different coordinates at a single moment if the map **F** couples the different degrees of freedom.

To summarise the last paragraph, a delay reconstruction of only one observable is a very peculiar map from A to $\mathbb{R}^m$. Since it mixes dynamics and geometry, the words "almost every C^1 map" in the generalised version of Whitney's theorem cited above do not say that almost every delay map is an embedding. In particular, the projection due to the measurement process might destroy this property. Indeed, a negative example reported in Sauer *et al.* (1991) is the dynamics on a limit cycle with a period equal to once or twice the sampling time Δt. It cannot be

---

[1] The box counting dimension is explained in Section 11.3.1.

reconstructed by a delay embedding: the orbit collapses to one or two points. Nevertheless, Takens could prove [Takens (1981)] that it is a generic property that a delay map of dimension $m = 2D + 1$ is an embedding of a compact manifold with dimension D if the measurement function $s : A \to \mathbb{R}$ is C^2 and if either the dynamics or the measurement function is generic in the sense that it couples all degrees of freedom. In the original version by Takens, D is the integer dimension of a smooth manifold, the phase space containing the attractor. Thus D can be much larger than the attractor dimension.

Sauer *et al.* (1991) were able to generalise the theorem, which they call the *Fractal Delay Embedding Prevalence Theorem*. Again let D_F be the box counting dimension of the (fractal) attractor. Then for almost every smooth measurement function s and sampling time $\Delta t > 0$ the delay map into $\mathbb{R}^m$ with $m > 2D_F$ is an embedding if there are no periodic orbits of the system with period Δt or $2\Delta t$ and only a finite number of periodic orbits with period $p\Delta t$, $p > 2$.

The theorem can be visualised in the following commutative diagram:

$$
\begin{array}{ccc}
\mathbf{x}_n \in \mathcal{A} \subset \Gamma & \overset{\mathbf{F}}{\mapsto} & \mathbf{x}_{n+1} \in \mathcal{A} \subset \Gamma \\
\downarrow h & & \downarrow h \\
s_n \in \mathbb{R} & & s_{n+1} \in \mathbb{R} \\
\downarrow e & & \downarrow e \\
\mathbf{s}_n \in \tilde{\mathcal{A}} \subset \mathbb{R}^m & \overset{\mathbf{G}}{\mapsto} & \mathbf{s}_{n+1} \in \tilde{\mathcal{A}} \subset \mathbb{R}^m
\end{array}
$$

Here e denotes the embedding procedure and can formally be written as a map on s_n by introducing a backward shift operator. Since the embedding map from $(\mathbf{x}_{n-m+1}, \dots, \mathbf{x}_n)$ onto $\mathbf{s}_n$ is invertible if $\mathbf{s}_n \in \tilde{\mathcal{A}}$, the dynamics $\mathbf{G}$ in the delay embedding space is uniquely determined by the dynamics $\mathbf{F}$ together with the measurement function and the embedding procedure.

Thus the main result of the embedding theorems is that it is not the dimension D of the underlying true state space that is important for the minimal dimension of the embedding space, but only the fractal dimension D_F of the support of the invariant measure generated by the dynamics in the true state space. In dissipative systems D_F can be much smaller than D. In fact, low dimensional motion was observed in hydrodynamic systems containing $O(10^{23})$ molecules. The simplest low dimensional dynamics are a steady state (a fixed point in the language of dynamical systems) and a steady oscillation (a limit cycle). Let us further remark that in favourable cases an attractor might already be reconstructed in spaces of dimension in between D_F and $2D_F$, as evidenced by Example 9.2.

Depending on the application, a reconstruction of the state space up to ambiguities on sets of measure zero may be tolerated. For example, for the determination of the correlation dimension, events of measure zero can be neglected and thus any embedding with a dimension larger than the (box counting) attractor dimension is sufficient; see Sauer & Yorke (1993).

We have neither reported the theorems in a mathematically rigorous language nor have we tried to outline their proof. Instead we have focused on explaining them in physicists' words. If you want to enter deeper into the mathematical aspects, we strongly recommend that you read the original "Embedology" paper by Sauer *et al.* (1991), and perhaps also the paper by Takens (1981).

9.2 The time lag

The time delay τ between successive elements in the delay vectors is not the subject of the embedding theorem (apart from the fact that certain specific values are forbidden). Therefore, from a mathematical point of view it is arbitrary, since the data items are assumed to have infinite precision. Also, it is assumed that there is an infinite number of them. In applications the proper choice of the delay time τ is quite important. If it is taken too small, there is almost no difference between the different elements of the delay vectors, such that all points are accumulated around the bisectrix of the embedding space. This is called redundancy in Casdagli *et al.* (1991a) and Gibson *et al.* (1992). If in addition the data are noisy, vectors formed like this are almost meaningless if the variation of the signal during the time covered by $m\tau$ is less than the noise level. (This gives a lower bound for τ. If σ_{noise}^2 is the variance of the noise and σ_{signal}^2 the variance of the data, τ has to be larger than the time when the normalised autocorrelation function decays to $1 - \sigma_{noise}^2/\sigma_{signal}^2$.)

If τ is very large, the different coordinates may be almost uncorrelated. In this case the reconstructed attractor may become very complicated, even if the underlying "true" attractor is simple. This is typical of chaotic systems, where the autocorrelation function decays fast (for *mixing* systems exponentially fast).

Unfortunately, since τ has no relevance in the mathematical framework,[2] there exists no rigorous way of determining its optimal value, and it is even unclear what properties the optimal value should have. In the literature, the issue of how to estimate τ has been emphasised greatly, and at least a dozen different methods have been suggested. All these methods yield optimal results for selected systems only, and perform just as average for others. Our attitude is therefore more pragmatic. Since all these methods yield values of similar magnitude, we are inclined to estimate τ just by a single preferred tool and to work with this estimate. In the particular application one can try to optimise the performance by a variation of τ. Anyway, the results should not depend too sensitively on τ since, otherwise, the invariance

---

[2] With the notable exception of Casdagli *et al.* (1991a), where the influence of τ on the quality of a phase space reconstruction is discussed rather rigorously. Unfortunately, this analysis does not lead to a practical method for determining an optimal value of τ from a time series.

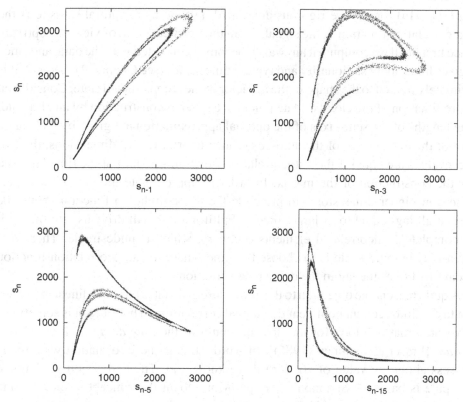

Figure 9.2 Two dimensional section through the three dimensional reconstruction of the NMR laser flow data. The larger the time delay the less are the two coordinates correlated (lags $\tau = 1, 3, 5, 15$).

of the properties of the attractor under smooth transformations is lacking, which suggests that one is not investigating a true attractor.

Example 9.3 (Delay embedding of the NMR laser flow data). In Fig. 9.2 we show two dimensional section through a three dimensional time delay embedding of an approximately 2.3 dimensional attractor (NMR laser data, Appendix B.2). The minimal τ is given by the sampling interval Δt ($\tau = 1$ in the notation of Section 3.3). In this example, for $\tau = 1$ the data are concentrated around the diagonal; for larger τ the attractor is reasonably well expanded, but finally becomes increasingly complicated. ☐

Now, what criteria for the choice of τ can be recommended? First of all, one can apply a geometrical argument. The attractor should be unfolded, i.e. the extension of the attractor in all space dimensions should be roughly the same. Statistics such as *fill factor* [Buzug & Pfister (1992)] or *displacement from diagonal* [Rosenstein *et al.* (1994)] are employed to evaluate this argument quantitatively [see Casdagli *et al.* (1991a) for a more rigorous approach], but the most natural statistic is the autocorrelation function of the signal. From the practical point of view it is optimal, since first of all we compute it anyway to become acquainted with the data, and since it gives hints about stationarity and typical time scales (see Chapter 2). Second, it is intimately related to the shape of the attractor in the reconstructed state. Consider an approximation of the cloud of data points in a given reconstruction by an ellipsoid. The lengths of the semi-axes of the optimal approximation are given by the square root of the eigenvalues of the auto-covariance matrix. In two dimensions, the two eigenvalues are equal if the autocorrelation function vanishes at the time lag used for the construction of the matrix. In order to approximate the data by a hyper-sphere in higher dimensions, in principle the autocorrelation function should be zero at all lags equal to or larger than τ. Fulfilling this will drive us into the limit of completely uncorrelated elements (large τ), which is undesirable. Therefore, a reasonable thumb rule is to choose the time where the autocorrelation function decays to $1/e$ as the lag in the delay reconstruction.

A quite reasonable objection to this procedure is that it is based on linear statistics, not taking into account nonlinear dynamical correlations. Therefore, it is sometimes advocated that one look for the first minimum of the *time delayed mutual information* [Fraser & Swinney (1986)]. In words, this is the information we already possess about the value of $s(t + \tau)$ if we know $s(t)$. The expression we have to compute is based on Shannon's entropy. More about the concept of entropy and information will be found in Section 11.4; here we just want to present the recipe for the computation of the mutual information.

On the interval explored by the data, we create a histogram of resolution ϵ for the probability distribution of the data. Denote by p_i the probability that the signal assumes a value inside the ith bin of the histogram, and let $p_{ij}(\tau)$ be the probability that $s(t)$ is in bin i and $s(t + \tau)$ is in bin j. Then the mutual information for time delay τ reads

$$I_\epsilon(\tau) = \sum_{i,j} p_{ij}(\tau) \ln p_{ij}(\tau) - 2 \sum_i p_i \ln p_i . \qquad (9.1)$$

It is customary to use a fixed resolution for binning the data. There are alternative algorithms using adaptive partitionings. In any case, we will have to do with a coarse grained version because the limit $\epsilon \to 0$ does not always exist, in particular not for fractal attractors with $D_f < 2$.[3] Since we are not interested in the absolute

---

[3] The discussion of ϵ-entropies on page 222*ff* shows the following. If the data distribution represents a continuous probability density in two dimensions, then the ϵ-dependence of the two terms in the time delayed mutual information cancel (both are given by $\ln \epsilon$), and what remains is the difference between continuous entropies in two and in one dimension. For invariant measures whose support is fractal in $m = 2$ dimensions, the continuous entropy needed here is undefined, see its definition in Eq. (11.29), and thus also limit $\epsilon \to 0$ of the mutual information is undefined.

values of $I(\tau)$ but only in its dependence on τ, coarse graining is not such a big issue here. The resolution should be selected such that a stable τ dependence can be read off. For small τ, $I_\epsilon(\tau)$ will be large. It will then decrease more or less rapidly. In the limit of large τ, $s(t)$ and $s(t + \tau)$ have nothing to do with each other and p_{ij} thus factorises to $p_i p_j$ and the mutual information becomes zero.

The first minimum of $I(\tau)$ marks the time lag where $s(t + \tau)$ adds maximal information to the knowledge we have from $s(t)$, or, in other words, the redundancy is least. To our knowledge there is no theoretical reason why there always has to be a minimum[4] and we suspect that in many cases the first one found is the first statistical fluctuation which leads to an increase of the estimated value of $I(\tau)$. Also, the arguments in favour of this choice for τ apply strictly to two dimensional embeddings only; see also the discussion in Grassberger *et al.* (1991). Nevertheless, when one finds that the minimum of the mutual information lies at considerably larger times than the $1/e$ decay of the autocorrelation function, it is worth optimising τ inside this interval. To this end, one looks for a maximal scaling range in a suitable nonlinear statistic, e.g. the correlation integral.

To speed up the computation of the mutual information, one can use the analogue expression of Eq. (9.1) based on the second-order Renyi entropies. They have the drawback that they do not strictly fulfil the additivity condition, and therefore the resulting mutual "information" is no more positive definite. This problem, however, is irrelevant when we just use it to find a suitable delay. Numerically, one has to compute the correlation sum Eq. (6.1) for embedding dimension $m = 2$ and fixed ϵ as a function of the time lag τ, $C(2, \epsilon, \tau)$. Although, as Eq. (11.23) states, the mutual information also involves the correlation sum in one dimension, due to its positivity and independence of τ, already the logarithm of $C(2, \epsilon, \tau)$ looks similar to the time delayed mutual information. It is argued in Prichard & Theiler (1995) that it supplies almost the same information with respect to its τ dependence.

To give you a feeling of why nonlinear statistics might be superior to linear ones when looking for a good embedding, consider the Lorenz system [Lorenz (1963)]. A typical trajectory stays some time on one wing of the well-known butterfly-like attractor, spiralling from the inside to its border before it jumps to the other wing. A record of the x- or y-variable thus shows alternations of oscillations around a positive mean value and oscillations around a negative mean, the total average being close to zero. The autocorrelation function thus decays smoothly and does not indicate that the average period of the motion on a single wing is of relevance. However, in a time delay embedding, a time lag of a quarter of this period yields the most reasonable reconstruction. One can detect this period either by computation

---

[4] For example, an AR(1) process, Section 12.1.1, does not have any.

of the autocorrelation function of the squared signal, or by computation of the time delayed mutual information for the original signal.

The literature contains many more suggestions on methods of how to determine an optimal time lag. Some of them have a nice heuristic justification. The reader who, after having acquired some experience with nonlinear time series analysis, feels the need for a better estimate, may try them, although, as we have said before, it is our impression that it might be more useful to optimise the lag with respect to a particular application such as dimension estimation. The *wavering product* of Liebert & Schuster (1989) is a proper ratio between near neighbour distances in different embeddings, and allows one to detect a reasonable unfolding of the attractor, i.e. the situation where overlaps due to projections disappear. Above we mentioned the geometric criteria *fill factor* and *displacement from diagonal* which are, however, related to the decay of autocorrelations.

In summary, we feel that the quest for the optimal time delay has been much over-emphasised in the literature. It should be clear that the best choice will depend on the purpose of the analysis. For predictions, e.g., it turns out that for longer prediction horizons, larger time lags are optimal. For noise reduction, small delays have been used with success. With today's computer power, good embeddings can be found by trial and error, comparing the results for predictions or estimation of invariants for a whole range of reasonable embedding parameters.

9.3 Filtered delay embeddings

Since a linear or at least a smooth transformation of the delay vectors does not change the validity of the embedding theorems, different manipulations of the delay vectors may be useful for a better representation of the data. This is of particular interest if the data are noisy and one wants to reduce the noise level implicitly. This comes at a price, however. Apart from requiring more computer storage than delay vectors, the embedding space will no longer be isotropic. The signal amplitude and properties may be different in different coordinate directions. If the dynamical ranges of the different components are very different in magnitude, it may be necessary to rescale them to equal variance or, equivalently, to use an anisotropic metric.

9.3.1 Derivative coordinates

When discussing a differential equation of higher order, a common method is to convert it into a set of differential equations of first order by the introduction of additional variables. In certain cases this procedure is invertible, such that one can derive from a dynamical system given by a set of differential equations of first order a single differential equation of higher order in one variable. When solving this equation as an initial value problem, one has to specify the values of all but the highest derivatives such that they define the state of the system uniquely.

From this point of view, the so-called *derivative coordinates* appear quite naturally. Numerically, one should form the adequate differences between successive observations, $\dot{s}(t) \simeq (s(t + \Delta t) - s(t - \Delta t))/2\Delta t$, $\ddot{s}(t) \simeq (s(t + \Delta t) + s(t - \Delta t) - 2s(t))/\Delta t^2$, etc., or use other suitable approximations. Then the state vector at time t is a vector of the signal and its derivatives at this time, $s(t) = (s(t), (s(t + \Delta t) - s(t - \Delta t))/2\Delta t, \ldots)$ of sufficiently many components. Instead of derivatives it might also be reasonable to introduce integrals of the signal, e.g. $I_1(s(t)) = \int_0^t dt' s(t')$, or to use a mixed representation.

Apart from the motivation given above, can we show that such coordinates really form an embedding? In fact we can, since the derivative coordinates are nothing but linear combinations of the elements of a delay embedding. Therefore, the dimension of the reconstructed state space in derivative coordinates has to be the same as in delay coordinates, i.e. $m > 2D_F$.

The advantage of derivative coordinates is their clear physical meaning. Their drawback lies in their sensitivity to noise. Let us assume that there is measurement noise $\eta(t)$ which is identically distributed with variance σ_{noise}^2 (zero mean) and which has a normalised autocorrelation function $c_{\text{noise}}(\tau)$. Suppose the data is recorded with a high sampling rate $1/\Delta t$, such that successive observations are strongly correlated. We claim that the first derivative is corrupted by a much larger noise level then the signal itself. Let $x(t)$ be the clean variable with autocorrelation function $c_{\text{signal}}(\tau)$ and variance $\sigma_{\text{signal}}^2 = \langle x(t)^2 \rangle - \langle x(t) \rangle^2$. The observed data $s(t) = x(t) + \eta(t)$ then has a relative noise level of $\sigma_{\text{noise}}/\sigma_{\text{signal}}$. If we rather take the numerical first derivative

$$\dot{s}(t) = \frac{1}{2\Delta t}(s(t + \Delta t) - s(t - \Delta t)) \qquad (9.2)$$

as an observable, then its signal component is

$$\dot{x}(t) = \frac{1}{2\Delta t}(x(t + \Delta t) - x(t - \Delta t)). \qquad (9.3)$$

Since because of stationarity the mean value of $\dot{x}$ is zero, it is straightforward to show that its variance is $\sigma_{\text{signal,derivative}}^2 = \sigma_{\text{signal}}^2(1 - c_{\text{signal}}(2\Delta t))/2\Delta t^2$. The corresponding noise part $\dot{s}(t) - \dot{x}(t)$ has a variance $\sigma_{\text{noise,derivative}}^2 = \sigma_{\text{noise}}^2(1 - c_{\text{noise}}(2\Delta t))/2\Delta t^2$. The relative noise level of the first derivative is thus

$$\frac{\sigma_{\text{noise,derivative}}}{\sigma_{\text{signal,derivative}}} = \frac{\sigma_{\text{noise}}}{\sigma_{\text{signal}}} \sqrt{\frac{1 - c_{\text{noise}}(2\Delta t)}{1 - c_{\text{signal}}(2\Delta t)}}, \qquad (9.4)$$

which can be much larger than the relative noise level of the original signal, if the autocorrelation of the signal decays considerably slower than the autocorrelation of the noise, a situation which is natural for flow data. Analogous considerations can be made for the higher derivatives. The noise levels can easily become intolerably large for a strongly correlated signal. Even more importantly, the noise amplitudes increase by taking higher derivatives. Thus, derivative coordinates usually have different noise levels. As you see, derivative coordinates are to be used with care, and usually at least require some low-pass filtering of the data. A particular estimator for derivatives which implicitly contains the optimal filter and hence can be recommended is the Savitsky–Golay filter. Details can be found in the corresponding chapter of the *Numerical Recipes*, Press *et al.* (1992).

If the data are map-like (e.g. when they were obtained from flow data by a Poincaré surface of section), the problem becomes less severe. Of course, the linear combinations of the signal at different times can no longer be interpreted as time derivatives. Thus, derivative coordinates of map data do not have any better physical meaning than ordinary delay coordinates.

9.3.2 Principal component analysis

Flow data with a high sampling rate contain high redundancy due to their strong linear correlations. When investigating the chaotic properties, which are represented only by the dynamics in the direction orthogonal to the tangent of the flow, the high sampling rate is often numerically expensive, since it exaggerates the amount of data covering a comparatively small time interval.

It can be a good idea to reduce the amount of data by increasing the sampling intervals and at the same time to reduce the noise level by exploiting the redundancy. The most straightforward approach would be a low pass filter and a re-sampling with the highest frequency passing the filter. For instance, one can compute the Fourier transform of the time series, skip the amplitudes and phases of all channels with a frequency larger than, say, $\frac{1}{4}$, and transform the remaining information into a time series of half-length and double sampling interval.

Another kind of data analysis which has appeared in the literature for decades in many variants with different names and was best studied in connection with signal processing offers a reasonable alternative. We like to use the name *Principal Component Analysis* (PCA), but it is sometimes referred to as *Karhunen–Loève transformation*, *Singular Value Decomposition*, or *Empirical Orthogonal Functions*. See Broomhead & King (1986) for a presentation in the nonlinear dynamics context. The idea is to characterise the time series by its most relevant components in a delay embedding space $\mathbb{R}^m$, where m is most probably larger than strictly necessary. The set of all delay vectors forms an irregular cloud in $\mathbb{R}^m$. Often, there are directions into which this cloud does not extend substantially, and the PCA allows for the computation of a series of one dimensional subspaces ordered according to their relevance to the data. Algorithmically, one has to compute the $m \times m$ covariance matrix[5]

$$C_{ij} = \langle (\mathbf{s})_i (\mathbf{s})_j \rangle = \frac{1}{N-m+1} \sum_{n=1}^{N-m+1} s_{n-m+i} s_{n-m+j} \,. \tag{9.5}$$

Since C_{ij} is a real symmetric matrix, its eigenvalues are real and its eigenvectors are orthogonal. The eigenvalues c_i are the squared lengths of the semi-axes of the (hyper-)ellipsoid which best fits the cloud of data points, and the corresponding eigenvectors give the directions of the axes. The most relevant directions in space are thus given by the vectors corresponding to the largest eigenvalues. If there are very small eigenvalues, the corresponding directions may be neglected.

For the PCA analysis, the data are represented by m dimensional vectors. On knowing the eigenvectors, one can transform the delay vectors into the eigenbasis, i.e. the components of the new vectors are the projections of the old ones onto the eigenvectors. If one decides that the m_0 most relevant eigenvectors are enough to describe the signal, one just truncates the new vectors after the m_0th component. Thus one has an m_0 dimensional embedding of the data. Since each of the m_0 dimensional vectors covers a time interval of m times the sampling interval, the effective time lag in this representation is m/m_0. One can show that in the limit $m \to N$ this transformation turns into a Fourier transform (see Broomhead & King (1986) for more details). For small m, however, the PCA selects the relevant structures in space rather than the relevant frequencies. Moreover, the spirit of the truncation of the basis is different. The only natural ordering of the eigenvectors is according to the magnitude of the eigenvalues. In a Fourier transform, one could also order the different channels according to their power instead of by frequency and perform a similar truncation. The reason why this is not usually done is related to a drawback of the PCA analysis. One cannot uniquely convert the set of m_0 dimensional vectors back into a scalar time series of length Nm_0/m, corresponding to the inverse Fourier transform after truncating the high frequencies.

Example 9.4 (NMR laser data). The original NMR laser data, Appendix B.2, are flow data, sampled exactly 15 times per period of the driving oscillation. Thus there are obviously 15 different possibilities of converting them into map data by a stroboscopic view, which is the optimal Poincaré section for periodically driven nonautonomous systems (compare Section 3.5). However, the different map-like

---

[5] The different names of the method basically refer to different ways of estimating the covariance matrix from a short time series. See Allen & Smith (1996) for a discussion. These differences are only important if the number of data points is not much larger than the embedding window m.

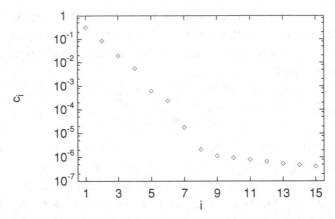

Figure 9.3 The eigenvalue spectrum of the covariance matrix for the NMR laser flow data.

series are not equivalently apt for further processing, since the signal-to-noise ratio is very different. Assuming that we have essentially independent measurement noise, the noise amplitude is the same on all elements of the time series. In certain stroboscopic views, however, the amplitude of the signal is very small, and thus the relative noise level is quite high. The optimal stroboscopic view can be found by PCA analysis.

We cut the 600 000 flow data into 40 000 segments of length 15 (i.e., one driving period) which are interpreted as vectors. Thus we do not use a true delay embedding, simply since it is our goal to obtain a stroboscopic view and therefore we want to compress every driving cycle into a single number. For the set of vectors we compute the covariance matrix and its eigenvalues and -vectors. The eigenvalue spectrum is shown in Fig. 9.3. Projecting the vectors onto a single selected eigenvector of the covariance matrix yields a new scalar time series. The projection onto the dominant direction leads to the largest variance of the data and thus to the smallest noise level (1.1% instead of 1.8% in the best stroboscopic view without PCA).[6] The resulting attractor is shown in Fig. 1.3. In Fig. 9.4 we show the projection onto the third and the fifth eigenvector. The signal-to-noise ratio is obviously worse, and the projections onto the seventh and higher components appear completely random.   □

Example 9.5 (Oversampled string data). The data of the vibrating string (Appendix B.3) are taken with an average sampling rate of 200 per cycle on the attractor. A reasonable time delay would be about one-quarter of the average period. Instead, one can perform principal component analysis on vectors with unit time lag

---

[6] To make things a bit more complicated, this procedure is not unique, and we maximise additionally over the position, where the first vector starts, in a window of length 15.

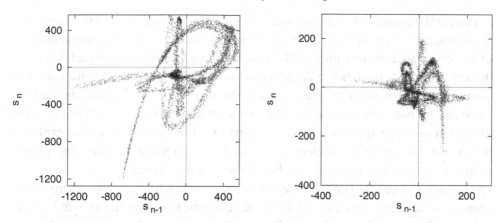

Figure 9.4 Projection of the NMR laser flow data onto the third (left hand panel) and fifth (right hand panel) principal component in a two dimensional delay representation. Compare with Fig. 3.6 for the projection onto the dominant component and with Fig. 9.2, lower right hand panel, for the stroboscopic view without PCA.

and length 50. Projection onto the first two components yields a two dimensional representation of the data. Since the data are rather noise free, in this case the result looks extremely similar to a delay embedding with delay between 25 and 50, but the clear advantage is that there is only a very weak dependence on the length of the embedding window. Projection onto the dominant eigenvectors corresponds to the moving average of the observations during the embedding window, i.e. all its components are identical. The second creates something like an average derivative, since its components vary linearly from positive to negative values as a function of their position inside the vector. ☐

PCA is essentially a linear method. The only nonlinear step is the determination and ordering of the eigenvectors and -values. The reconstruction itself is just a linear combination, and if you are familiar with linear filters you will already have realised that in fact each component of the new vectors is obtained from the time series by a FIR filter (compare Section 12.1.1). Correspondingly, the representation is optimal only in the least squares sense, which is not always what we need for further nonlinear investigations. Comparing the NMR laser attractor in the PCA optimised stroboscopic view, Fig. 3.6, to the attractor in the best view found by selecting only every 15th element of the original series shows that it is somewhat more folded. This may have negative consequences for certain applications. For example, self-similarity of the attractor is destroyed on the large length scales. In estimates of the correlation dimension the scaling range is therefore shifted towards the smaller length scales, and although one gains due to the suppression of noise, one sometimes loses even more due to the distortion on the large scales.

Principal component analysis was also used for dimension estimates; see Broomhead *et al.* (1987). The claim was that the number of significant eigenvalues of the covariance matrix mirrors the dimension of the subspace which contains the attractor. This implies the definition of a *noise floor*. All eigenvalues smaller than this are considered to reflect mere noise directions. Numerical simulations show that for many model systems there are more eigenvalues larger than the noise floor than are expected from the known dimension of the attractor. One explanation is that the manifolds which contain attractors are not usually linear subspaces. A low dimensional surface may be bent in such a way inside a high dimensional space that it can be fitted only by a hyper-ellipsoid which has far more macroscopic semi-axes then the covered manifold has dimensions. For an application of PCA as an embedding procedure in the calculation of the correlation dimension, see Albano *et al.* (1988).

Instead of a global PCA one can collect all points within the neighbourhood of some reference point and thus perform the same analysis locally. The criterion of the last paragraph now applies, since manifolds locally are (hyper-)planes. The dimension determined this way is called *local intrinsic dimension* in Hediger *et al.* (1990). Local PCA can also improve local linear predictions; see Sauer (1993).

9.4   Fluctuating time intervals

In certain experimental set-ups it is not possible to make measurements periodically in time. One example is the measurement of the velocity of a fluid by laser-Doppler-anemometry. Tracer particles with a high reflectivity are added to the fluid of interest. Two laser beams are directed onto the fluid (which has to be in a transparent container) such that they cross somewhere inside the fluid. Each time a tracer particle passes this crossing point, beams are reflected and the frequency shift of the reflected beams gives information about the local velocity. Using two beams allows for a high spatial resolution; performing the experiment with only one beam means losing this information. Since the passage times of tracer particles through the crossing point of the two beams fluctuates around some mean value, the resulting time series consists of values which are not spaced uniformly in time. For such data the embedding theorems do not apply, and the attempt nevertheless to use the time delay embedding leads to a very messy attractor which does not yield the desired insight into the data structure. The only way out is to resample the data or to use the data to perform a Poincaré section.

Resampling is possible if the information about the observation times is stored. If the average sampling rate is very high, one can just reduce the data set and use as new values some appropriate average over all observations within the new sampling

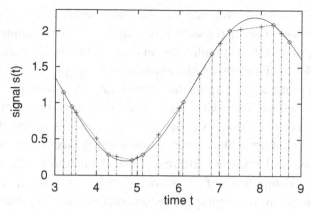

Figure 9.5 At irregularly distributed times observations (◇) of the continuous signal (solid line) are made. After linear interpolation, the new time series, equidistant in time, consists of the +. The differences between each + and the true signal (the corresponding point on the solid line) can be interpreted as measurement noise.

interval. This averaging acts as a low-pass filter, and one might cut off important parts of the signal if the average sampling rate of the original data is not high enough. A good resampling is more difficult and makes use of a nonlinear noise reduction scheme described in Chapter 10. The new sampling rate in this case can be quite small, even smaller than the average sampling rate of the original data. As a first step, one interpolates the signal linearly between the single observations. The new observations are the values of this polygonal curve at the correct times (see Fig. 9.5). This new time series is now noisy with respect to the true underlying data, and the noise (i.e. the deviations of the true values from the results of the linear interpolation) can be reduced by applying a nonlinear noise reduction algorithm to these data. There are two limitations which prevent the resulting signal from being identical to what one would receive if one observed the continuous signal with the desired sampling. The first is that the noise is correlated to the signal, since on average the errors are larger where the second time derivative of the signal is large. This can reduce the performance of the noise reduction scheme. Second, and more important, what we called noise above contains some systematics. Interpolated points tend to be closer to the centre than their "correct" values. This will lead to a slight reduction of the dynamical range of the signal. It can be avoided by using spline interpolations, for example.

9.5 Multichannel measurements

So far we have considered only the situation where a univariate time series is available. Then, phase space is reconstructed using (linear combinations of) delay vectors. Nevertheless, Whitney's embedding theorem ensures us that a set of different variables obtained at the same time may form an embedding as well. Thus one could think of performing multichannel measurements, i.e. measuring a set of different variables simultaneously. Guckenheimer & Buzyna (1983) report on a hydrodynamic experiment where more than 1000 thermistors were placed at different positions in a container. The data analysis was performed for selected sub-sets of these variables. For dimension estimates, e.g., they reconstructed state spaces up to dimension 27, combining the output of different selected thermistors to vectors. Alternatively, one could measure different physical quantities at the same position, such as different components of the velocity vector. Of course, if such a set of variables is insufficient for a complete reconstruction of the system's state space, one can augment the different measurements by time lagged copies to yield a mixed time delay and multivariate embedding. In the following we want to discuss some aspects to be considered when designing experiments or when evaluating vector valued time series.

9.5.1 Equivalent variables at different positions

In many situations it will be most convenient to measure similar physical quantities simultaneously at different positions in a spatially extended system, like in the above hydrodynamic example. This is like rotating the delay reconstruction into a spatial direction. Thus one faces exactly the same problems. One has to choose the embedding dimension and the spatial distance δ between the positions of the measurements, which corresponds to the delay τ. The latter again attempts to find the best trade-off between redundancy and decorrelation. Obviously, the smaller δ the stronger the different variables tend to be correlated, whereas when they are too far away they may seem to be completely independent. Like for the time delay, the correlation function or the mutual information between two simultaneously measured variables as a function of their spatial distance yields the required information, since the same arguments as given in Section 9.2 hold. Thus one could start with recording the signal of only two channels, computing their correlations as a function of the distance of the measuring devices, and using the estimate for the optimal distance for the design of the real experiment. The dimension m of the state space reconstructed in this way again has to fulfil $m > 2D_F$. However, homogeneous spatially extended systems have some properties which are distinct from other systems with many degrees of freedom. It has been conjectured, and demonstrated for a number of models, that systems exhibiting spatio–temporal chaos have attractors which are not finite dimensional but possess a finite dimension-density. This means that the dimension determined for a sub-system is proportional to its spatial extent. Therefore, the attractor of the (infinite) full system is infinite dimensional and cannot be reconstructed by a finite set of coordinates.

9.5.2 Variables with different physical meanings

An alternative approach can be taken by simultaneously measuring several observables with different physical meanings. In a thermal convection experiment, these could in fact be the variables of the Lorenz model [Lorenz (1963)]. In this special case this would simplify the analysis of the results tremendously, since these variables span the original state space of the model equations. Very often, however, macroscopically meaningful variables may be quite complicated nonlinear functions of the microscopic variables. In this space the attractor might be more folded than in a delay embedding of only one coordinate. A serious difficulty with multivariate data lies in the fact that the different observables most often have different dynamical ranges, or even different physical units which cannot be easily scaled relative to each other. A work-around is to scale each variable to have the same variance.

If it turns out that a given number of variables is not enough for a reconstruction of the state space, it might be difficult to measure another one, since this demands a completely different experimental device. Therefore very often one will have to involve an additional delay reconstruction. For example for a nine dimensional space one could use three different observables, each at three different times. This combination does not induce conceptual problems, but it tends to make algorithms somewhat more complicated technically. For examples of mixed multivariate and time delay algorithms see Hegger & Schreiber (1992), and Prichard & Theiler (1994), and the TISEAN routines [Hegger *et al.* (1999)].

9.5.3 Distributed systems

In extended systems with very many degrees of freedom, multichannel measurements may yield only poor insight. One could imagine that distant parts of a large system are coupled only very weakly, such that one could hope to describe each part as low dimensional – subject to some small perturbation – and thus apply the methods discussed in this book. But if this is the case, it is quite probable that different macroscopic observables will mainly describe different subsystems and will thus be almost uncorrelated. Physiological data from living beings may serve as examples, for instance the data set B from the SFI time series competition. To study a disease called *sleep apnoea* (breathing stops for some time during sleep), Rigney *et al.* (1993) simultaneously measured the breath rate, the oxygen concentration in the blood, the heart rate and nine other observables. Of course, there is a connection between the breath rate and the oxygen concentration, but the heart rate already depends on so many other degrees of freedom in the body that using it as an additional signal increases the dimensionality of the sub-system under observation more than it helps to understand the breath data. Another prominent situation in medical research where routinely multichannel data sets are produced are EEG recordings, either from arrays of electrodes on the scalp or even, in epilepsy research, from arrays of implanted electrodes.

In Chapter 14 we will discuss some information theoretical concepts which may help to select those variables of a multidimensional time series which are most suited to span the state space of a selected subsystem. Here, a subsystem is defined in a pragmatic way by the choice of a variable. When we are interested in the phenomenon of sleep apnoea, then the breath rate clearly is representative of the dynamical subsystem we are interested in. In order to find those other variables which supply additional information about this system, we may, e.g., ask and quantify numerically, which variables improve the predictability of the breath rate fluctuations. When looking for determinism, all the variables which do not improve the predictability are worthless and hence should not be included in our state space reconstruction.

9.6   Embedding of interspike intervals

In many situations the recorded signal itself is quite uninteresting and the relevant information is contained in the time intervals between certain characteristic events. The most popular example of data of this type are the R–R intervals obtained from ECG time series, which are the times between successive heart beats. Also think of series of intervals between the surfacing of a sea mammal in order to breathe, or intervals between earthquakes.

Now, consider as an observable the time interval in between two successive events. The peculiarity of this kind of "time" series is that, usually, the "time" index of each event is the event (e.g. beat) number, and not the true time any more. If one insists on using the true time, the observations are not uniformly spaced on this time axis, and the information stored in the signal, namely the length of the last interval, is redundant with the information of the position in time. An alternative could consist of a uniform resampling of the data as described in Section 9.4 and a conversion of the observable into a rate, the heart rate for instance in the ECG case. However, this practice is also questionable. The fact that a heart rate is only defined over times longer than one heart beat can cause artefacts in a subsequent analysis of the resampled data. In the power spectrum, for example, we will find aliasing of the spurious high-frequency contribution into the lower frequencies. The resampling procedure can also cause additional predictability in heart rate time series which can be mistaken for nonlinear determinism.

There are many settings where it seems attractive to analyse the interspike intervals directly, without resampling and even without assuming a nonuniform sampling. In view of the material presented in this book the important question is whether one can embed these data in the standard way if they were created by a deterministic source. This is in fact questionable, since the time between certain events is not usually an observable based on phase space coordinates; we cannot usually write it as a phase space function $s(\mathbf{x})$, and the ideas behind the time delay embedding (Section 9.1.2) do not apply immediately.

In the last paragraph we used the expression "usually", since in fact there are situations where such a relation can be established and the embedding theorems can be shown to be valid. One thus has to specify the process behind the emergence of the spikes.

One common way in which nature creates spikes is an *integrate-and-fire* process. Consider some nonlinear unit which accumulates its input, and when this value reaches a certain threshold, the unit "discharges" by emission of a spike. It is generally supposed that this process holds for the firing of neurons in the nervous system, and one can imagine that, among the examples given above, the surfacing of sea mammals, and perhaps earthquakes, can also be modelled by such a process. Sauer (1994) and (1995) shows that if in such a situation the integrated signal is deterministic, then the embedding theorems are valid and a time delay embedding of the interspike intervals yields a faithful reconstruction of state space. The integrated value of some observable in phase space is itself a valid signal for time delay embedding, and determining the times when its value assumes integer multiples of a certain value means wrapping the data on a torus and performing a Poincaré section (see Fig. 9.6).

More generally, the times between successive passages of a continuous trajectory through a *Poincaré surface of section* are deterministically related to the properties of the motion in between. An individual time interval is given by the length of the path from one intersection to the next divided by the average velocity of the phase space vector on this path. Therefore, it is plausible that the sequence of time intervals obtained from a surface of section allows the reconstruction of the deterministic motion using a time delay embedding.

Example 9.6 (Laser with feedback). The time series of the CO_2 laser intensities described in Appendix B.10 is reduced to map-like data by a suitable Poincaré surface of section. The attractor reconstructed from the new series is shown in the left hand panel of Fig. 9.7. We have also recorded the series of time intervals between successive intersections of the Poincaré surface. A two dimensional embedding of these data yields the attractor in the right hand panel, which is characterised by the same invariants. □

It is obvious that, very often, spikes cannot be traced back to an integrate-and-fire process, and, even if this is possible, the integrated signal may not be deterministic at all. As an alternative scenario, assume some strongly nonlinear filter which enhances

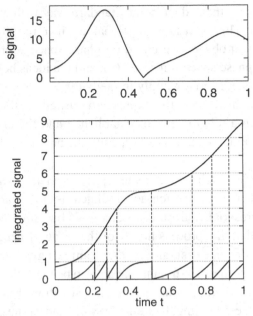

Figure 9.6 Integrate-and-fire process. The signal shown in the upper panel is integrated. Each time when the integral assumes an integer, a spike is emitted and the integration process starts anew. The time intervals between spikes form a new time series.

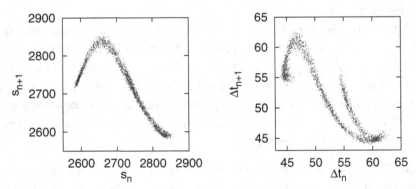

Figure 9.7 Poincaré map and return map of the corresponding time intervals. For details see Example 9.6.

local maxima of the signal to spikes. We have shown earlier that collecting local maxima of a time series can be interpreted as performing a Poincaré section in a plane of zero time derivatives of the signal, so that such interspike intervals can again be embedded. The only problem in such an approach is that there may be events which are somewhere between spike or no spike, i.e. there is some freedom of interpretation. And missing a point (or adding an erroneous point) in such a

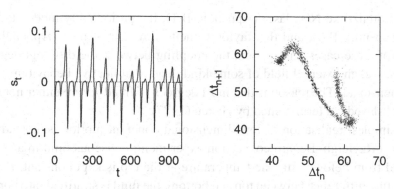

Figure 9.8 The spike train created by enhancing the extrema of a "harmless" time series and the delay plot of the interspike intervals (Example 9.7).

situation creates false points on the reconstructed attractor. Generally, this leads to time intervals which are either close to zero (one spike is interpreted as two) or about twice the average interspike interval (one spike is missed). Such points can sometimes be located on the reconstructed attractor, since they are disconnected from the main data cloud. Time series of interbeat intervals often have to be hand-edited for missing beats.

Example 9.7 (Laser with feedback). The data from Example 9.6 (see also Appendix B.10) are transformed into a spike train by computing the third power of their normalised deviation from the mean value. A part of the resulting signal is shown in Fig. 9.8. The time delay embedding of the series of interspike intervals obtained from this signal yields the attractor shown in the right hand panel of this diagram. Compare it to the structures of Fig. 9.7. The objects look topologically and geometrically the same. □

Summarising, as long as we can interpret the interspike intervals in some way as the time intervals in some kind of Poincaré section, we can assume that they can be used to reconstruct the main features of the attractor. See also Hegger & Kantz (1997).

9.7 High dimensional chaos and the limitations of the time delay embedding

Many systems are composed of a huge number of internal microscopic degrees of freedom, but nevertheless produce signals which are found to be low dimensional. In such cases, it is an important consequence of the embedding theorems that the embedding dimension is determined by the attractor dimension and not by the phase space dimension. Among the experimental examples of this book, these are the Lorenz-like NH$_3$ laser (Appendix B.1), the periodically modulated NMR laser (Appendix B.2), and the Taylor–Couette flow experiment (Appendix B.4). In each of these cases, dissipation, the coupling between the different degrees of freedom, and an external field of some kind, lead to collective behaviour which is low dimensional. The reason is that most degrees of freedom are either not excited at all or "slaved", a term coined by Haken (1977).

The simplest realisation of low dimensional coherent motion is a steady state. Imagine a Rayleigh–Bénard convection experiment: a viscous fluid in a container is heated from below while the temperature at the top is kept constant. For small temperature difference between top and bottom, the fluid is stratified into horizontal layers with constant temperature. Heat is exchanged via thermal conduction. When this state is stable, an arbitrary observable yields a constant signal, as long as it does not resolve the thermal noise of the molecules of the liquid. For larger temperature gradients the laminar Rayleigh–Bénard convection sets in, where the liquid rotates in convection rolls inside the container. Heat is then carried by the moving liquid. At every point in real space, the velocity is time independent, such that again an arbitrary observable is constant. These two types of motion can be described by the Lorenz equations (see Exercise 3.2) at parameter values where the attractor is a stable fixed point. When for even larger gradients the convective motion becomes unstable, the Lorenz system (and also the corresponding laser system, the data set in Appendix B.1) show low dimensional chaos. In certain geometries which allow for a single convection roll only, the Lorenz equations are still approximately valid. In an extended fluid however, adjacent rolls interact with each other leading to weak turbulence. In such a case the coherence between degrees of freedom at different positions in space is lost and the whole motion becomes high dimensional.

Rayleigh–Bénard convection is an example where the high dimensionality is related to the spatial extension of the system. Spatially extended dynamical systems are however not the most general class of high dimensional systems. If the *aspect ratio* is large, that is, if edge effects can be neglected, the dynamics of the system (but not necessarily the solutions) becomes effectively translation invariant. This very strong symmetry may lead to a simplified description despite the large number of degrees of freedom.

The most general high dimensional system would be described by many coupled equations, each of its own structure. This would describe the collection of different nonlinear dynamical units which are interacting with each other. An experimental example of a high dimensional but probably still deterministic system is the Nd:YAG laser described in Appendix B.12, see Gills *et al.* (1992) for a study of the chaotic dynamics. When pumping is increased, more and more laser modes become active. These modes are themselves described by nonlinear equations of motion and additionally interact in a nonlinear way and lead to high dimensional, chaotic behaviour.

This section is devoted to the discussion of the possibility to resolve intermediate to high dimensional deterministic motion from measured signals. All our experimental examples where we can numerically verify that nonlinear *deterministic* dynamics are dominantly responsible for the observed signals possess attractors of dimension between two and three when regarded as flows. (In a Poincaré surface of section the dimension is between one and two). That means that they are of the lowest dimension chaotic signals can have. Thus you might already suspect that there are difficulties with higher dimensional systems. In fact, when scanning the literature, you will find rather few experimental observations of *hyper-chaos*, i.e. of systems with more than one positive Lyapunov exponent [Rössler (1979)]. In an experiment on the catalytic reaction of CO and O_2 on a platinum surface, a second positive Lyapunov exponent was measured [Kruel *et al.* (1993)], but despite the great care of the authors the results are not completely convincing due to the high noise level on the data which alternatively could be a signature of a very high dimensional component of the signal, since the core of the experiment is a spatially extended object. Experiments have been made with an externally driven NMR laser [Stoop & Meier (1988)] and a p-Ge semiconductor [Stoop *et al.* (1989)]. These are very nice experiments, but they are special in the sense that the experimental devices are very close to nonlinear oscillatory electric circuits and therefore similar to analogue computers that solve predefined ODEs.

To explain this lack of higher dimensional experimental attractors we can offer two possible explanations: either typical systems in nature possess either exactly one or very many positive Lyapunov exponents, or the reason is that systems with a higher than three dimensional attractor are very difficult to analyse. Since the first explanation is not very plausible, we have to think about the limits of our analysis tools with respect to the dimensionality of the underlying dynamics. The problems we will encounter will be directly related to the issue of phase space reconstruction and determinism.

All numerical algorithms to compute nonlinear quantities from time series exploit neighbourhood statistics in some way or another. The dimension is, loosely speaking, the rate at which a point on the attractor loses its neighbours if we decrease the radius of the neighbourhood. The entropy correspondingly is the loss rate if we increase the embedding dimension at a fixed length scale, and the maximal Lyapunov exponent can be determined by the increase of distances between neighbours under forward iteration. All these quantities follow exponential or power laws and therefore require high spatial resolution to access the scaling regime.

In the first place, analysing high dimensional attractors is a problem of access to length scales and therefore a problem of sparse data: if we assume a uniform distribution of N data points on a D_F dimensional attractor, the average interpoint distance is $\epsilon_i = N^{-1/D_F}$. Most of the statistical tools[7] work only on length scales which are larger than ϵ_i. For a safe indentification of scaling behaviour we have to cover a range in length scale of one to two octaves for dimension and entropy estimates, and of a factor $e^{5\lambda}$ in nearest neighbour distances for estimation of the maximal Lyapunov exponent *inside* the scaling regime. If we take into account the violation of scaling on the large scales, we see that we need extremely many data points to recover high dimensional chaos. Arguments like these lead to several slightly different estimates about the relation between the attractor D_F and the number N of points required for a faithful determination of D_F. Assuming for simplicity that $\epsilon_i = 1/20$ is a typical value, then this yields

$$ N_{\min} > \left(\frac{1}{\epsilon_i}\right)^{-D_F} = 20^{D_F} . \tag{9.6} $$

When we compute the correlation dimension, we would require to have about 100 pairs left at the length scale ϵ_i. If the scaling behaviour persisted up to the largest scales, we had $N(\epsilon) \approx \frac{N^2}{2}\epsilon^{D_F}$ pairs at scale ϵ. Together this leads to the weaker bound

$$ N_{\min} > 10^{\frac{D_2+2}{2}} . \tag{9.7} $$

In other words, if a time series of length N represents deterministic dynamics on a set whose dimension is larger than $D_{\max} \approx 2\log_{10} N - 2$, then its deterministic properties cannot be identified any more by a dimension analysis. In Section 11.4.4 we will come back to the fact that on the large length scales deterministic and stochastic systems are indistinguishable.

Using these estimates, however, we would expect to have good empirical evidence for the existence of up to five dimensional chaotic attractors, since time series of length $N = 10^5$ and even much longer are quite easily available nowadays. When performing numerical simulations of e.g. the "hyper-chaotic" Rössler system [Rössler (1979), dimension slightly above three, two positive Lyapunov exponents], or the "hyper-chaotic" version of the Hénon map [Baier & Klein (1990)], one can in fact compute all the invariants with reasonable effort, in agreement with the limitations expressed above. However, when the same analysis is attempted on the basis of a scalar time series we encounter additional difficulties. It is mainly the reconstruction of the attractor from univariate data which induces distortions relative to the original state space and thus complicates the analysis.

---

[7] Compare Chapter 6: the correlation dimension is a friendly statistics in the sense that it is sensitive to the regions on the attractor where the density of points is largest. Because of this and the fact that the finite sample correlation sum is an unbiased estimator of the correlation integral, the smallest accessible length scale is given by the closest points, rather than the *average* distance of nearest neighbours.

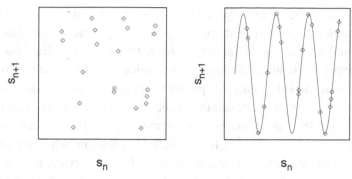

Figure 9.9 The few points in the left panel hardly allow to detect the deterministic structure underlying them. The solid line represents a cut through the surface defined by $s_{n+1} = F(\mathbf{s}_n)$. Note that in this example the average inter-point distance is already smaller than the average width of a fold of the surface.

Let $\mathbf{x}_n$ be a state vector of the system in its "natural" state space[8] at time n, and $\mathbf{s}_n = (s_{n-m+1}, \ldots, s_n)$ a delay vector. The map from $\mathbf{x}_n$ to $\mathbf{x}_{n+1}$ generally is nontrivial and nonlinear in every component. The map from $\mathbf{s}_n$ to $\mathbf{s}_{n+1}$ is a simple shift in all but the last component of $\mathbf{s}_{n+1}$, which is a complicated function of all components of $\mathbf{s}_n$. When passing from the original state space to the delay embedding space, one shuffles all the nonlinearities of the time evolution of the different components of $\mathbf{x}_n$ onto the time evolution of the last component of $\mathbf{s}_n$. This function is therefore expected to be much more complicated, and, when regarded as a surface in $\mathbb{R}^{m+1}$, it is much more folded than any of the components of $\mathbf{x}_{n+1} = \mathbf{F}(\mathbf{x}_n)$. In order to reveal the deterministic structures in the data, we have to be able to resolve the folds of the surface. Consequently, the resolution we need is determined by the width of the smallest fold (see Fig. 9.9).

For a more precise reasoning which yields also a quantitative statement about the complications introduced by the time delay embedding, we need the concept of ϵ-entropies to be introduced in Section 11.4.4. If you have the feeling that you cannot make any sense out of the following paragraphs, return here after having read Chapter 11. To relate the critical length scale to characteristics of the system, let us introduce some coarse-graining through an ϵ-partition, where we choose ϵ such that we cover the dynamical range of the data by A partition elements. We can convert the series of real numbers into a symbol sequence composed of A symbols (the alphabet) and compute the entropy of this sequence (Section 11.4) per sampling interval τ, which is $h\tau$ (h then is the entropy per unit time). It is trivially limited by $h\tau \leq \ln A$. Obviously, if A is too small, we cannot compute

---

[8] Which representation of a dynamical system is most natural is generally not well defined, but in most cases some set of coordinates can be obtained which at least look more natural than delay coordinates. Usually that representation is considered most natural which achieves determinism with the smallest number of variables.

the true entropy of the process. We need at least $e^{h\tau}$ partition elements in order to resolve the underlying "grammar" of the symbolic dynamics, which amounts to a spatial resolution of $\epsilon = e^{-h\tau}$ (assuming normalised data). If the dimension of a chaotic system is high, we also expect a large value of h. In particular, for spatially extended systems h will often grow proportional to the system size. Note that the resolution $\epsilon < e^{-h\tau}$ was estimated under the assumption that we can use natural coordinates. For delay coordinates, this estimate has to be replaced by the much stronger condition $\epsilon < e^{-h(m-1)\tau}$, for the following reason: when we suppose that we need m_0 dimensions for the reconstruction of a state space, then there will be false neighbours in every lower dimensional projection, as argued in Section 3.3.1. Hence, in order to separate different states from each other, we need to separate them at least in the two dimensional subspace spanned by the first and the last component of a delay vector. In this projection, we see an effective entropy given by $(m-1)\tau h$, the entropy produced by the chaotic dynamics within the time interval spanned by the embedding vector.[9]

In summary, the full deterministic structure underlying the data can be seen only on length scales smaller than $\epsilon \propto e^{-h(m-1)\tau}$. What can we expect to observe on the larger scales? Fig. 9.9 evidences that on this level of resolution the data seem to stem from a random process. In fact, we can construct surrogates as in Section 7.1 and compare them to the data. We will find out that above the critical length scale data and surrogates look equivalent. If the critical ϵ is smaller than the noise level, there is no way to distinguish such a signal from a truly random process. To illustrate these considerations we construct a somewhat artificial example based on experimental data.

Example 9.8 (Composed NMR laser system.). The active medium of the NMR laser (Appendix B.2) is an extended ruby crystal. Let us assume that inside the crystal two domains are formed which due to some process get desynchronised. In such a case the resulting output power is the sum of the individual powers of the two subsystems. We create a time series of this situation by adding the original NMR laser series (after noise reduction, in the stroboscopic view; the periodic external drive acts synchronously on both subsystems) to its time-shifted version, $\tilde{s}_n = s_n + s_{n+n_0}$, where the time shift n_0 is arbitrary but large. In Fig. 9.10 we show the correlation dimension of the signal $\tilde{s}$, the dimension obtained for surrogates (see Section 7.1) of $\tilde{s}$, and the correlation dimension of the original signal s. Since the scalar distribution of $\tilde{s}$ differs strongly from that of s, the two curves $d(\epsilon)$ cannot be related to each other on large scales. But, although it is difficult to describe the changes quantitatively, we see that a simple doubling of the complexity of the signal

---

[9] The reasoning here does not apply if we proceed to $m' > m_0$, since then the distinction between different states is already given in the lower dimensional space m_0, so that even if we are unable to distinguish two state vectors in their first and m'th component, they are already distinguished (at least, when we use a maximum norm).

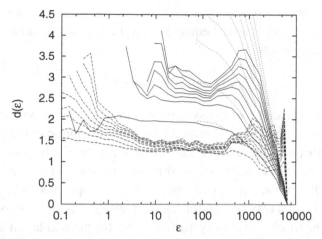

Figure 9.10 The correlation dimension of the NMR laser data (broken), the "double NMR laser" data (continuous), and their phase randomised surrogates (dotted) as a function of ϵ. The curves of the surrogates are truncated at $\epsilon < 100$.

(and of the attractor dimension) reduces the scaling range drastically. Since both the entropy per time step and the required embedding dimension have doubled, the upper end of the scaling range has shrunk by a factor of four, whereas the smallest accessible ϵ is about a factor of one hundred larger than in the original data. □

To summarise, the negative result of these considerations is that we do not only have the problem that the average inter-point distance increases if we distribute a given number of points on higher and higher dimensional objects, but that at the same time, for systems with positive entropy, we have to proceed towards smaller and smaller length scales in order to observe determinism in a time delay embedding space. This affects the estimation of all quantities which are related to the deterministic properties of the motion, i.e. dimensions, entropies, and Lyapunov exponents. The only hope might lie in the reconstruction of the underlying deterministic dynamics.

The important aspect, however, is that a good deal of the problems is related to the reconstruction of the attractor from a scalar observable. When we possess multichannel observations, we should be able to study high dimensional systems much better, in particular if the different variables are sufficient to reconstruct the attractor.

9.8　Embedding for systems with time delayed feedback

A very particular class of systems which in principle can create attractors of arbitrarily high dimension despite a rather simple structure are systems with time delayed feedback. In continuous time, these are *delay-differential equations* or *difference-differential equations*, in discrete time they are sometimes called *delayed maps*. A delay-differential equation in a d dimensional configuration space with a single delay reads

$$\dot{\mathbf{x}}(t) = \mathbf{f}(\mathbf{x}(t), \mathbf{x}(t - T)) , \tag{9.8}$$

where T is the time delay of the system (not to be confused with the time lag τ of our time delay embedding). Hence, the temporal evolution of the present state $\mathbf{x}$ depends not only on the present state but also on some state in the past, $\mathbf{x}(t - T)$. The state space of this system is the infinite dimensional space of functions on the time interval $[-T, 0]$, since $\mathbf{x}(t)$ in this full interval has to be specified as initial condition. See Hale (1977) for a textbook on this class of systems. A popular example in one dimension is the Mackey–Glass system, a model for the regulation of the number of red blood cells in the body. In many experiments, the feedback of a signal back into the system occurs with some time lag, sometimes even purposefully.

For a start, one may choose to ignore the peculiar structure of the system. Whenever attractors of these systems have finite, even if large, dimension, Takens's embedding theorem applies. However, if the attractor dimension is beyond, say, 5, then for practical problems outlined above the analysis in a time delay embedding space will most likely not reveal the deterministic properties of the system. However, the particular structure of systems with time delayed feedback offer a different embedding technique, which is a straightforward modification of the Takens embedding.

Assume that we do not want to reconstruct the high dimensional attractor, but only the low dimensional space $(\mathbf{x}(t), \mathbf{x}(t - T))$ which defines the input vector of Eq. (9.8). This would allow us to verify the deterministic structure, since first of all the temporal derivative should be found to be a unique function of the vectors in this space. Moreover, we can use this input for predictions and for modelling. If the full vector $\mathbf{x}$ were observable, we only would have to identify the correct time lag T. In higher dimensional settings where this is not the case, we employ Takens's theorem for the reconstruction of the actual and of the delayed input vector. It then turns out that the reconstructed space we are working in is given by vectors

$$\mathbf{s}_n = (s_n, s_{n-1}, \dots, s_{n-m+1}, \quad s_{n-l}, \dots, s_{n-l-k+1}) , \tag{9.9}$$

i.e., these vectors are concatenations of two embedding windows. The first one with an embedding dimension m is supposed to yield a reconstruction of $\mathbf{x}(t = n\Delta t)$, the second one of dimension k is delayed by $l = T/\Delta t$ with respect to the first one (where Δt is the sampling interval) and provides a reconstruction of the variables which are fed back into the system. We distinguish between m and k, since often not all variables are fed back, so that in many applications $k < m$.

It is natural to assume that the time lag T of the feedback loop is unknown. The time lag l in number of sampling intervals is easily identified by minimising the prediction error with respect to l. It is straightforward to introduce the time lag τ of the time delay embedding, if the sampling interval Δt is very short. In Hegger *et al.* (1998) the example of a CO_2-laser with scalar feedback loop is treated. Theoretical and practical considerations including pointers to further literature are thoroughly discussed in Bünner *et al.* (2000 a and b).

Further reading

Classical articles about embedding are of course Takens (1981) and Sauer *et al.* (1991), which proves many important extensions of the theorems. Issues arising with finite data are discussed in Casdagli *et al.* (1991a), Ding *et al.* (1993), and in Malinetskii *et al.* (1993). References for the singular value decomposition (or PCA) are Broomhead & King (1986) and Vautard *et al.* (1992). A very interesting discussion of different applications of the PCA method can be found in Allen & Smith (1996).

For the discussion about the signature of dimension densities in time series of spatially extended systems see Grassberger (1989), Torcini *et al.* (1991), Tsimring (1993), and Bauer *et al.* (1993). Pattern formation and spatiotemporal chaos is thoroughly reviewed by Cross & Hohenberg (1993), and there exists a recommendable collection of articles on space-time patterns in the SFI series edited by Cladis & Palffy-Muhoray (1994).

Exercises

9.1 Revisit Exercise 3.2. Increase the sampling rate and add a small amount of noise to your Lorenz data. Use a low-pass filter to increase the sampling time Δt and to smooth the data. Convince yourself by comparing the delay plots that you have reduced noise.

9.2 In Exercise 3.2 we asked you to use the x-coordinate as the scalar observable. Create different time series using each of $s^{(1)} = x$, $s^{(2)} = y$, $s^{(3)} = z$, and $s^{(4)} = x^2 + y^2 + z^2$ as scalar observables, and then reconstruct the attractor.

9.3 Instead of a delay reconstruction, use the scalar signal and its numerical time derivative to plot the attractor. Do this for both a clean and a noisy Lorenz time series. Compare the delay and derivative reconstruction for the noisy data, for example, by plotting a short part of the trajectory with lines.

Chapter 10

Chaotic data and noise

All experimental data are to some extent contaminated by noise. That this is an undesirable feature is commonplace. (By definition, *noise* is the unwanted part of the data.) But how bad is noise really? The answer is as usual: it depends. The nature of the system emitting the signal and the nature of the noise determine whether the noise can be separated from the clean signal, at least to some extent. This done, the amount of noise introduces limits on how well a given analysing task (prediction, etc.) can be carried out.

In order to focus the discussion in this chapter on the influence of the noise, we will assume throughout that the data are otherwise considerably well behaved. By this we mean that the signal would be predictable to some extent by exploiting an underlying deterministic rule – were it not for the noise. This is the case for data sets which can be embedded in a low dimensional phase space, which are stationary and which are not too short. Violation of any one of these requirements leads to further complications which will not be addressed in this chapter.

10.1 Measurement noise and dynamical noise

When talking about noise in a data set we have to make an important distinction between terms. *Measurement noise* refers to the corruption of observations by errors which are independent of the dynamics. The dynamics satisfy $\mathbf{x}_{n+1} = \mathbf{F}(\mathbf{x}_n)$, but we measure scalars $s_n = s(\mathbf{x}_n) + \eta_n$, where $s(\mathbf{x})$ is a smooth function that maps points on the attractor to real numbers, and the η_n are random numbers. (Even in multichannel measurements, $\mathbf{x}_n + \eta_n$ is not generally recorded, but different scalar variables corresponding to different measurement functions $s^{(j)}(\mathbf{x})$ are.) The series $\{\eta_n\}$ is referred to as the *measurement noise*.

Example 10.1 (Analogue/digital converter). To store data electronically, they have to be converted into digital numbers. Even if the measurements have infinite

174

precision, our digital data contain measurement noise of an amplitude which is half the resolution of the digital numbers. Thus, 8-bit data have a noise amplitude of at least 0.2%. Note that the error due to digitisation does not have the properties we usually expect from random noise (the distribution is uniform on a bounded interval, the "noise contributions" are identical for identical measurements, and the noise is not independent in time and correlates with the signal). □

Dynamical noise, in contrast, is a feedback process wherein the system is perturbed by a small random amount at each time step:

$$\mathbf{x}_{n+1} = \mathbf{F}(\mathbf{x}_n + \eta_n). \tag{10.1}$$

Dynamical and measurement noise are two notions that may not be distinguishable *a posteriori* based on the data only. Both descriptions can be consistent to some extent with the same signal. For strongly chaotic systems which are everywhere expanding (more precisely, for *Axiom A* systems), measurement and dynamical noise can be mapped onto each other, as has been proved by Bowen & Ruelle (1975). Interesting material for the non-hyperbolic case is contained in Grebogi *et al.* (1990). However, this equivalence does not help very much in practice and the effects of the noise on the predictability and other properties of the data will depend on the nature of the noise. In these cases we will thus specify which kind of noise we are assuming. Since the effect of adding a noise term into the map **F** can be compensated by a small change of some control parameters in **F**, conversely a small stochastic fluctuation of control parameters of **F** can always be replaced by noise in its argument, even if the statistical properties of the noises may change. Hence, dynamical noise as a notion includes the very realistic situation that also all system parameters suffer from some random perturbations. Generally, dynamical noise induces much greater problems in data processing than does additive noise, since in the latter case a nearby clean trajectory of the underlying deterministic system always exists.

Furthermore, what one interprets to be dynamical noise may sometimes be a higher dimensional deterministic part of the dynamics with small amplitude. Even if this is not the case, dynamical noise may be essential for the observed dynamics because transitions to qualitatively different behaviour (bifurcations) can be induced or delayed by dynamical noise.

10.2 Effects of noise

We have encountered the effects of noise in otherwise deterministic data on several occasions already. The noise level limited the performance of the simple prediction algorithm, Section 4.2, and we will see in Section 12.5 that this is not a flaw of this

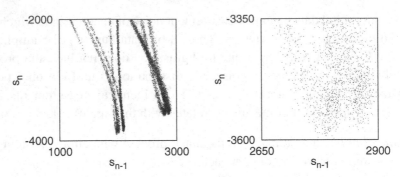

Figure 10.1 Enlargements of the phase portrait of the NMR laser data, Fig. 3.6. The resolution corresponds to the middle left and the lower right panels of Fig. 6.1.

particular algorithm but that noise is one of the most prominent limiting factors for the predictability of deterministic systems. The range of length scales through which we could observe exponential divergence of nearby trajectories is bounded from below by the noise in the data (see Section 5.3). A dramatic consequence of the noise is the breakdown of self-similarity of the strange attractor underlying the system.

Example 10.2 (Broken self-similarity of NMR laser attractor). In Example 6.1 we demonstrated the self-similarity of the attractor underlying the NMR laser data, Appendix B.2. The noise in the data we used there had, however, already been suppressed by the nonlinear noise reduction scheme that we will introduce later in this chapter. Now look what little self-similarity we find in the unprocessed data, Fig. 10.1 and remember that this is a remarkably clean data set with around 1.1% noise only.   □

Self-similarity is reflected by the correlation sum and we have already shown in Example 6.2 how the power law scaling of $C(\epsilon)$ is broken when noise is present. We studied the effect of measurement errors more carefully for the (optimistic) case of Gaussian uncorrelated noise in Section 6.7, leading us to the conclusion that the maximally tolerable noise level for the estimation of the correlation dimension is about 2%. This is a remarkable result since the correlation algorithm is the least sensitive to noise of all the methods for determining the dimension of a strange set given by a finite set of data points. The case is no better for Lyapunov exponents and for entropies, for which detailed bounds are given in Schreiber & Kantz (1996).

This sensitivity to noise is the price we have to pay for the invariance property of these quantities. The definitions of all of them involve the limit of small length scales because it is only then that the quantity becomes independent of the details of the measurement technique, data processing and the phase space reconstruction method. Moreover, the theorems which justify the use of delay (or other) embedding vectors recovered from a scalar measurement as a replacement for the "original" dynamical variables are themselves strictly valid only for data of infinite resolution. Which noise level can be considered small enough for a practical application of the embedding method depends in a very complicated way on the details of the underlying system and the measurement. The only guideline we can give with confidence is that whenever a result depends sensitively on the details of the embedding procedure, we should be very careful when interpreting that result.

There are many useful concepts which are not invariant. For instance, the *mean* depends on the calibration of the measurement apparatus, the *variance* is scale dependent and also the *autocorrelation function* and the *power spectrum* change under smooth coordinate transformations. This is not always a serious problem. Most changes are trivial and can easily be accounted for in subsequent analysis. The variance, for example, is always given with respect to some physical units. The dependence of *coarse-grained* quantities – such as estimated Lyapunov exponents from finite resolution data – on the details of the analysis can be much more intricate and it is often impossible to distinguish some spurious effects of coordinate changes from nontrivial changes due to different dynamics. Thus we do not recommend the use of "effective" dimensions at fixed length scales without a clear scaling region when we want to interpret the result as the attractor dimension. In Section 7.2 we discussed how non-invariant quantities may be used as relative measures. But even when such quantities are only used to compare data from similar situations (such as electrocardiograms (ECGs) of a patient doing exercise and at rest) one has to proceed very carefully. Obvious differences, such as the fact that the ECG taken during exercise is more noisy due to the sweat on the patient's skin, affect coarse-grained quantities as well. In the long run, the reputation of nonlinear science will not benefit from publications which diagnose a "higher complexity of the heart during exercise" just because of sweat on the skin.

After the preceding discussion it should be clear that noise is an omnipresent phenomenon that we cannot simply ignore and that we have to do something about. The important first step is to be aware of the problem and to recognise noise by its effects on the data analysis. Let us repeat the most characteristic examples: self-similarity is visibly broken, a phase space reconstruction appears as high dimensional on small length scales, nearby trajectories diverge diffusively (following a power law) rather than exponentially, the prediction error is found to be bounded from below no matter which prediction method is used and to how many digits the data are recorded. If we find that the noise level is only moderate and we have found hints that there is a strong deterministic component in the signal, then we can attempt a second step, the separation of the deterministic signal from the noise.

Once we know how to achieve this separation we have to verify the results. In Sections 10.3.4/5 we will discuss how we can check the consistency of the data processing. The separation into a deterministic signal and noise can only be justified if the resulting signal is in fact deterministic and the suppressed part of the data has the characteristic properties of additive measurement noise.

10.3 Nonlinear noise reduction

We have already proposed a simple nonlinear noise reduction scheme back in Section 4.5. It is not a coincidence that this was done in the context of nonlinear prediction since prediction and noise reduction are closely related. Let us assume that the data follows a deterministic evolution rule but is measured only with some uncertainty η:

$$s_n = x_n + \eta_n, \qquad x_{n+1} = F(x_{n-m+1}, \ldots, x_n). \tag{10.2}$$

For convenience, we assume an evolution in delay coordinates and set the delay time to unity. Both for prediction and noise reduction, we somehow have to estimate the dynamical equations F which express the fact that the dynamics are deterministic. A forecast of a future value is then obtained by applying the estimated map $\hat{F}$ to the last data points s_N available, yielding a prediction for the following measurement $\hat{s}_{N+1}$. For the purpose of noise reduction we need the information about the dynamical structure to construct a cleaned data sequence $\hat{x}_n$, $n = 1, \ldots, N$ which is close to what we measured but more consistent with the estimated dynamics. That is,

$$\hat{x}_{n+1} = \hat{F}(\hat{x}_{n-m+1}, \ldots, \hat{x}_n) + \eta'_n, \tag{10.3}$$

where $\langle \eta'^2 \rangle$ is the remaining discrepancy from the dynamical equations which should be smaller than the variance of the noise η in Eq. (10.2).

Those who have never tried it out tend to think that a good estimate of $\hat{F}$ is in fact *all* you need in order to be able to reduce noise. The simplest scheme you might suggest is to throw away every measured point and replace it by a prediction based on the previous values, i.e. $\hat{x}_{n+1} = \hat{F}(s_{n-m+1}, \ldots, s_n)$. Unfortunately, the $s_{n-m+1}, \ldots, s_n$ themselves are all contaminated by noise. If F is a chaotic map this error will on average be enhanced and $\hat{x}_{n+1}$ will be more noisy than before. The amount of "noise amplification" is given by the average of the largest eigenvalue of the Jacobian matrix of F; typical values for chaotic dynamics range from just above unity for densely sampled flows to ≈ 2.6 for the Hénon map or even larger. Thus a second ingredient is necessary for successful noise reduction which is an algorithm to adjust the trajectory $\hat{x}_n$, $n = 1, \ldots, N$ to the estimated dynamics $\hat{F}$.

The different noise reduction methods which have been proposed in the literature differ in the way the dynamics are approximated, how the trajectory is adjusted and how the approximation and the adjustment steps are linked to each other. Apart from algorithms which are restricted to the uncommon case where the dynamical equations are known, most methods will reduce noise by a similar amount and their performance will not differ dramatically. Instead, the major criteria for the preferred algorithm will be robustness, ease of use and implementation, as well as the resources needed (computer time and memory). We freely admit that we have not implemented and compared all the algorithms which have been proposed so far, and consequently the presentation given here is to some extent subjective. Conversely, we have found both the simple algorithm (Section 4.5) and the locally projective scheme that we will introduce below (Section 10.3.2) to be reliable and effective on a broad variety of data sets, including artificial and real examples.

10.3.1 *Noise reduction by gradient descent*

As an introductory and instructive example let us assume for the moment that we know the dynamical equations of motion F or that we have a very good approximation $\hat{F}$. For simplicity we concentrate on the one dimensional case; the generalisation to higher dimensional maps is straightforward. As we have argued already, the replacement $\hat{x}_{n+1} = \hat{F}(s_n)$ does not reduce noise. Instead, we can evaluate the consistency criterion, Eq. (10.3), which constitutes a measure for the remaining error η'_n in a cleaned trajectory. We could now boldly require all η'_n to become zero for the desired trajectory and numerically solve Eq. (10.3) for the $\hat{x}_n$. This turns out to be a tough numerical problem since for chaotic maps the equations become effectively ill-conditioned. Nevertheless, some of the difficulties can be overcome when the exact dynamics are known; see Hammel (1990) and Farmer & Sidorowich (1991). We do not follow this path since we will not usually have enough faith in the approximate map $\hat{F}$ in order to require the cleaned data to follow it *exactly*.

It is more realistic not to set all the η'_n to zero but, rather, to minimise the remaining mean squared error

$$e^2 = \sum_{n=1}^{N-1} \eta'^2_n = \sum_{n=1}^{N-1} (\hat{x}_{n+1} - \hat{F}(\hat{x}_n))^2 . \tag{10.4}$$

The simplest way to solve this nonlinear minimisation problem numerically is by gradient descent:

$$\hat{x}_n = s_n - \frac{\alpha}{2} \frac{\partial e^2}{\partial s_n}$$

$$= (1 - \alpha)s_n + \alpha[\hat{F}(s_{n-1}) + \hat{F}'(s_n)(s_{n+1} - \hat{F}(s_n))] , \tag{10.5}$$

where the step size α is a small constant number. Observe the two special cases which are $\hat{x}_1 = s_1 + \hat{F}'(s_1)(s_2 - \hat{F}(s_1))$ and $\hat{x}_N = \hat{F}(s_{N-1})$. The adjustment step, Eq. (10.5), is usually repeated many times.

This very intuitive algorithm, which is due to Davies (1992), explicitly shows the two distinct steps of approximating the dynamics and adjusting the trajectory to it. Note that in this second step information from the past and from the future of the signal naturally enters the estimate $\hat{x}_n$. In practice this method is mainly useful when a very good global approximation to the dynamics is available, together with its derivatives. Unfortunately, noise limits the ability to construct accurate global models. Global function fits to experimental data can be full of surprises and require some expertise. As a general purpose algorithm, the locally linear approach that we will take below proves more reliable and yields superior results if enough data is available for the local approximations.

Let us remark that the rms error Eq. (10.4) can be minimised immediately by solving a linear system of equations if $\hat{F}$ was estimated in the form of locally linear maps. Again, since these maps are only approximations we do not want to go all the way down to the minimum in one step. Rather, we take a small step in the right direction and fit new and better maps for $\hat{F}$ before we take the next step. Up to technical details related to the ill-conditioning of the above matrix problem, this is what Kostelich & Yorke (1988) proposed.

10.3.2 Local projective noise reduction

A classical way to approximate the dynamical equations underlying a deterministic signal is by the use of local linear maps. All we have to assume is that the map $\hat{\mathbf{F}}$ (or the differential equation) is a smooth function of the coordinates, that is, it is at least piecewise differentiable. Then we can linearise $\hat{\mathbf{F}}$ locally in the vicinity of each point $\mathbf{x}_n$:

$$\hat{\mathbf{F}}(\mathbf{x}) = \hat{\mathbf{F}}(\mathbf{x}_n) + \mathbf{J}_0(\mathbf{x} - \mathbf{x}_n) + O(\|\mathbf{x} - \mathbf{x}_n\|^2)$$
$$\approx \mathbf{J}_n\mathbf{x} + \mathbf{b}_n . \tag{10.6}$$

The matrix $\mathbf{J}_n$ is the Jacobian of $\hat{\mathbf{F}}$ in $\mathbf{x}_n$. Eckmann & Ruelle (1985) suggested that the above linear approximation could be determined from an experimental time series by a least squares fit. Eckmann *et al.* (1986) used the linear fits to the dynamics to estimate Lyapunov exponents from time series. Locally linear maps were later used for time series prediction by Farmer & Sidorowich (1987).

Above, we have argued that if we have a description of the dynamics in an explicit, forward-in-time form we can perform nonlinear noise reduction only with the help of a separate trajectory adjustment step. This is necessary because of the sensitive dependence on initial conditions in chaotic systems. Let us now develop a formulation which avoids this instability and which thus leads to a one-step algorithm. The basic idea is to correct each delay vector s_n individually by solving a minimisation problem which renders the corrected vector more consistent with the dynamics. Stability is achieved by *avoiding corrections to the earliest and latest coordinates* of every single delay vector. Such dangerous corrections are discouraged by choosing an appropriate metric in phase space. Let us now collect the ingredients we need for the local projective correction scheme.

In order to write the dynamics such that they are neither forward nor backward in time, we express the linearised relationship Eq. (10.6) in a neutral, implicit way, using delay coordinates. An $m - 1$ dimensional map $x_n = F(x_{n-m+1}, \ldots, x_{n-1})$ can be written implicitly as $\tilde{F}(x_{n-m+1}, \ldots, x_n) \equiv \tilde{F}(\mathbf{x}_n) = 0$, or linearised as

$$\mathbf{a}^{(n)} \cdot \mathbf{R}(\mathbf{x}_n - \overline{\mathbf{x}}^{(n)}) = 0 + O(\|\mathbf{x}_n - \overline{\mathbf{x}}^{(n)}\|^2). \tag{10.7}$$

Here, $\overline{\mathbf{x}}^{(n)} = |\mathcal{U}_n|^{-1} \sum_{n' \in \mathcal{U}_n} \mathbf{x}_{n'}$ is the centre of mass of the delay vectors in a small neighbourhood $\mathcal{U}_n$ of $\mathbf{x}_n$. Further, we have taken the chance to introduce a diagonal weight matrix $\mathbf{R}$ which will allow us to focus the noise reduction on the most stable middle coordinates of the delay vectors. This is achieved by choosing R_{11} and R_{mm} large and all other diagonal entries $R_{ii} = 1$. For $\mathbf{R} = \mathbf{1}$ we would obtain orthogonal projections.[1] This means that there is one direction in phase space, $\mathbf{a}^{(n)}$, into which the attractor does not extend – to linear approximation. Of course, for a noisy sequence s_n, the above relationship will not be valid exactly but only up to some error related to the noise:

$$\mathbf{a}^{(n)} \cdot \mathbf{R}(\mathbf{s}_n - \overline{\mathbf{s}}^{(n)}) = \eta_n. \tag{10.8}$$

In the following we will suppress the notation (n) but keep in mind that this linear equation is valid only locally and that the direction $\mathbf{a}^{(n)}$, as well as the centre of mass $\overline{\mathbf{s}}^{(n)}$, depends on the position in phase space.

It is now important to observe that we do not usually know the correct dimensionality $m - 1$ of the dynamics. Moreover, for general maps the transformation to delay coordinates will increase the minimal value of m that we need to write the dynamics deterministically. Thus the common situation is that we use a delay reconstruction in m dimensions while the dynamics confine the trajectories to a lower dimensional manifold of dimension m_0. In this case we can find up to $Q = m - m_0$ mutually independent sub-spaces $\mathbf{a}^q$, $q = 1, \ldots, Q$ fulfilling Eq. (10.7), respectively Eq. (10.8). For obvious reasons we will call the linear space spanned by these Q vectors the *nullspace* at point $\mathbf{x}_n$. Since the noise-free attractor does not

---

[1] Local orthogonal projections have been proposed for nonlinear noise reduction independently by Cawley & Hsu (1992) and Sauer (1992).

extend to this space, the component of s_n that we find in it must be due to noise. The locally projective noise reduction algorithm that we are about to propose tries to identify this nullspace and then removes the corresponding component of s_n.

When we assume that the nullspace is Q dimensional, we have to find Q orthonormal vectors a^q such that the local projection onto these vectors is minimal. If we use the notation $z_n = R(s_n - \bar{s})$, the projection of z_n onto the nullspace (assuming normalised vectors a^q) is $\sum_{q=1}^{Q} a^q \cdot (a^q \cdot z_n)$ and we require $\sum_{n' \in \mathcal{U}_n} [\sum_{q=1}^{Q} a^q \cdot (a^q \cdot z_{n'})]^2$ to be minimal for the correct choice of the set of a^q. If we introduce the constraint that the a^q have unit length by means of Lagrange multipliers λ^q and use the fact that the a^q are orthogonal,[2] $a^q \cdot a^{q'} = 0$, $q \neq q'$, we have to minimise the Lagrangian

$$L = \sum_{n' \in \mathcal{U}_n} \left[\sum_{q=1}^{Q} a^q \cdot (a^q \cdot z_{n'}) \right]^2 - \sum_{q=1}^{Q} \lambda^q (a^q \cdot a^q - 1) \qquad (10.9)$$

with respect to a^q and λ^q. This can be done separately for each q and yields

$$C a^q - \lambda^q a^q = 0, \qquad q = 1, \dots, Q, \qquad (10.10)$$

where C is the $m \times m$ covariance matrix of the vectors $z_{n'}$ within the neighbourhood $\mathcal{U}_n$,

$$C_{ij} = \sum_{n' \in \mathcal{U}_n} (z_{n'})_i (z_{n'})_j . \qquad (10.11)$$

Of course, the solutions of Eq. (10.10) are nothing but the orthogonal eigenvectors a^q and eigenvalues λ^q of C. These can be readily determined using standard software. The global minimum is given by the eigenvectors to the Q smallest eigenvalues. The noise component of the vector z_n is thus removed by replacing it by

$$\hat{z}_n = z_n - \sum_{q=1}^{Q} a^q \cdot (a^q \cdot z_n) . \qquad (10.12)$$

Finally, we write the result in terms of the original delay vectors s_n:

$$\hat{s}_n = s_n - \Delta s_n = s_n - R^{-1} \sum_{q=1}^{Q} a^q \cdot [a^q \cdot R(s_n - \bar{s})] . \qquad (10.13)$$

The projective approach is illustrated in Fig. 10.2. If points within a manifold of dimension m_0 are contaminated by noise they will be spread around the manifold in all available directions by an amount corresponding to the noise amplitude. We

---

[2] We could also require orthogonality by additional Lagrange multipliers. This would complicate the algebra but would lead to the same result, since the a turn out to be eigenvectors of the real symmetric matric C, and hence are orthogonal to each other anyway.

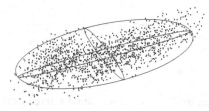 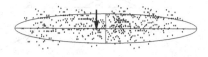

Figure 10.2 Schematic diagram of the local projective noise reduction process. Consider a cluster of points on a two dimensional surface which are corrupted by noise. The noise spreads the points above and below the surface so that in three dimensional space they form a three dimensional object of small but finite thickness (left hand panel). We can approximate the distribution of points by an ellipsoid and thus identify the direction of smallest extent. Then we project the points onto the plane orthogonal to it. The right hand panel shows a slice through the ellipsoid of the left hand panel to illustrate this projection (arrow).

can now approximate this cloud of points by an ellipsoid which is described by a quadratic form given by the local covariance matrix $\mathbf{C}$. The eigenvectors and eigenvalues give the directions and squared magnitudes of the principal axes of the best-fitting ellipsoid. In the noise-free case, the Q smallest principal directions would have zero eigenvalues, while the remaining directions locally span the manifold. In the presence of noise we identify the Q directions with the smallest eigenvalues as directions dominated by the noise and project onto the hypersurface spanned by the m_0 largest principal directions.

This projection is done for each embedding vector s_n separately, yielding a set of corrected vectors in embedding space. Since each element of the scalar time series occurs as a component of m different embedding vectors, we finally have as many different suggested corrections for each data value. We take their average, again suppressing corrections to first and last components by appropriate weights. Due to this averaging procedure, in embedding space the corrected vectors do not lie precisely on the local sub-spaces defined by Eq. (10.7) but are only moved towards them. Furthermore, all points in the neighbourhoods change a little such that the covariance matrices have new eigenvectors after the correction has been applied. Thus one has to repeat the correction procedure several times to find convergence.

10.3.3 *Implementation of locally projective noise reduction*

Before we become more specific as to how to implement the above algorithm and, in particular, how to choose the parameters involved, let us introduce an improvement that we will routinely use in order to obtain the best results available. The refinement, which was inspired by a suggestion by Sauer (1992), is particularly useful if we want to further suppress noise which is already quite small. In this case the remaining

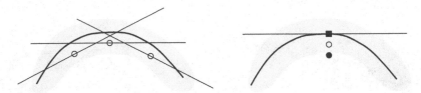

Figure 10.3 Consider a cloud of points around a one dimensional curve. Left panel: the local linear approximations (straight lines) are not tangents but secants and all the centres of mass (circles) of different neighbourhoods are shifted inward with respect to the curvature. Right panel: a tangent approximation is obtained by shifting the centre of mass outward with respect to the curvature. The filled circle denotes the average of the centres of mass of adjacent neighbourhoods, the square is the corrected centre of mass, $2\bar{s} - \bar{\bar{s}}$.

deviation from the original attractor is dominated by the curvature error resulting from the linear approximations. Here the main problem is that the hypersurfaces on which we project are not tangent to the manifold but "secant". We illustrate the situation in Fig. 10.3. The reason is that the centre of mass $\bar{s}$ is systematically shifted inward. The linear approximations can be rendered tangent by eliminating the shift of the centre of mass. Observe that the local average $\bar{\bar{s}}$ of the centre of mass vectors of adjacent neighbourhoods is shifted by roughly twice the amount compared to a single vector $\bar{s}$. Thus $2\bar{s} - \bar{\bar{s}}$ is a good estimate of the touching point of the tangent hypersurface. Technically this means that we have to sweep through the data twice. First we compute the centres of mass for all neighbourhoods and in the second sweep we compute the corrections. If computational time (or memory) is a critical issue we can omit this refinement at the price of a small, smooth distortion of the attractor which should not affect invariants such as dimensions, etc.

Now let us summarise the nonlinear noise reduction algorithm. First we embed the time series in an m dimensional phase space using delay coordinates. For each embedding vector s_n find a neighbourhood of diameter at least ϵ_0 which contains at least k_0 points. For efficiency it is advisable to use a fast neighbour search algorithm.

We go through the data set twice. In the first sweep we only compute the centres of mass of the neighbourhoods $\bar{s}$ for each embedding vector. In the second sweep we use these vectors to compute the corrected centre of mass $\bar{s}' = 2\bar{s} - \bar{\bar{s}}$. With the current estimate $\bar{s}$ or $\bar{s}'$ we build the vectors z_n and the covariance matrix C. The diagonal weight matrix R is set to be

$$R_{ii} = \begin{cases} 10^3, & i = 1 \quad \text{or} \quad i = m \\ 1, & \text{otherwise}. \end{cases} \tag{10.14}$$

Now we determine the eigenvalues and eigenvectors a^q of C in increasing order and compute the corrections Δs_n. Remember that the nth element s_n of the time series appears as a component of the delay vectors $s_n, \ldots, s_{n+m-1}$. Therefore its correction is the average of the corresponding components of the above corrections $\Delta s_n, \ldots, \Delta s_{n+m-1}$. We use weights $w_i = \sum_{j=1}^{m} R_{jj}/R_{ii}$ with the same $\mathbf{R}$ as above for the averaging.

When all data points have been handled, the time series has been replaced by the corrected one. The procedure can be repeated for best results. Our experience with data from known dynamical systems shows that about three to eight iterations are a good tradeoff between accuracy and time consumption.

There are three parameters of the algorithm which have to be chosen individually for each data set. The embedding dimension m should not be chosen much smaller than twice the dimension m_0 for which the noise-free dynamics in delay coordinates become deterministic (and which is $2D_F + 1$ when D_F is the box counting dimension of the attractor, see Chapter 9). For flows, especially if the sampling rate is high, m can be much larger because of the redundancy in consecutive data points.

The assumed dimension of the nullspace has to be chosen such that the remaining sub-space has a dimension m_0 as small as possible but large enough to accommodate the attractor, i.e. it should be about the supposed dimension of the attractor.

The optimal neighbourhood size is governed by two effects. Too small neighbourhoods are insufficient for stable linear approximations, both if too few points are available and if all the directions are dominated by noise. Too large neighbourhoods induce strong curvature effects and can lead to a distortion of the attractor. Consequently, the diameter of a neighbourhood must not be smaller than the noise level. In addition we usually require at least 50 points within each neighbourhood. Both settings can be reduced for subsequent iterations, but remember that an ellipsoid in m dimensions cannot be spanned with less than $m + 1$ points.

The implementation available in the TISEAN package is discussed in Section A.8. The routine implements both improvements, the cutoff for spuriously large correction vectors and the shift of the centre of mass. These refinements require the temporary storage of one m dimensional vector per data item, which can be a nuisance in particular for long, oversampled data sets where we have to choose m quite large (say $m = 15$). It is possible to cut memory consumption dramatically by omitting the centre of mass correction. If the cutoff is omitted as well – which we do not recommend – the whole noise reduction procedure can be done in one sweep in order to cut the computation time by half.

Example 10.3 (Nonlinear noise reduction on NMR laser data). We have already used the NMR laser data set after nonlinear noise reduction in Example 6.1 to illustrate the concept of a self-similar set. Let us now reveal how the filtering was done. We have computed the correlation sum for the unprocessed data (Example 6.2). From this we obtain two important pieces of information. First, we

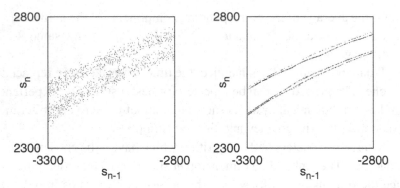

Figure 10.4 Another detail of the phase portrait of the NMR laser data before and after noise reduction. Note the linear enlargement factor of 14 relative to the total phase portrait in Fig. 1.3.

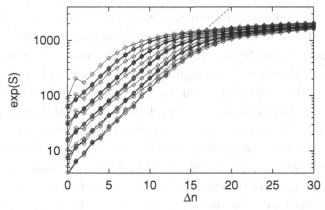

Figure 10.5 Same as Fig. 5.3 but with data after noise reduction. Indeed, initially close trajectories diverge exponentially over almost two decades in length scale.

find approximate scaling behaviour in a small range suggesting a low dimensional attractor of a dimension below or around two. Thus we choose an embedding in $m = 7$ dimensions (slightly more than $2D + 1$) and leave room for a local $m_0 = 2$ dimensional manifold. Second, we obtain the noise level using the method described in Section 6.7. Due to deviations from a Gaussian distribution we can only give the estimate that the noise has an amplitude between 20 and 25 units (see also Fig. 10.6). This is, however, good enough to determine the parameters of the noise reduction algorithm. We choose the radius of the neighbourhoods in the first iteration to be 25 units and require at least 50 neighbours for each neighbourhood.

To illustrate the effect of the nonlinear noise reduction, we show details of the phase portrait in Fig. 10.4. We also repeat the analysis carried out in Example 5.3 in order to estimate the largest Lyapunov exponent. Figure 10.5 is obtained exactly as Fig. 5.3, except that the cleaned data were used. Since the fast initial divergence due to the noise is suppressed, a larger region of exponential divergence can be seen, in particular in the bottom curves. It is highly nontrivial that this behaviour is correctly restored. □

10.3.4 How much noise is taken out?

A fundamental question which arises when nonlinear noise reduction is performed is how we can verify and quantify whether noise has actually been reduced in the data. The answer depends partly on the assumptions one makes about the nature of the noise in the input signal. The previous algorithm has been designed to be effective when the data are additively composed of a deterministic part plus random noise. Thus we can measure the success of a noise reduction algorithm by testing whether the signal has become more deterministic and the subtracted noise is random. In the following we outline methods for testing for this property.

Two error measures have been used for the development of algorithms. When the noise-free signal is known, the error in the input and output is given by the rms distance between the data s_n and the original noise-free signal s_n^0:

$$e_0 = \sqrt{\frac{1}{N} \sum_{n=1}^{N} (s_n - s_n^0)^2}. \qquad (10.15)$$

A similar quantity $\hat{e}_0$ can be computed for the cleaned data $\hat{x}_n$. If $\hat{e}_0 < e_0$ then the noise has been reduced. A slightly weaker error measure can be computed when the exact dynamical evolution equation $\mathbf{x}_n^0 = \mathbf{F}(\mathbf{x}_{n-1}^0)$ is known. It measures the deviation from deterministic behaviour according to the equation

$$e_{\mathrm{dyn}} = \sqrt{\frac{1}{N} \sum_{n=1}^{N} (\mathbf{x}_n - \mathbf{f}(\mathbf{x}_{n-1}))^2}. \qquad (10.16)$$

As before, the dynamical error can be compared before and after the noise reduction procedure has been applied.

These are useful measures when the original trajectory or the "true" dynamics are known. However, these choices are not unique. For example, the output from a noise reduction scheme might be considered as a cleaner set of data for a slightly different value of a parameter in the dynamical equations. It is also possible for a noise reduction procedure to shift the data slightly in a systematic way (for instance when the noise has nonzero mean), resulting in large values for e_{dyn} and e_0. Nevertheless, the data might still be regarded as a less noisy realisation of the dynamics in slightly different coordinates. The misinterpretations in these and other examples are the consequence of the predominance of computer generated data in the science of time series analysis. We are brought up with the prejudice that every dynamical phenomenon is generated by some hidden "original" equations of motion and if we knew these we could in principle understand everything. For computer generated data we feel that the equations *we* used to obtain the signal are privileged as the underlying *true* dynamics. In reality there are usually many ways of generating the same data which might not be distinguishable *a posteriori*.

Example 10.4 Here is a simple example for a one-parameter family of models which describe the same data equally well [Čenys (1993)]:

$$x_{n+1} = f(x_n) + \alpha(x_n - f(x_{n-1})). \tag{10.17}$$

A data set generated with $\alpha = 0$ will also be a valid trajectory of any of the two dimensional maps, Eq. (10.17), for arbitrary α since the expression in the bracket vanishes identically. It might turn out, however, that the trajectory is not a typical one for some values of α. □

The main disadvantage of quantities such as e_0 and e_{dyn} is that in an experimental situation neither a noise-free sample of data nor any exact dynamical equations can be known. Thus we must employ criteria which are accessible given only the data. Figure 10.4 shows an example where the processed data "looks less noisy", but we need a more objective, quantitative criterion. In Section 6.7 we have learned how to determine the noise level of a predominantly deterministic data set. With this technique we can estimate the amplitude of the noise before and after noise reduction in order to verify successful cleaning. Note that we cannot expect errors which remain after noise reduction to be Gaussian. Noise is often suppressed to such a degree that the typical increase of $d(\epsilon)$ for small ϵ is not visible since we are in the regime where sparseness of points prevails. In this case we can give an upper bound on the remaining noise level. For data sets which allow the determination of the noise level from the correlation sum, this is the method of choice for the verification of successful noise reduction.

Example 10.5 (Scaling of NMR laser data before and after noise reduction). We determine the difference between the scaling functions $d(m, \epsilon)$ for consecutive values of the embedding dimension m for the NMR laser data (Appendix B.2) before and after the above filtering has been applied (Fig. 10.6). For details of the noise reduction see the text of Example 10.3. As explained in the text of Fig. 10.6, noise could be suppressed by almost two orders of magnitude. □

If a noise reduction procedure successfully removes most of the noise, then the processed data set should have better short-term predictability than the raw data set. In other words, the data should be more self-consistent. This can be quantified by

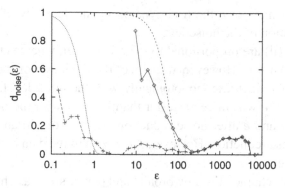

Figure 10.6 Signature of noise in NMR laser data before and after nonlinear filtering. We show $d_{\text{noise}}(\epsilon) = d(m = 7, \epsilon) - d(m = 6, \epsilon)$. In the scaling range, $d_{\text{noise}}(\epsilon)$ is close to zero as expected. The curves increase (ultimately towards 1) around the noise level. We also show the theoretical curves, Eq. (6.7), for Gaussian noise of amplitudes 25 and 0.25 units respectively. Although the noise does not seem to follow a Gaussian distribution (even before noise reduction) the diagram suggests a suppression of the noise level by almost two orders of magnitude!

computing the *out-of-sample* prediction error with some nonlinear predictor before and after noise reduction. In the simplest case, the data are split into two parts, one of which must be long enough to fit a nonlinear mapping $\hat{\mathbf{F}}(\mathbf{s})$. (One can use the same maps as in the noise reduction itself, but this does not have to be the case.) The prediction error

$$e = \sqrt{\frac{1}{N_p} \sum_{n=N-N_p+1}^{N} (\mathbf{s}_n - \hat{\mathbf{F}}(\mathbf{s}_{n-1}))^2} \tag{10.18}$$

is computed over the remaining N_p data points. It is essential to take the out-of-sample error, because one can always find a mapping which yields a zero in-sample prediction error simply by interpolating between the data points. If the data set is too short to be split into two sufficiently long parts, then *take-one-out* statistics can be used to obtain an out-of-sample error; see e.g. Grassberger *et al.* (1991) and Efron (1982).

Three effects contribute to the one-step prediction error in low dimensional chaotic systems. (i) The observed value $\mathbf{s}_n$ is contaminated by noise, inducing an error proportional to and of the order of the noise level. (ii) The data values used as arguments in the prediction function are contaminated by noise, leading to an error which is approximately proportional to the noise level but is enhanced due to the sensitivity on the initial conditions. The proportionality constant is determined by the largest eigenvalue of the Jacobian of $\hat{\mathbf{F}}$. (iii) The fitted prediction function is only an approximation to the true dynamics. The accuracy of the fit can be a complicated function of the noise level.

Effects (i) and (ii) are proportional to the noise level, whereas the size of effect (iii) is mostly unknown. However, we expect the deviation of the predictor from the true dynamics to increase monotonically with the noise level. In the presence of all three effects, e will increase faster than linearly with the noise level. If we compare e before and $\hat{e}$ after noise reduction we only obtain an upper bound on the amount of noise reduction. However, if $\hat{e} < e$ then the data have been rendered more self-consistent.

Chaotic data are characterised by broad band power spectra. Thus it is generally not possible to reduce noise in such data by using methods based on power spectra. However, there are cases where signal and noise can be readily distinguished in *some* part of the spectrum, for example at high frequencies or around a dominant frequency. Since nonlinear noise reduction methods should suppress noise throughout the spectrum, their success should be visible in these parts of the power spectrum.

Let us take densely sampled flow data as an example. High-frequency noise can be removed with a simple low-pass filter. Nonlinear noise reduction methods tend to suppress the high frequencies as well. However, some of the high frequencies are part of the dynamics. Nonlinear noise reduction methods can also remove some of the lower-frequency noise that is not part of the dynamics. Such distinctions cannot be made with a low-pass or Wiener filter.

Example 10.6 (The vowel [u:] spoken by a human). Isolated phonemes of articulated human voice are to a very good approximation limit cycle dynamics. Within running speech, their initial and final parts are distorted by the concatenation to preceeding and following phonemes. When spoken in an isolated and elongated way, they supply pretty stationary data segments. We use about 2s of [u:] (like in English "shoe") (see Fig. 9.1 on page 145 for the time series), which is contaminated by nearly white noise from the recording by the built-in microphone of a laptop computer. In Fig. 10.7 we show the power spectra of the original recording and after noise reduction using the local projective algorithm (routine `ghkss` of TISEAN). Evidently about ten harmonics whose power is far below the spectral noise level are reconstructed by the nonlinear method. This is the somewhat surprising consequence of the fact that the nonlinear algorithm uses other than spectral information to distinguish between signal and noise.

We do not recommend the power spectrum as a measure of the amount of noise reduction in chaotic time series data, because it is essentially a linear method and is therefore insensitive in the case of broad band spectra. As an extreme case,

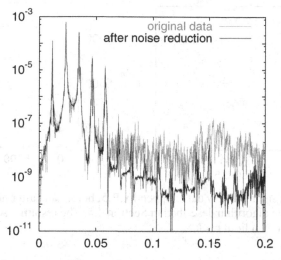

Figure 10.7 The power spectrum of our noisy recording of the vowel [u:] (grey) and after local projective noise reduction (black).

the nonlinear noise reduction scheme can even reduce in-band noise, i.e. noise with exactly the power spectrum of the clean data. As we discuss below, however, the power spectrum is a useful characterisation of the corrections applied to the input data. Most nonlinear methods have limiting cases where they behave like linear filters (for example, when very large neighbourhoods are used to fit local linear models or when inappropriate basis functions are used to fit global models). Successful nonlinear noise reduction should ensure that we do not pick one of these cases.

10.3.5 Consistency tests

So far we have only described tests which can be applied to the cleaned signal in order to check whether it is likely to be low dimensional deterministic chaos. If the input data consists of uncorrelated noise added to a deterministic signal, then the corrections applied to the data should resemble a random process. One can exploit all that is known about the sources of noise in the data. In most cases, it has only short correlation times. The spectrum will not always be flat, but if it has unexpected features, then one must consider whether the method used to process the data is appropriate.

For example, it is a good idea to check the distribution of the corrections to the data. Measurement errors are expected to have a normal distribution, whereas discretisation errors are uniformly distributed in $[-0.5, 0.5]$ if the data are stored as integers.

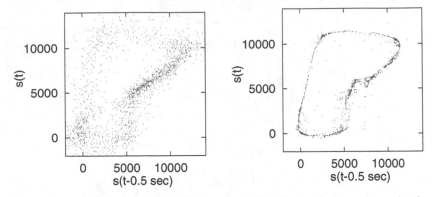

Figure 10.8 Human air flow data, Appendix B.5, before and after noise reduction using the simple algorithm described in Section 4.5. The resulting sequence looks like the motion on a limit cycle.

In addition, one should look for cross-correlations between the signal and the corrections. They should be small if the noise reduction procedure has successfully removed random measurement errors. Significant cross-correlations can arise for different reasons, including some threshold in the measurement device; fluctuations in the scale of measurement; varying noise levels in different regions of the phase space; and corruption of the data due to an inappropriate noise reduction method.

Example 10.7 (Noise reduction on air flow data). The human air flow data, Appendix B.5, see Example 1.2, shows significant nonlinearity when compared to surrogate data. Nevertheless, the correlation integral fails to show a scaling region, perhaps due to a high noise level. Now let us make the working hypothesis that the data consist of a deterministic component plus measurement noise. From the phase portrait, left hand panel in Fig. 10.8, we learn that if this is the case, the noise must be quite large, say a few hundred units. Since only 2048 data samples are available if we require the sequence to be approximately stationary, we use the robust simple noise reduction scheme described in Section 4.5 to separate the data into signal and noise. The right hand panel of Fig. 10.8 shows the result of three iterations of the locally constant approximations, using embedding dimension 7 and an initial neighbourhood size of 1500 units. Indeed, we obtain a low dimensional signal, a limit cycle. The correction has an amplitude of 574 units, which is about 15% of the total variance.

As a cross-check for our working hypothesis, we create a surrogate data set with the same properties as the "noise" subtracted by the noise reduction algorithm, using the algorithm described in Section 7.1.2. We add this surrogate to the cleaned series. If the assumption of additive independent measurement noise were true, the resulting phase portrait (Fig. 10.9) should look the same as that of the original noisy

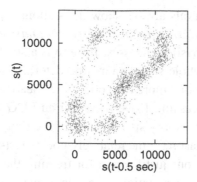

Figure 10.9 Same data as in Fig. 10.8. A surrogate of the subtracted "noise" was made and added back onto the cleaned data. There is a significant difference from the original noisy data.

data. Superficially, this is the case, but we find some significant differences. We conclude that motion on a limit cycle is a good approximate description, but the dynamics are probably subject to both dynamical and measurement noise.

10.4 An application: foetal ECG extraction

Throughout this chapter we have considered the case where the noise is the undesired contamination of the time series that we want to get rid of. However, there are other situations where it is useful to think of a data set as a superposition of different signals which can possibly be separated by a generalised filtering technique. As an example, consider the problem of measuring the electrocardiogram (ECG) of a foetus prior to birth. It is desirable to measure the electrochemical activity noninvasively, that is, without having to use an electrode placed within the uterus. Depending on the location of the electrodes on the skin of the mother, it is possible to measure an ECG signal where the foetal component is not completely hidden by the maternal ECG. But even in an abdominal ECG, the maternal component will be dominant. Current methods of extracting the foetal component from such an ECG are usually either based on multiple lead measurements which allow for an approximate cancellation of the maternal ECG or on the subtraction of a time averaged maternal cycle from a single lead ECG.

Here we want to treat the maternal ECG as the signal and the foetal component as the noise. On this base we apply nonlinear noise reduction in order to separate both parts. In order to do so, we first have to justify why we treat the ECG as a low dimensional deterministic dynamical system – which is what the nonlinear methods were designed for. As you probably know, the human heartbeat is not completely periodic, except in the case of certain severe diseases. The nature of the fluctuations of the interbeat intervals has been subject to considerable dispute. Long records of interbeat intervals always show nonstationarity, but at least in healthy subjects the fluctuations *during approximately stationary episodes* seem to be well described by linear stochastic processes. If there is a stochastic component in the interbeat intervals, the whole signal cannot be deterministic and, indeed, we find that ECGs are hard to predict more than one cycle ahead, while predictions *during* one cycle are pretty successful. If we represent an ECG in delay coordinates, as we did in Fig. 3.4, the trajectory spends most of the time near a lower dimension manifold, which is the reason for the relatively good short-time predictability. This empirical observation is our justification for treating the ECG as approximately low dimensional. As an interpretation we propose that the stochastic aspect of the ECG mainly concerns the exact position of the trajectory on this manifold which is largely unaffected by the noise reduction procedure.

Local projections onto this approximate manifold have been used in Schreiber & Kaplan (1995) to suppress measurement noise in ECGs. The same algorithm can be used to extract the foetal component from an abdominal ECG. Here we show a simple example where the noise is much smaller than the foetal component. Schreiber & Kaplan (1996a, b) also investigate the case where the foetal component is hidden under measurement noise. Two separate projection steps lead to a good separation of foetal ECG and noise. For the projections we can simply use the noise reduction algorithm described in Section 10.3.2 (see also Appendix A.8). In order to save time and memory we leave out the curvature correction step when separating the maternal signal from the rest: some distortion of the maternal component is of no particular concern here. The delay window has to be long enough to cover some reasonable fraction of one ECG cycle. With the high sampling rates (100–500 Hz) typical for ECGs this leads to high embedding dimensions. Do not be surprised to use a 50 dimensional embedding. The neighbourhoods have to be large enough to cover the whole foetal signal and can be determined by inspection. We project down to one or two dimensions (again, the maternal signal might be somewhat distorted without harm). We have found reasonable results using recordings which were only a few (maternal) cycles long. The smaller the foetal component the more data you will need.

We describe this application to show that although we have been very careful about claims of determinism, in particular in the context of Lyapunov and dimension estimates, we think that nonlinear techniques can also be very useful in situations where determinism could not be established. We feel that people have already spent too much of their time trying to find an answer to the question "is it chaos or is it noise?" Often it is much more fruitful to ask which is the most *useful* approach for a given experiment. That a data set *is* stochastic to some degree does not mean that we have to use stochastic methods exclusively. In particular, we are not obliged to

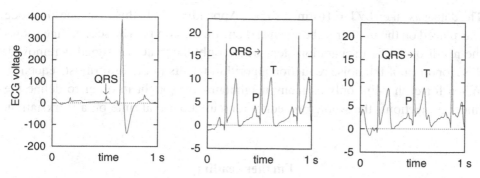

Figure 10.10 Detail of the input signal (left hand panel), the (artificially added) foetal component (middle panel), and the reconstructed foetal component (right hand panel). We also identified some clinically relevant features of the foetal part: the P-wave indicates the depolarisation of the atrium. The QRS complex reflects the depolarisation and the T-wave the repolarisation of the ventricle.

give up when no 100% appropriate method is available, as in this case, where the heart rate fluctuations are not yet fully understood.

Example 10.8 (Foetal ECG). We took two ordinary ECG recordings (Appendix B.7) and superposed them in a way which mimics an abdominal ECG. In this example, the foetal component is about 2.5 times faster and 20 times smaller in amplitude than the maternal part. The reconstruction of such a small foetal ECG, Fig. 10.10, is only possible in an almost noise-free situation like this. In a realistic situation, the foetal component of the recording will be hidden in a more or less strong noise background. In such a case, the foetal signal has to be separated first from the maternal ECG and then, by a second noise reduction step, from random fluctuations. In Fig. 10.11 we show a real ECG recording of a 16-week-old foetus.

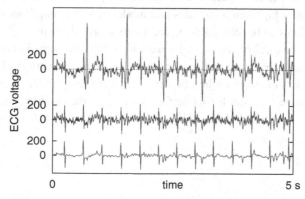

Figure 10.11 Upper: a foetal ECG recording using cervical electrodes. Middle: the maternal component removed. Lower: random fluctuations removed.

The data was taken by J. F. Hofmeister; see Appendix B.7. In this case, one electrode was placed on the mother's abdomen and one on the cervix. The second line shows the result of the first separation step, where only the maternal signal is removed. A second run of the noise reduction algorithm results in the lowermost sequence. When for each point only as many neighbours are sought in order to define the linear projections, the whole procedure runs in about real time on a workstation.

□

Further reading

Many of the nonlinear noise reduction algorithms which are currently used are reviewed in Kostelich & Schreiber (1993). Classical articles are Farmer & Sidorowich (1988), Kostelich & Yorke (1988), and Hammel (1990). More recent references are Davies (1992) and (1994), Schreiber & Grassberger (1991), Cawley & Hsu (1992), and Sauer (1992). A comparative study of the three latter papers is given in Grassberger *et al.* (1993).

Exercises

10.1 Check, how the simplest available symmetric low pass filter, $\hat{s}_n = \frac{1}{4}(s_{n+1} + 2s_n + s_{n-1})$, affects the power spectrum.

10.2 Noise reduction in numerically generated data. Choose your preferred numerical time discrete model system (i.e., a map) and create a scalar time series. Add a few percent of uncorrelated measurement noise. Apply a simple low pass filter (e.g., the program low121 from TISEAN) and compare orignial and final data. It should be evident that map-like data can never be improved by passing them through a linear filter.

10.3 Repeat the numerical experiment for highly sampled flow data. You should see that for sufficiently high sampling rates and moderate noise amplitudes, the usage of a low-pass filter can be reasonable, if one wants to avoid too much thinking.

10.4 Now apply the local projective noise reduction as implemented in the TISEAN routines ghkss or project or the simple nonlinear noise reduction scheme (nrlazy) to the data sets of the preceding two exercises. Try to optimise the parameters. Use as an objective criterion for its performance, e.g., the e_0 from Eq. (10.15), or the length scale ϵ, at which the correlation dimension starts to reflect the noise regime.

Chapter 11

More about invariant quantities

In the first part of this book we introduced two important concepts to characterise deterministic chaos: the maximal Lyapunov exponent and the correlation dimension. We stressed that one of the main reasons for their relevance is the invariance under smooth transformations of the state space. Irrespective of the details of the measurement process and of the reconstruction of the state space, they will always assume the same values. Of course, this is strictly true only for ideal, noise-free and infinitely long time series, but a good algorithm applied to an approximately noise-free and sufficiently long data set should yield results which are robust against small changes in the parameters of the algorithm.

The maximal Lyapunov exponent and the correlation dimension are only two members of a large family of invariants, singled out mainly because they are the two quantities which can best be computed from experimental data. In this chapter we want to introduce a more complete set of invariants which characterises the stability of trajectories and the geometrical and information theoretical properties of the invariant measure on an attractor. These are the spectrum of Lyapunov exponents and the generalised dimensions and entropies. These quantities possess interesting interrelations, the Kaplan–Yorke formula and Pesin's identity. Since these relations provide cross-checks of the numerical estimates, they are of considerable importance for a consistent time series analysis in terms of nonlinear statistics.

11.1 Ergodicity and strange attractors

Since every given dynamical system may provide quite a variety of different solutions depending on the chosen initial conditions (e.g., there are infinitely many different unstable periodic solutions inside a chaotic attractor), reasonable characteristic quantities cannot just represent a system, but they represent a particular solution of a given system. This sounds very discouraging, since if every initial

197

condition would produce its own value of, e.g., the maximal Lyapunov exponent, then these numbers were completely irreproducible, at least for chaotic solutions due to sensitive dependence on initial conditions. The right way out of this problem is given by the observation that there exist *invariant subsets* in the phase space, i.e. sets which are mapped onto themselves, when each of its elements are mapped forward in time by the dynamics. So instead of characterising the orbits emerging from all possible initial conditions, one characterises only their long time behaviour, after discarding transients, which takes place on one of the possibly many invariant indecomposable subsets.[1] Inside each set, there is again a kind of universality due to the fact that on the set we have at least one *invariant measure*, i.e., a probability distribution to find an orbit on the invariant set inside a measurable subset. Again invariance means that when I distribute a huge number of points according to this measure on the invariant set and map them forward in time, the measure is reproduced exactly. The natural invariant measure of a given set is the one which will be produced by almost all semi-infinite trajectories which asymptotically reside inside this invariant set. So in the end we need quantities to characterise invariant measures. It is evident that a time series represents a single solution of a dynamical system only, but that we will usually assume that it is a typical solution which generates the natural invariant measure on an attractive invariant subset of the phase space.

Since in time series analysis the measurement function involves sometimes unknown nonlinearities (and almost always some rescaling), and in addition we are working in a reconstructed phase space instead of the original one, we have access only to a transformed version of the dynamical system. Hence, it would be nice when we can characterise such a time series by quantities which are invariant under smooth coordinate transformations, such that someone else, who uses a different measurement function or a different phase space reconstruction will find the same values.

Example 11.1 (Dependence of the covariance matrix on the time lag). In Fig. 9.2 (page 149) we showed embeddings of the NMR laser data for different time lags. It is evident that the two by two covariance matrix for these four different distributions will be different, hence, covariances are not invariant under smooth coordinate transforms (evidently, they are not invariant under rescaling). □

In fact, as the title of this whole chapter indicates, those properties of invariant measures which are suitable to characterise chaos and fractality are at the same time invariant under smooth coordinate transforms. This invariance is a consequence

---

[1] Indecomposable means that the semi-infinite trajectory emerging from almost every element of this subset is dense in the subset.

from the fact that these quantities characterise either scaling behaviour or tangent space properties of the dynamics on recurrent solutions on bounded subsets.

But let us first define some terminology. If the time average along a semi-infinite trajectory is identical to the spatial average according to a probability distribution (here: according to the invariant measure) we call this dynamics ergodic on the set supporting the measure. Hence speaking about ergodicity first requires to identify an invariant set, which is the image of itself under time evolution. There are many invariant sets of a dynamical system, for example any unstable periodic orbit, but in an experimental situation we are mainly interested in stable, attracting objects, or attractors. Non-attracting sets are unstable with respect to even infinitesimal perturbations and are thus not observed in general.[2]

In order to observe a stable attractor we need dissipation in the system. To be useful for the study of ergodic properties, an attractor has to be irreducible: it must not consist of more than one invariant sub-set. A trajectory emerging from a typical initial condition will eventually settle on such an attractor. We can thus define a *natural measure* on the attractor: $\mu(x)dx$ is the average time a typical trajectory spends in the phase space element dx. Ergodicity is guaranteed when this measure is independent of the initial condition. Phase space averages taken with respect to $\mu(x)dx$ are then equal to time averages taken over a typical trajectory.

Attractors of dissipative chaotic systems generally have a very complicated, *strange* geometry. They are *fractals* in the sense of Mandelbrot (1985). The natural invariant measure has a fractal set as its support. Such measures are also called *fractal*. For any fractal set there exists one measure which is "most uniformly" spread over the whole set: the Hausdorff measure [Hausdorff (1918)]. If the natural measure is not absolutely continuous with respect to the Hausdorff measure, we call it *nonuniform* or *multi-fractal*. The natural measure of a dynamical system is in general multi-fractal.

11.2 Lyapunov exponents II

In the first approach given in Chapter 5 we restricted our interest to the maximal Lyapunov exponent. It is the inverse of a time scale and quantifies the exponential rate by which two typical nearby trajectories diverge in time. In many situations the computation of this exponent only is completely justified, as we shall discuss at the end of this section. However, when a dynamical system is defined as a

---

[2] We are not very specific when saying that an attractor should attract. In the literature there exist several definitions, differing on the question as to whether one should require a whole neighbourhood around the attractor (the *basin of attraction*) from which trajectories are attracted with probability 1, or whether one should only demand that there exists a set of initial conditions with positive measure leading to the attractor [Milnor (1985)]. In fact, there exist simple maps which possess attractors in the second sense which are not embedded in a basin of attraction with nonempty interior [Pikovsky & Grassberger (1991), Alexander *et al.* (1992)].

mathematical object in a given state space, as in Eq. (3.1) or Eq. (3.2), there exist as many different Lyapunov exponents as there are space dimensions. Typically, some of them are negative, indicating that there are also directions in space along which trajectories converge exponentially fast towards each other. The sum of all Lyapunov exponents is the rate of phase space contraction or expansion. This should be zero (for energy conserving systems) or negative. In a physically meaningful system it cannot be positive. In a complete data analysis we would like to determine all of the Lyapunov exponents, although the "all" will turn out to be questionable for various reasons when our knowledge about the system is based on a scalar time series only.

11.2.1 The spectrum of Lyapunov exponents and invariant manifolds

Chaotic systems have the property of sensitive dependence on initial conditions, which means that *infinitesimally* close vectors in state space give rise to two trajectories which separate exponentially fast. The time evolution of infinitesimal differences, however, is completely described by the time evolution in tangent space, that is, by the linearised dynamics. Let $\mathbf{x}$ and $\mathbf{y}$ be two such nearby trajectories in m dimensional state space. Considering the dynamical system as a map, the time evolution of their distance is

$$\mathbf{y}_{n+1} - \mathbf{x}_{n+1} = \mathbf{F}(\mathbf{y}_n) - \mathbf{F}(\mathbf{x}_n)$$
$$= \mathbf{J}_n(\mathbf{y}_n - \mathbf{x}_n) + O(\|\mathbf{y}_n - \mathbf{x}_n\|^2), \qquad (11.1)$$

where we have expanded $\mathbf{F}(\mathbf{y}_n)$ around $\mathbf{x}_n$ and $\mathbf{J}_n = \mathbf{J}(\mathbf{x}_n)$ is the $m \times m$ Jacobian matrix of $\mathbf{F}$ at $\mathbf{x}$. Once the perturbation $\delta_n = \mathbf{y}_n - \mathbf{x}_n$ is given, we can thus compute its modulus one time step later. Let $\mathbf{e}_i$ be an eigenvector of $\mathbf{J}$ and let Λ_i be its eigenvalue. Decomposing δ_n into these vectors with coefficients β_i, we find $\delta_{n+1} = \sum \beta_i \Lambda_i \mathbf{e}_i$. In particular, if δ_n is parallel to one of the eigenvectors $\mathbf{e}_i$, it is either stretched or compressed by the factor Λ_i. We thus find m different local stretching factors and a decomposition of the state space into m linear subspaces.

Up to now the analysis has been local in time and thus also in space. It is obvious that at an arbitrary other point of the state space we will find different eigenvectors of the Jacobian and different eigenvalues since the Jacobian itself is in general position dependent. In order to characterise the system as a whole one has to introduce global objects and quantities. As in the case of the maximal Lyapunov exponent, we introduce a proper average over the different local stretching factors. The Lyapunov exponent λ_i is defined as the asymptotic rate at which vectors from some suitable subspace grow in modulus. In order to find the largest one we want to maximise the modulus of the Nth iterate of a tangent space vector which initially was normalised, i.e. solve

$$\left(\prod_{n=1}^{N} \mathbf{J}_n \mathbf{u}\right)^{\dagger} \cdot \left(\prod_{n=1}^{N} \mathbf{J}_n \mathbf{u}\right) - \lambda(\mathbf{u}^{\dagger}\mathbf{u} - 1) \overset{!}{=} \max, \tag{11.2}$$

where† denotes the transpose. The solution of this problem is given by a system of linear equations in the coefficients of $\mathbf{u}$. When you write it down you will immediately see that this system is solved by the normalised eigenvectors of $(\prod_{n=1}^{N} \mathbf{J}_n)^{\dagger}(\prod_{n=1}^{N} \mathbf{J}_n)$. The eigenvalues of this real symmetric matrix are real. Their logarithms divided by the number of iterations N yield, in the limit $N \to \infty$, the different Lyapunov exponents, so that we do not just find the largest one. The *multiplicative ergodic theorem* by Oseledec (1968) guarantees that this set of eigenvalues normalised by N converges in the limit $N \to \infty$ and is the same for almost all initial conditions in the invariant set. This is not obvious since the multiplication of matrices is non-commutative. In summary the Lyapunov exponents are given by

$$\lambda_i = \lim_{N \to \infty} \frac{1}{2N} \ln |\Lambda_i^{(N)}|, \tag{11.3}$$

where Λ_i is defined by

$$\left(\prod_{n=1}^{N} \mathbf{J}_n\right)^{\dagger} \left(\prod_{n=1}^{N} \mathbf{J}_n\right) \mathbf{u}_i^{(N)} = \Lambda_i^{(N)} \mathbf{u}_i^{(N)}. \tag{11.4}$$

We use the convention that for each finite N the eigenvalues are ordered according to their magnitude, starting with the largest. For every N, the eigenvectors of this symmetrised matrix are orthogonal to each other and have not a direct interpretation as directions in tangent space.

For one dimensional maps the Jacobian is a real number and the above definition reduces to

$$\lambda = \lim_{N \to \infty} \frac{1}{N} \sum_{n=1}^{N} \ln |f'(x_n)|. \tag{11.5}$$

In this case the existence and uniqueness of λ is established by the usual ergodic theorem, and for its numerical computation the time ordering of $f'(x_n)$ can be ignored.

The set of the m different exponents is often called the *Lyapunov spectrum*. For all initial conditions except the set of measure zero which does not lead to the natural invariant measure, the spectrum is the same. Furthermore, Lyapunov exponents are invariant under smooth transformations of the state space. To see this consider the invertible and differentiable map $\mathbf{g}$ from the phase space Γ onto itself be our transformation. Define the dynamical system $\tilde{\mathbf{F}}$ by $\tilde{\mathbf{F}}(\tilde{\mathbf{x}}) = \mathbf{g} \circ \mathbf{F} \circ \mathbf{g}^{-1}(\tilde{\mathbf{x}})$.

Then, following the chain rule, the product of Jacobians of $\tilde{\mathbf{F}}$ along a trajectory is $\prod_{n=1}^{N} \tilde{\mathbf{J}}_n = \mathbf{J}_N^{(\mathbf{g})} \prod_{n=1}^{N} \mathbf{J}_n \mathbf{J}_1^{(\mathbf{g}^{-1})}$, since all inside the product all factors of Jacobians of the transformation cancel each other to unit matrices. Inserting this in Eq. (11.4) yields, in the limit $N \to \infty$, the same eigenvalues and thus the same Lyapunov spectrum since we can neglect the outmost factors $\mathbf{J}^{(\mathbf{g})}$ and $\mathbf{J}^{(\mathbf{g}^{-1})}$ of the phase space transformation $\mathbf{g}$.

If the dynamic $\mathbf{F}$ is invertible, every invariant measure of $\mathbf{F}$ is also invariant under the time reversed dynamic $\mathbf{F}^{-1}$. One can again compute the Lyapunov exponents and will find that they all just exchange their signs. However, one has to be careful with the interpretation of this fact, since the *natural measure* of the forward-in-time dynamics is generally not the natural measure of the time reversed system.[3]

A theoretically very interesting question is which global objects in state space correspond to the different local directions given by the eigenvectors of $\mathbf{J}_n$. In fact, likewise there is no direct relation between the eigenvalues of a single Jacobian and the Lyapunov exponents; these directions themselves do not possess much relevance. But using $\mathbf{x}_0$ as the starting point for an infinite product of Jacobians yields a basis of eigenvectors in the tangent space at $\mathbf{x}_0$, the vectors $\mathbf{u}_i^{(\infty)}$ of Eq. (11.4). They describe the directions in which perturbations increase exponentially fast over time with the corresponding Lyapunov exponent. If one denotes by $E_{\mathbf{x}}^{(i)}$ the sub-space formed by all but the first $i-1$ eigenvectors, then it is easy to verify that

$$\lambda^{(i)} = \lim_{N \to \infty} \frac{1}{N} \ln \| \prod_{n=1}^{N} \mathbf{J}(\mathbf{x}_n)\mathbf{u} \| \Leftrightarrow \mathbf{u} \in E_{\mathbf{x}}^{(i)} \qquad (11.6)$$

with probability 1. Below we will use this expression for the numerical determination of the Lyapunov spectrum.

After having defined local directions, the set of $\mathbf{u}_i^{(\infty)}$, we can ask how they align to global objects. It turns out that they are tangential to global invariant manifolds, the *stable* and *unstable* manifolds. Given a point $\mathbf{x}$, these manifolds are the set of points $\mathbf{y}$ whose images (pre-images) approach the images (pre-images) of $\mathbf{x}$ exponentially fast.

11.2.2 Flows versus maps

For continuous time dynamical systems one can adopt the above derivation by just considering the map formed by integrating the dynamics over a finite time τ, Eq. (3.3). In a numerical simulation of differential equations this corresponds to

---

[3] Attractors turn into repellers under time inversion. There can be even more invariant measures which are different from the *natural* measure. Examples are measures defined on invariant sub-sets of the attractor, such as sets of periodic points.

the use of an integration routine with a finite step width, and in experimental flow data the time interval between successive measurements can be interpreted as τ. In a flow which remains in a finite region of state space without converging to a fixed point, one Lyapunov exponent has to vanish [Benettin *et al.* (1976), Haken (1983)]. This exponent corresponds to the direction tangent to the flow. Since two points with a spacing in this direction only can always be considered to be parts of the same trajectory at different times, this means that the distance covered by the flow within a fixed time interval (i.e. the generalised velocity in state space) will never diverge nor decrease to zero. A direction corresponding to a Lyapunov exponent zero is called *marginally stable*. Thus we can now make the seemingly arbitrary distinction between flows and maps employed in Chapter 3 more precise. We call data, flow data if they contain this marginally stable direction tangent to the trajectory. Reducing such data to map data by a Poincaré surface of section removes this direction from the state space. Generally, map-like data have no Lyapunov exponents identical to zero, and if they have, this corresponds to symmetries and related conserved quantities. For further reading on Lyapunov exponents and invariant manifolds we recommend Eckmann & Ruelle (1985).

11.2.3 Tangent space method

For the computation of the maximal Lyapunov exponent we have relied on the information supplied by the separation of nearby trajectories in the reconstructed state space. In principle, one can compute the whole spectrum of exponents with this method, making use of the fact that a typical m_0 dimensional hypersurface grows over time at a rate given by the sum of the m_0 largest Lyapunov exponents. In Section 5.3 we mentioned the method by Wolf *et al.* (1985), who suggested looking for the two nearest neighbours of a given point and computing the increase in the areas of the triangles formed by the images of these three points. Since on average an area will grow with the sum of the two maximal Lyapunov exponents, one can estimate the second largest one if the maximal one has been computed before. Similarly, by using three dimensional volumes, one could compute the third exponent, and so on. Although this is a very intuitive concept, in practice this method fails due to its lack of robustness and is thus not used much.

In numerical simulations of dynamical systems, where the time evolution is known explicitly, one computes Lyapunov exponents by a numerical implementation of Eq. (11.6). When iterating the trajectory $\mathbf{x}_{n+1} = \mathbf{F}(\mathbf{x}_n)$ one computes the Jacobians $\mathbf{J}_n$ and applies them in time order to each vector $\mathbf{u}_i$ out of a set of m initially orthonormal vectors. With probability 1 they are all elements of $E^{(1)}$, such that they grow with λ_1 and thus tend to align. Well before they become degenerate, one applies an ortho-normalisation procedure to them. Obviously, the first vector

(which is only normalised) will point to the most unstable direction after sufficiently many time steps. Its average increase in length thus yields λ_1. Subtracting
its contribution from all other vectors means constructing the sub-space $E^{(2)}$ such
that all remaining vectors contain a contribution of the second Lyapunov exponent.
Normalising the vector $\mathbf{u}_2$ gives the stretching according to λ_2, and projecting it
out of all other vectors leaves the sub-space $E^{(3)}$ as the remainder. Thus the successive projections lead to the sequence of sub-spaces $E^{(i)}$, and the norm of $\mathbf{u}_i$ gives
the corresponding stretching factor. Taking the logarithm and averaging over time
yields the desired exponents. Note that in numerical simulations, this is the standard
way by which Lyapunov exponents are computed. It goes back to Benettin *et al.*
(1978).

Once we have embedded the time series data in a state space, we can try to derive
the local Jacobians from the data and employ the same method. The Jacobian is
the linear part of the dynamics in a local approximation. Thus when we want to
determine the Jacobian at the point $\mathbf{s}_0$, we can collect the set $\mathcal{U}$ of all neighbours of
$\mathbf{s}_0$ and minimise

$$e^2 = \frac{1}{|\mathcal{U}|} \sum_{\mathcal{U}} \|\mathbf{s}_{n+1} - \hat{\mathbf{J}}\mathbf{s}_n - \mathbf{b}\|^2 \tag{11.7}$$

with respect to the elements of the matrix $\hat{\mathbf{J}}$ and the offset vector $\mathbf{b}$. If, on the one
hand, the size of the neighbourhood is small enough such that this linear relation is
justified and, on the other hand, there are enough neighbours such that $\hat{\mathbf{J}}$ and $\mathbf{b}$ are
statistically well determined, then $\hat{\mathbf{J}}$ is an estimate of the Jacobian of the unknown
dynamics at $\mathbf{s}_0$. (This is exactly what we shall call a local linear fit of the dynamics
in Section 12.3.2.) In particular, when the state space is a delay embedding space,
almost all elements of the Jacobian are zero, apart from those on the lower off-
diagonal (which are unity) and apart from the first row which contains a nontrivial
vector $\mathbf{j}$ describing $s_{n+1} = F(\mathbf{s}_n)$ in linear approximation.

The method we sketched above was first suggested by Eckmann & Ruelle (1985)
and applied to data by Sano & Sawada (1985) and Eckmann *et al.* (1986). A
more recent reference containing many technical details is Darbyshire & Broomhead (1996). The results that one obtains depend strongly on the quality of the
data.

Before we point out some fundamental problems let us discuss some modifications of this method. As we shall argue in Section 12.3.2, the size of the neighbourhoods must be larger than the noise level in order to recover the dynamics and
not to fit the peculiarities of the noise. This requirement is in competition with the
requirement for small neighbourhoods in order to avoid artefacts from the ignored
curvature. Moreover, on parts of the attractor where the natural measure is "thin",

one might have difficulties finding enough neighbours. In a nutshell, there are many situations in which the local linear approach yields a very bad reconstruction of the deterministic dynamics underlying the data. In such cases global nonlinear fits may be preferable. Again, one finds the Jacobians by inserting the observed data into the matrix of derivatives of the fitted dynamics, and applies them in time ordering (see Section 12.4 for details about global nonlinear fits).

11.2.4 Spurious exponents

When embedding the data in an m dimensional space, the complete spectrum of exponents consists of m values. Let us think about their meaning. The embedding theorems of Chapter 9 state that one generally has to choose an embedding space which has about twice the dimensionality of the chaotic attractor and may even have a larger dimension than the "true" underlying state space. Thus among the m Lyapunov exponents in the reconstructed state space there may be some which are not defined in the true state space and which are consequently called *spurious*. Moreover, according to the usual assumptions, a fractal attractor of dimension D_F generated by a dynamical system can be locally embedded in a smooth manifold with a dimension equal to the next integer m_0 larger than D_F. Therefore, local neighbourhoods of the data cannot supply information for more than the first m_0 Lyapunov exponents. In favourable cases a global fit of the dynamics may contain information about additional space dimensions.

Example 11.2 (Motion on a limit cycle). Consider the following flow:

$$\dot{x} = -a(x^2 + y^2 - 1)x - \omega y,$$
$$\dot{y} = -a(x^2 + y^2 - 1)y - \omega x. \tag{11.8}$$

It can be interpreted as an over-damped motion (i.e. where the force is proportional to the velocity instead of the acceleration) in a quartic potential with a circular constant forcing. For positive a every trajectory eventually settles down on a one dimensional limit cycle (the unit circle) and moves along it with angular frequency ω. Thus a typical scalar time series would consist of the values $s(t) = \sin \omega t$. We need a two dimensional delay embedding space for a unique reconstruction of this kind of one dimensional motion. We can easily compute the Lyapunov exponent corresponding to the tangent direction of the "attractor" (which is zero). But from observations made after transients have died out, we cannot obtain information about the dynamics perpendicular to this. Knowing the equations of motion, we can rewrite them in polar coordinates. The time evolution of the radius is given by $\dot{r} = -a(r^2 - 1)r$, and small deviations δ from the stationary solution $r(t) = 1$

evolve according to $\dot{\delta} = -a\delta$. Thus the second exponent[4] is $\lambda_2 = -a$ and thus depends on a parameter whose value cannot be inferred from the motion on the limit cycle. In particular, the orbit is also a solution when $a < 0$ in the same potential. If that is the case, however, it is an unstable orbit (and therefore transient). The Lyapunov exponent is then positive.   □

Example 11.3 (Ulam map).   The chaotic Ulam map $x_{n+1} = 1 - 2x_n^2$ is a special case of the well-known logistic equation (see also Exercise 2.2). Making use of the transformation $y_n = -\cos \pi x_n$, one can map it to the *tent map* $y_{n+1} = 1 - 2|y_n - \frac{1}{2}|$, which is piecewise linear. Applying Eq. (11.5), one easily finds the Lyapunov exponent $\lambda = \ln 2$. The attractors of both the Ulam map and the tent map are one dimensional, and in a two dimensional representation, the pairs (x_{n-1}, x_n) of the Ulam map form a parabola. In this two dimensional state space, the Lyapunov exponent corresponding to the tangential direction of the attractor is $\lambda = \ln 2$ (the attractor is the unstable manifold), whereas the one corresponding to the direction perpendicular to the attractor is undefined and in principle should be $-\infty$, since any point not belonging to the attractor would be mapped exactly onto it in only one iteration (since the outcome does not depend on x_{n-1}). Computing the two exponents using local linear fits for a time series of length 10 000, we always find a positive exponent close to $\ln 2$, but a negative one ranging from -0.1 to almost zero, depending on the neighbourhood size we require.   □

To repeat, the straightforward application of the Jacobian based algorithm always yields m exponents in an m dimensional delay embedding space, irrespective of the dimension of the underlying "true" state space. Although the two preceding examples are very simple, they illustrate that within the set of m exponents several may be meaningless. Thus we either have to find out which ones are spurious, or we have to introduce a sophisticated trick in order to reduce the dimensionality of the Jacobians and to avoid spurious exponents.

There exists a conceptually convincing suggestion by Stoop & Parisi (1991) which allows for the computation of the relevant Lyapunov exponents only. First, the time series is embedded in the required m dimensional embedding space, but after having selected the neighbourhood the fit for the linear dynamics is restricted to the minimal sub-space necessary for a local embedding of the attractor. There are some technicalities, especially for map data, thus we do not discuss this method further.

---

[4] For first-order ordinary differential equations $\dot{\mathbf{x}} = \mathbf{f}(\mathbf{x})$ the Lyapunov exponents are given by the eigenvalues of the *monodromy matrix* $\mathbf{M} = D\mathbf{f}$. Heuristically we can understand this very easily. The simplest integration scheme (Euler scheme) yields the map $\mathbf{x}_{n+1} = \mathbf{x}_n + \Delta t \mathbf{f}(\mathbf{x}_n)$. Its Jacobian is $\mathbf{J} = \mathbf{1} + \Delta t \mathbf{M}$, and for small Δt the logarithms of its eigenvalues are in lowest order the eigenvalues of $\Delta t \mathbf{M}$ which have to be normalised by Δt to obtain the Lyapunov exponents.

Example 11.4 (Ulam map). Let us embed a trajectory of the Ulam map $x_{n+1} = 1 - 2x_n^2$ in two dimensions, and determine the parameters of the more general model equation $x_{n+1} = 1 + ax_n^2 + bx_{n-1}$. For noise-free data the outcome will be $a = -2$ and $b = 0$, which leads to exponents $\lambda_1 = \ln 2$ and $\lambda_2 = -\infty$. On the other hand, when allowing for a quartic polynomial in the two variables, the result will be an arbitrary linear combination $x_{n+1} = \alpha(1 - 2x_n^2) + (1 - \alpha)(-1 + 8x_{n-1}^2 - 8x_{n-1}^4)$, depending on the details of the fit. For $\alpha = 0$ the two Lyapunov exponents are $\lambda_1 = \lambda_2 = \ln 2$. This result implies an additional instability in the direction perpendicular to the attractor. In fact, this instability is easy to understand. Each trajectory of $x_{n+1} = 1 - 2x_n^2$ is also an exact solution of the map $x_{n+1} = -1 + 8x_{n-1}^2 - 8x_{n-1}^4$, but not vice versa. When iterating the latter map, the slightest perturbation will kick the trajectory off the attractor. Finally, performing a generic quartic fit can yield even more complicated results. □

Now we can formulate a recipe to detect the spurious exponents. We have said that the values we find for them depend on the realisation of the noise, on the embedding and on the details of the fit. Thus they change when we add further numerical noise to the data, when we change the delay in the delay reconstruction or when we change the details of the fitting procedure. In contrast, the true exponents should be relatively robust against this. Otherwise we could not determine them at all, since there exists no prescription for the details of their computation. It is exactly this undeterminedness of spurious exponents which makes them worthless for us, since otherwise they could possibly supply some additional information about the dynamics on the reconstructed attractor which goes beyond the knowledge of the system in its original phase space.

Apart from computing the Lyapunov spectrum in different ways there exists a powerful test to estimate the reliability of those exponents which have been identified as non-spurious. One uses the fitted dynamic $\hat{F}$ as an artificial dynamical system and simply begins to iterate it. An arbitrary point on the observed attractor serves as the initial condition. This *bootstrapping* (compare Chapter 12) works both for local linear and global nonlinear models. If the trajectory escapes to infinity after only a few steps, then the observed data are unstable under the fitted dynamics, which means that there has to be at least one positive exponent which is spurious. One should doubt all results obtained using this dynamics, in particular the Lyapunov exponents. One has to repeat the fit of the dynamics with different parameters or with a different model. In a better case, the iteration yields a similar attractor which is essentially a deformed version of the original one. Now, all positive exponents are true ones, but their magnitudes could be wrong. Only if the observed attractor is reproduced fairly well by the model system can one be sure that the exponents are also well determined. This discussion will be continued in Section 12.4.4.

Example 11.5 (NMR laser data: global and local fits). We fit the dynamics both by radial basis functions and by a multivariate polynomial in a three dimensional delay embedding space (NMR laser data, Appendix B.2). Details of the radial basis function fit and the attractor obtained when iterating it will be presented in Example 12.8. The attractors formed by the iteration of a seventh- and eighth-order polynomial look indistinguishably the same. The Lyapunov exponents we obtain from the first fit are $\lambda_1 = 0.272$, $\lambda_2 = -0.64$, and $\lambda_3 = -1.31$. Using polynomials, we find $\lambda_1 = 0.27$, $\lambda_2 = -0.64$, and $\lambda_3 = -1.15$. (All values vary by about a few percent depending on details such as size of the training set, width and number of the radial basis functions. The average one-step prediction error on the data after noise reduction is smaller than 5×10^{-3} for all reasonable cases.) In Table 11.1 we present these exponents together with those obtained by local linear fits in 2–7 embedding dimensions, as a function of the number of neighbours k. We can learn from this that the local linear method is quite sensitive to the dimension of the embedding space and can yield quite ambiguous results. More generally, the dependence on the structure of the model introduces some uncertainty about the exponents obtained by the tangent space method. Our estimate for the maximal one is $\lambda_+ = 0.3 \pm 0.05$. Since the attractor dimension is slightly above 1, there can exist only a single positive exponent; all others (in the ($m > 3$) dimensional models) have to be spurious.

How many non-spurious negative exponents do we expect to find? The data were originally flow data with an attractor dimension between 2 and 3. Thus the minimal number of equations for the underlying dynamical system is three, and one needs at least two dimensions for the Poincaré map. Thus, apart from the exponent $\lambda_0 = 0$, which we lost by passing over to the Poincaré map, we expect at least one negative non-spurious exponent. However, it turns out that we need one additional dimension in order to reconstruct the dynamics uniquely by a delay embedding. The corresponding additional exponent might be spurious.

The first negative exponent of a chaotic system always corresponds to a degree of freedom which is active and thus contributes to the Kaplan–Yorke estimate of the attractor dimension (see Section 11.5.2). It should therefore come out quite clearly in the analysis, and in fact, as Table 11.1 shows, we can identify it in all the three dimensional models: $\lambda_2 = -0.6 \pm 0.05$. Finally, the most negative exponent of all the approaches is also very robust, and we are inclined to conclude that the exponent $\lambda_3 = -1.2 \pm 0.1$ is not spurious. However, based on the data analysis, this is a bit vague, since we possess no criterion apart from its robustness in favour of its presence. Fortunately, the NMR laser system of Flepp *et al.* (1991) is sufficiently simple that a set of differential equations can be given where all terms are justified from the knowledge of the underlying physical process. The data are reproduced extremely well with a periodically driven system of three first-order equations.

Table 11.1 *Lyapunov exponents of the NMR laser data. m is the embedding dimension and k is the number of neighbours used in the local linear fits. Those exponents which are clearly spurious are shown with grey shade*

m	k	λ_{dubious}	λ_+	λ_{dubious}				λ_-
3	20		0.32			−0.40		−1.13
	40		0.30			−0.51		−1.21
	160		2.28			−0.68		−1.31
4	20		0.34		−0.03	−0.49		−1.08
	40		0.31		−0.01	−0.52		−1.12
	160		0.29		−0.03	−0.69		−1.35
5	20		0.36	0.16	−0.20	−0.57		−1.11
	40		0.35	0.14	−0.21	−0.59		−1.14
	160		0.31	0.13	−0.35	−0.77		−1.34
6	20	0.39	0.24	−0.02	−0.26	−0.58		−1.09
	40	0.41	0.25	−0.02	−0.27	−0.64		−1.18
	160	0.38	0.25	−0.17	−0.44	−0.83		−1.34
7	20	0.42	0.27	0.08	−0.09	−0.28	−0.57	−1.06
	40	0.45	0.27	0.11	−0.12	−0.33	−0.65	−1.16
	160	0.40	0.25	−0.02	−0.24	−0.50	−0.85	−1.38
3	global[a]		0.272			−0.64		−1.31
3	global[b]		0.273			−0.64		−1.15

[a] radial basis functions
[b] polynomial

The system is non-autonomous and the phase space is four dimensional, hence two negative exponents are non-spurious. The error bars given above are mere estimates and reflect mainly systematic errors. Statistical errors are errors induced by using different parts of the data, and these errors are small compared to the deviations between the exponents gained from different models.   □

We do not want to leave the discussion about spurious exponents without mentioning a suggestion for distinguishing between spurious and non-spurious Lyapunov exponents, due to Parlitz (1992). Under time inversion all Lyapunov exponents of a system simply change their signs, whereas what happens to the spurious exponents is undetermined. Thus Parlitz feeds a time series twice into the same algorithm, once as usual and once reversing the time ordering of the values. Comparing the resulting two Lyapunov spectra, he concludes that all values which occur in both spectra but with opposite signs are the non-spurious exponents. When trying to reproduce these results with local linear fits and global models we failed in the sense that even those exponents which are most surely not spurious are not reproduced well enough to be distinguished from the spurious ones.

Table 11.2 *Estimated invariants of NMR laser data*

Quantity	Value	Uncertainty	Remarks
λ_1	0.3	0.01	See Example 5.3
λ_2	−0.6	0.05	See Example 11.5
λ_3	−1.2	0.1	See Example 11.5
D_1	1.45	0.1	See Example 11.7
D_2	1.5	0.1	See Example 6.2
D_{KY}	1.5	0.17	$= 1 + 0.3/0.6$
h_{KS}	0.3	0.05	See Example 11.14
h_2	0.3	0.05	See Example 11.13

For the data set underlying Table 11.1 we find after time inversion, e.g., in a five dimensional embedding requiring 20 neighbours, the exponents −0.82, −0.38, −0.18, 0.14, 1.04, to be compared to 0.36, −0.16, −0.20, −0.57, −1.11. This would indicate that the exponent −0.16 might be non-spurious, whereas −0.57 does not have a counterpart in the reversed spectrum. The global models and also the dimension estimates (see Table 11.2) suggest that nearby −0.57 is the negative exponent closest to zero. Parlitz uses a local interpolation of the dynamics with radial basis functions (which includes the possibility of over-fitting), such that we do not know under which conditions the expected change of signs really occurs. Thus although at first sight this method appears very promising, its usefulness is not yet clear to us. In the TISEAN routine `lyap_spec` we have included an option to compute the exponents on the time reversed version of the input data so that you can try it on your data set.

Almost all of the experimental data sets where chaos can be verified in this book possess only one positive Lyapunov exponent. This is so common in experiments that for models with more than one positive Lyapunov exponent the name *hyper-chaos* was coined [Rössler (1979)]. If, furthermore, the positive Lyapunov exponent is smaller than the least negative one (which is also quite common), then the Kaplan–Yorke formula (see Section 11.5.2) tells us that the attractor dimension should be less than two (for flows, this holds for the Poincaré section). The converse is also true: if the attractor (more precisely, its Poincaré section) looks between one and two dimensional, we can say without any further computations that the system has one positive Lyapunov exponent only and one or several negative ones. On such an attractor, the invariant measure locally is a direct product of a continuous line, which represents the unstable manifold, and a Cantor set on the stable (attracting) manifold. The most relevant information about the instability of such a system can readily be extracted by the application of the algorithm introduced in Chapter 5.

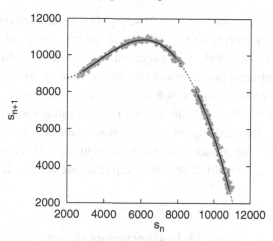

Figure 11.1 Hatched dots: return map of the chaotic string data (Appendix B.3) in the Poincaré surface of section (497 points). Dashed line: the graph of the fitted polynomial $s_{n+1} = \sum_{j=0}^{4} a_j s_n^j$. Solid line on this graph: the attractor formed by 2000 iterates of this map.

11.2.5 Almost two dimensional flows

For flows on an attractor which in the Poincaré plane looks almost one dimensional, the method of the last section degenerates to a much simpler procedure. For such data one can represent the dynamics by a one dimensional map on some interval of $\mathbb{R}$. Often it is enough to just consider the first component of every delay vector representing a point in the surface of section, and to plot it versus the corresponding component of its predecessor over time. If one is lucky, the points thus obtained align to the graph of a single-valued function. If not, they still form a one dimensional object, and one has to find a suitable function of the components of each delay vector (or another surface of section) which yields a single valued graph. This graph defines the map by which the time evolution of the new scalar observable is characterised. We can thus compute its Lyapunov exponent with the help of Eq. (11.5), that is, we have to compute

$$\lambda_+ = \langle \ln | \hat{F}'(s) | \rangle_{\{s\}}, \tag{11.9}$$

where $\langle \ldots \rangle_{\{s\}}$ denotes the average over the collection of points. The derivative $\hat{F}'(s)$ has to be estimated from the data. This method was successfully employed in one of the pioneering works on nonlinear time series analysis [Packard *et al.* (1980)].

Example 11.6 (Data of a vibrating string). The attractor (after nonlinear noise reduction) is represented as dots in Fig. 11.1. Since this attractor looks like the graph of a single valued function, we model it by a fourth-order polynomial, $s_{n+1} = \sum_{j=0}^{4} a_j s_n^j$. Iterating this function, we can compute the only Lyapunov exponent as described above, and find $\lambda_+ \approx 0.214$. Applying the method of Chapter 5 to the experimental time series after nonlinear noise reduction (in the Poincaré surface of section) yields $\lambda_+ \approx 0.21$, in excellent agreement. Dividing this number by the average time between two intersections of the time-continuous trajectory with the Poincaré surface of section (which is about $\tau_P \approx 181$) yields the positive Lyapunov exponent of the flow in units of one per sampling interval, $\lambda_+^{(f)} \approx 0.0012$. Since we know that the next exponent of the flow is exactly zero, we should determine the exponents with an accuracy of more than 10^{-3} if we work directly with the flow data. This again illustrates that map data are much nicer to handle than flow data. $\square$

11.3   Dimensions II

The dynamical side of chaos manifests itself in the sensitive dependence of the evolution of a system on its initial conditions. This strange behaviour over time of a deterministically chaotic system has its counterpart in the geometry of its attractor. We have demonstrated that the inherent instability of chaotic motion leads to an exponential stretching of infinitesimal distances. Considering an arbitrary tiny part of the attractor as a set of initial conditions, we understand that after some time its images will cover the whole attractor. (This property is called *mixing* and leads to an exponential decay of the autocorrelation function.) An immediate consequence is that a typical chaotic attractor has to be continuous along the unstable direction.[5] Otherwise every small scale structure would be spread out by the stretching and eventually grow as large as the attractor itself. This would be in contradiction with the required invariance.

On the other hand, due to dissipation, there are contracting directions in space. For these directions we can apply the same argument as for the unstable ones, but with a different conclusion. Assume two initial conditions which are a small distance apart and whose difference vector points to a stable (i.e. contracting) direction. Then this difference will shrink over the course of time and will eventually disappear. Thus one could think that in the long-term limit the attractor would be zero dimensional in these directions. In fact this can be the case, but one generally observes a more complicated structure. The above argument ignores the folding process. We can only have a finite attractor and at the same time exponential divergence of initial conditions if, due to the nonlinearity, the attractor is folded in space and finally mapped onto itself. This process partially "fills up" the stable direction. Together with the dissipation this leads to *self-similarity*, to a *scale invariance* of the invariant set. If a piece of a strange attractor is magnified it will resemble itself. Thus,

---

[5] As we argued in Section 11.2, there may be more than just one unstable direction.

the attractor will show structure on all length scales and will be self-similar in a statistical sense.

11.3.1 Generalised dimensions, multi-fractals

Self-similarity of *sets* is characterised by the Hausdorff dimension, although the box counting dimension is much more convenient to compute. It presents an upper bound on the Hausdorff dimension from which it is known to differ only for some constructed examples. Consider a point set located in $\mathbb{R}^m$. If we cover it with a regular grid of boxes of length ϵ and call $M(\epsilon)$ the number of boxes which contain at least one point, then for a self-similar set

$$M(\epsilon) \propto \epsilon^{-D_F}, \qquad \epsilon \to 0. \tag{11.10}$$

D_F is then called the *box counting* or *capacity* dimension.[6]

Now, as we have said, natural measures generated by dynamical systems are usually not homogeneous. Thus it is rather unphysical to characterise this measure just by the dimension of its *support*, i.e. the set it lives on. We want to give more weight to those parts which are visited more frequently and thus contain larger fractions of the measure. It is, however, not immediately obvious exactly how we want to weight the regions of different densities when we define a dimension of a measure. In fact there is a one-parameter family of dimensions [Renyi (1971)], the *generalised* or Renyi dimensions which differ in the way regions with different densities are weighted.

If we want to study a fractal measure μ defined in phase space, let us denote by $p_\epsilon(\mathbf{x}) = \int_{\mathcal{P}_\epsilon(\mathbf{x})} d\mu(\mathbf{x})$ the probability of finding a typical trajectory in a ball[7] $\mathcal{U}_\epsilon(\mathbf{x})$ of radius ϵ around $\mathbf{x}$. We can then define a generalised correlation integral:

$$C_q(\epsilon) = \int_{\mathbf{x}} p(\mathbf{x})_\epsilon^{q-1} d\mu(\mathbf{x}) \equiv \left\langle p_\epsilon^{q-1} \right\rangle_\mu. \tag{11.11}$$

Now we have for a self-similar set

$$C_q(\epsilon) \propto \epsilon^{(q-1)D_q}, \qquad \epsilon \to 0. \tag{11.12}$$

Thus we can determine the generalised dimensions D_q to be

$$D_q = \lim_{\epsilon \to 0} \frac{1}{q-1} \frac{\ln C_q(\epsilon)}{\ln \epsilon}. \tag{11.13}$$

---

[6] Let us remark at this point that, however innocent the definition of the box counting dimension looks, it is unsuitable as an algorithm for extracting dimensions from experimental data, unless extremely many high-resolution data are available. Although the problem of establishing a grid of boxes in higher embedding dimensions can be overcome using appropriate data management, edge effects due to the finite size of the attractor are severe. This can only be overcome when very small length scales are accessible.

[7] Note that when we compute inter-point distances with the sup norm, such a ball is actually a cube.

The particular case $q = 1$ can be obtained by the de l'Hospital rule and yields the definition of the *information dimension*:

$$D_1 = \lim_{\epsilon \to 0} \frac{\langle \ln p_\epsilon \rangle_\mu}{\ln \epsilon}. \tag{11.14}$$

It gets its name from the fact that $\langle \ln p_\epsilon \rangle_\mu$ is the average (Shannon) information needed to specify a point $\mathbf{x}$ with accuracy ϵ. The information dimension specifies how this amount of information scales with the resolution ϵ. Since the Shannon information is the only proper information measure which fulfils the requirement of additivity, D_1 is the most interesting of all the D_q. The generalised dimension D_0 is equivalent to the capacity dimension D_F of the support of the measure.

If the D_q indeed depend on q, we call the measure *multi-fractal*. D_q is a non-increasing function of q and we often compute D_2 as a lower bound of the desired D_1. The reason is that, from a limited amount of data, D_2 is by far the easiest to estimate. As we have seen in Chapter 6, the algorithm for estimating D_2 from a time series is indeed very simple. Nevertheless, whenever it is feasible due to the data quality, we should try to estimate D_1 as well. A significant discrepancy is an indicator for possible intermittency effects. The other generalised dimensions are mostly interesting in order to check for multi-fractality. If D_q strongly depends on q, the use of D_2 as an approximation to D_1 must be questioned.

The generalised dimensions D_q can be obtained from high-quality time series data if care is taken. For $q \neq 2$ we have to take systematic errors due to finite statistics into account. To understand why this is the case, let us rewrite Eq. (11.11) using the Heaviside step function, $\Theta(x) = 0$ if $x \leq 0$ and $\Theta(x) = 1$ for $x > 0$:

$$C_q(\epsilon) = \int_{\mathbf{x}} d\mu(\mathbf{x}) \left[\int_{\mathbf{y}} d\mu(\mathbf{y}) \Theta(\epsilon - \|\mathbf{x} - \mathbf{y}\|) \right]^{q-1}. \tag{11.15}$$

From a finite sequence of points $\mathbf{x}_n$ we estimate

$$\hat{C}_q(\epsilon) = \frac{1}{N(N-1)^{(q-1)}} \sum_{i=1}^{N} \left[\sum_{i \neq j} \Theta(\epsilon - \|\mathbf{x}_i - \mathbf{x}_j\|) \right]^{q-1}. \tag{11.16}$$

For $q \neq 2$ we have to take the sum in brackets (and its fluctuations) to nontrivial powers and thus obtain a biased estimate of C_q. The most serious problem occurs for $q \leq 1$. For some centres the sum can be zero since they do not have any points at all in their neighbourhoods. The leading finite sample corrections which have been published in the literature [Grassberger (1985) and (1988)] offer a partial solution since they remove the singularity. A much better approach for $q \leq 1$ is a "fixed mass" algorithm, which we will describe below for the case $q = 1$. For the case $q < 1$ we have to refer the insistent reader to the literature [Badii & Politi (1985), van de Water & Schram (1988)].

For $q > 1$ we can use an algorithm which is very similar to that for the order-two correlation integral we gave in Section 6.3. This amounts to the evaluation of Eq. (11.16). The μ-average in Eq. (11.11) is performed by averaging over a collection of centre points chosen from the attractor without bias (the sum over i). For each centre $\mathbf{x}_i$, $p(\mathbf{x})_\epsilon$ is then estimated by the fraction of attractor points closer than ϵ in phase space. We must not count $\mathbf{x}_i$ itself and must also avoid spurious temporal correlations by excluding points closer in time than some $t_{\min} = \Delta t n_{\min}$. Thus we have

$$C_q(\epsilon) = \frac{1}{\alpha} \sum_{i=n_{\min}}^{N-n_{\min}} \left[\sum_{|j-i|<n_{\min}}^{N} \Theta(\epsilon - \|\mathbf{x}_i - \mathbf{x}_j\|) \right]^{q-1}, \qquad (11.17)$$

where $\alpha = (N - 2n_{\min})(N - 2n_{\min} - 1)^{(q-1)}$ yields the correct normalisation. Again we have to look for a scaling range at finite length scales, preferably by looking for a plateau of $D_q(\epsilon) = d \ln C_q(\epsilon)/d \ln \epsilon$.

In all of the above formulas, the state vectors $\mathbf{x}$ will be replaced by delay vectors $\mathbf{s}$ when working with scalar time series data. In this case, the embedding dimension m enters as an additional and very relevant parameter, which we will explicitly include into our notation as $C_q(m, \epsilon)$. It is evident that a data set embedded in m dimensions cannot show a scaling behaviour corresponding to dimensions $D_q(m) > m$, hence as long as $m < D_q$ the results thus obtained should either be trivial (namely, $D_q(m) = m$) or artefacts. As for the correlation dimension, we routinely compute $C_q(m, \epsilon)$ for all m ranging from unity to some reasonable value between 5 and 10 and compare the scaling behaviour in the different embedding dimensions. Only if $D_q(m)$ approaches a constant for m larger than this constant, good evidence for a finite dimension is given, since all generalised dimensions of an embedded set are independent of the dimension of the embedding space, if the latter is larger than the minimum of $2D_F$.

11.3.2 Information dimension from a time series

We defined the generalised dimensions D_q via a correlation integral $C_q(m, \epsilon)$ which was determined as an average over the content of neighbourhoods of equal radius ϵ. Alternatively, one can construct neighbourhoods with variable radii but which all contain the same fraction of the total measure or the same fixed "mass". One can then ask for the scaling behaviour of the average radii of these neighbourhoods as a function of the mass contained in each of them. One finds that [Badii & Politi

(1985), Grassberger (1985)]:

$$\left\langle \epsilon_p^{(1-q)D_q} \right\rangle \propto p^{1-q}, \qquad p \to 0. \tag{11.18}$$

The mass p of the neighbourhoods is estimated by $p \approx k/N$, where k is the number of points in a neighbourhood (excluding the centre) and N is the total number of points. The neighbourhood size ϵ_p of each single point of the time series has to be computed, such that it contains exactly k neighbouring points. We will use a finite sample correction to this estimate below. Only the limit $q \to 1$ of Eq. (11.18) can be solved explicitly for D_q:

$$D_1 = \lim_{p \to 0} \frac{\ln p}{\langle \ln \epsilon_p \rangle}. \tag{11.19}$$

For all other values of q it is a nuisance that D_q appears within the sum. Basically one investigates the scaling of $\langle \epsilon_p^\beta \rangle$ for a range of values of β and subsequently maps them to D_q. Since we do not find the computation of D_q for $q < 1$ particularly interesting and recommend to compute D_q for $q > 1$ using the correlation sum we leave out the details here.

When Eq. (11.19) is applied to a scalar time series, we have again an explicit dependence on the embedding dimension m. It is evident that the neighbourhood radius $\epsilon_p(m)$ containing exactly k neighbours can only increase with m. However, the scaling of $\frac{\ln p}{\langle \epsilon_p(m) \rangle}$ in p should be independent of m as soon as m is large enough. Of course, first of all a possible scaling range has to be established by plotting $D_1(m, p) = d \ln p / d \langle \ln \epsilon_p(m) \rangle$, which should exhibit a plateau when scaling is present. The mass p can be varied either by varying the number k of points within the neighbourhoods or by changing N', the total number of points considered. In Grassberger (1985), the leading finite sample correction for p is given, yielding the improved estimator $\ln p \approx \psi(k) - \ln N'$. Here, $\psi(x) = d/dx \ln \Gamma(x)$ is the *digamma function*.

In order to design an efficient algorithm which takes advantage of a fast neighbour search algorithm, variation of p is achieved in the following way. For the small values of p we consider all available points $N' = N$. The mass p is varied by increasing k from 1 to, say, 100. Above $p = 100/N$, we only use $N' < N$ points of the time series to increase p. The number of centre points over which the average is performed is chosen large enough to suppress statistical fluctuations. Again, temporal correlations are taken care of by discarding neighbours closer in time than $t_{\min}$. An implementation of this algorithm is given by d1 of the TISEAN package, which is discussed in Appendix A.6.2. The CPU time consumption is considerably larger than for the correlation dimension, but an analysis such as that in Example 11.7 can still be carried out in a few minutes on a workstation.

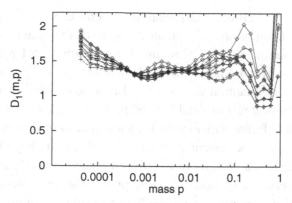

Figure 11.2 Local slopes $D_1(m, p)$ versus mass p for embedding dimension $m = 2, \ldots, 7$ (counted from below, NMR laser data). Scaling is clearly visible.

Example 11.7 (Information dimension of the NMR laser data). In analogy to the analysis of Example 6.2, we embed the NMR laser attractor (Appendix B.2, this time after nonlinear noise reduction, Example 10.3) in 2–7 dimensions. We compute $\langle \ln \epsilon_p(m) \rangle$ over 2000 reference points, over a range of p between $1/N$ and 1. In Fig. 11.2 we plot $D_1(m, p) = d \ln p / d \langle \ln \epsilon_p(m) \rangle$ versus the mass, $\ln p \approx \psi(k) - \ln N'$. A scaling range can be discerned which yields an estimate $D_1 = 1.45 \pm 0.1$. There is some ambiguity in the lower end of the scaling region. Keeping in mind that multi-fractality would yield $D_1 \geq D_2$, we attribute the difference between D_1 and D_2 (we found $D_2 = 1.5 \pm 0.1$ in Example 6.2) to the estimation errors, statistical and systematical. In a previous analysis of this data set [Kantz *et al.* (1993)], the generalised dimensions D_q for $q \geq 1$ were found to be compatible with the null assumption of no q-dependence. Thus, multi-fractality could not be established. $\square$

11.4 Entropies

11.4.1 Chaos and the flow of information

The concept of the entropy is fundamental for the study of statistical mechanics and thermodynamics. Entropy is a thermodynamic quantity describing the amount of disorder in the system. One can generalise this concept to characterise the amount of information stored in more general probability distributions. This is in part what *information theory* is concerned with. The theory has been developed since the 1940s and 1950s; the main contributions came from Shannon, Renyi, and Kolmogorov.

Information theory provides an important approach to time series analysis. The observation of a system is regarded as a source of information, a stream of numbers which can be considered as a transmitted message. If these numbers are distributed according to some probability distribution, and transitions between different numbers occur with well-defined probabilities, one can ask such questions as: "How much do I learn on average about the state of the system when I perform exactly one measurement?", or "How much information do I possess about the future observations when I have observed the entire past?". Information theory supplies concepts for quantitative answers.[8] Let us give a few examples. When I know that a system is at rest in a stable fixed point, a single observation suffices to determine the whole future with exactly the same precision with which I made my observation. For a periodic system an observation of one single period is enough to know all about the signal. When my system is a computer and it produces random numbers in the interval [0,1], I am not able to predict the next outcome even with an infinite number of previous observations. When I want to minimise the prediction error, Eq. (4.1), I should just use the mean value of my previous observations as a prediction.

11.4.2 Entropies of a static distribution

A series of entropy-like quantities, the order-q Renyi entropies, characterise the amount of information which is needed in order to specify the value of an observable with a certain precision when we know the probability density $d\mu/d\mathbf{x}$ that the observable has the value $\mathbf{x}$. As an alternative to the formulation via generalised correlation integrals in Section 11.3.1, let us cover the space in which the $\mathbf{x}$ live with *disjoint* boxes $\mathcal{P}_j$ of side length $\leq \epsilon$.[9] Let $p_j = \int_{\mathcal{P}_j} d\mu(\mathbf{x})$ be the fraction of the measure contained in the jth box. Then one defines

$$\tilde{H}_q(\mathcal{P}_\epsilon) = \frac{1}{1-q} \ln \sum_j p_j^q \qquad (11.20)$$

to be the order-q *Renyi entropy* for the partition $\mathcal{P}_\epsilon$. The case where $q = 1$ can be evaluated by the de l'Hospital rule,

$$\tilde{H}_1(\mathcal{P}_\epsilon) = - \sum_j p_j \ln p_j . \qquad (11.21)$$

$\tilde{H}_1$ is called the *Shannon entropy* [Shannon & Weaver (1949)]. In contrast to the other Renyi entropies it is additive, i.e. if the probabilities can be factorised into

---

[8] The concepts we discuss in this book are all part of *probabilistic* information theory. Thus we can address questions about *average* uncertainties, etc. The computational effort necessary to actually obtain this information is not the subject of this theory. Consult the literature on *algorithmic* information theory, e.g. Chaitin (1987), if you are interested in such questions. We should also point out that here we are only dealing with information about the *state* of the system, not its modelling. The information about parameters entering a model is described by the Fisher information [Fisher (1925); see also Kullback (1959)].

[9] The set of these boxes is called a *partition* $\mathcal{P}_\epsilon$. The neighbourhoods $\mathcal{P}_\epsilon(\mathbf{x})$ that we used to define $p_\epsilon(\mathbf{x})$ in Eq. (11.11) constitute a non-disjoint *covering* of space.

independent factors, the entropy of the joint process is the sum of the entropies of the independent processes.

Example 11.8 (Renyi entropies of a uniform distribution). Consider the probability density $\mu(x) = 1$ for $x \in [0, 1]$ and 0 elsewhere. Partitioning the unit interval into N intervals of length $\epsilon = 1/N$ yields $\tilde{H}_q(\epsilon) = (1 - q)^{-1} \ln(N\epsilon^q) = -\ln \epsilon = \ln N$. Thus all order-$q$ entropies are the same, which is a consequence of the homogeneity of the probability distribution. Its numerical value is the logarithm of the number of partition elements, which means that the better we resolve the real numbers by the partition the more information we gain. $\square$

Example 11.9 (Renyi entropies of the invariant measure of a periodic orbit). Let $\mu(x)$ be the measure created by an orbit of period l of a map on an interval of the real line (e.g. of the *logistic equation* $x_{n+1} = 1 - ax_n^2$). The invariant measure thus consists of p δ-functions at s_j, $j = 1, \dots, l$, the points of the orbit. Therefore, out of every arbitrarily large number of partition elements, at most l can possess nonzero probabilities, which are all identical to $p = 1/l$. Thus all Renyi entropies are $\tilde{H}_q = \ln l$. $\square$

The entropies introduced in this section have already occurred in the definitions of the order-q dimensions in Section 11.3.1, where, however, non-disjoint coverings were used. The dimensions are the scaling exponents of the Renyi entropies computed for equally sized partition elements as functions of ϵ, in the limit as $\epsilon \to 0$. Thus in Example 11.8 all $D_q = 1$ and in Example 11.9, $D_q = 0$, as expected.

A recipe for the computation of the *mutual information* has been given already in Section 9.2. There we were interested in the mutual information between a process and a time shifted copy of itself as a function of the time lag τ. More generally, the mutual information, or *redundancy*, can be defined for any m variables $x^{(1)}, \dots, x^{(m)}$ and their corresponding distributions. The case $m = 2$, for example, reads:

$$I^{(xy)} = \sum_{i,j} p_{i,j}^{(xy)} \ln p_{i,j}^{(xy)} - \sum_i p_i^{(x)} \ln p_i^{(x)} - \sum_j p_j^{(y)} \ln p_j^{(y)}$$

$$= \sum_{i,j} p_{i,j}^{(xy)} \ln \frac{p_{i,j}^{(xy)}}{p_i^{(x)} p_j^{(y)}} . \tag{11.22}$$

The $p_{i,j}^{(xy)}$ are joint probabilities, i.e., if $p_i^{(x)}$ is the probability that x is found in box i and $p_j^{(y)}$ is the probability that y is in box j, then $p_{i,j}^{(xy)}$ is the probability that, simultaneously, x is found in box i *and* y is in box j. For systems with continuous variables, the Shannon entropy diverges when the partition is refined further and further (as in Example 11.8). In the mutual information, however, only ratios of joint and ordinary probabilities appear as arguments of the logarithm in Eq. (11.22).

Hence the divergent term cancels and the value of $I^{(xy)}$ approaches a finite limit which is independent of the partitions underlying the $p_i^{(x)}$ and $p_j^{(y)}$. The latter only holds if the joint distribution is non-singular. It is not true for example if x and y are connected by a functional relation $y = f(x)$ or for coordinates of points on a fractal set. On the one hand, these are theoretically the kind of processes we are interested in. But then, any fractal or exact relationship will be blurred to some extent by noise. In practical applications we will consider coarse grained versions of $I^{(xy)}$ and observe relative changes only.

The obvious generalisation of Eq. (11.22) consists of replacing the logarithm by powers with exponents $q - 1$. Then the equality between the first and the second lines of Eq. (11.22) does not hold any more. While the second line in Eq. (11.22) still assumes the value zero for independent processes, as desired, the right hand side of the first line is very attractive computationally when $q = 2$. If we return to the case where the second variable is a time lagged copy of the first, the joint distribution is nothing but the distribution in two dimensional embedding space. Then the first form of the order-two redundancy reads at resolution ϵ:

$$I_2(\epsilon, \tau) = \ln \sum_{i,j} p_{i,j}^2 - 2 \ln \sum_i p_i^2$$
$$= \ln C_2(2, \epsilon, \tau) - 2 \ln C_2(1, \epsilon). \qquad (11.23)$$

The usual order-two correlation integrals $C_2(m, \epsilon, \tau)$ are easily estimated from the data (see Section 6.3). The asymptotic, invariant value $I_2(\tau)$ can be obtained by looking for scaling regions of $C_2(2, \epsilon, \tau)$ and $C_2(1, \epsilon)$. Note that $C_2(1, \epsilon)$ obviously does not depend on τ. Further material and applications of redundancies can be found in Paluš (1995) and in Prichard & Theiler (1995).

11.4.3 The Kolmogorov–Sinai entropy

The definitions of the last section gain even more relevance when they are applied to transition probabilities. So let us again introduce a partition $\mathcal{P}_\epsilon$ on the dynamical range of the observable. For simplicity, we directly consider the case of scalar observables which is typical of a time series situation, although the generalisation to vector valued observations is straightforward. For scalar observables the partition elements become intervals I_j of the real line. Let us introduce the joint probability $p_{i_1,i_2,\ldots,i_m}$ that at an arbitrary time n the observable falls into the interval I_{i_1}, at time $n + 1$ it falls into interval I_{i_2}, and so on. Then one defines *block entropies of block size m*:

$$H_q(m, \mathcal{P}_\epsilon) = \frac{1}{1 - q} \ln \sum_{i_1, i_2, \ldots, i_m} p_{i_1, i_2, \ldots, i_m}^q, \qquad (11.24)$$

The order-q entropies are then

$$h_q = \sup_{\mathcal{P}} \lim_{m \to \infty} \frac{1}{m} H_q(m, \mathcal{P}_\epsilon) \quad \text{or equivalently} \quad h_q = \sup_{\mathcal{P}} \lim_{m \to \infty} h_q(m, \mathcal{P}_\epsilon),$$

(11.25)

where

$$h_q(m, \mathcal{P}_\epsilon) := H_q(m+1, \mathcal{P}_\epsilon) - H_q(m, \mathcal{P}_\epsilon), \quad h_q(0, \mathcal{P}_\epsilon) := H_q(0, \mathcal{P}_\epsilon).$$

(11.26)

In Eq. (11.25) the supremum $\sup_{\mathcal{P}}$ indicates that one has to maximise over all possible partitions $\mathcal{P}$ and usually implies the limit $\epsilon \to 0$. In the strict sense only h_1 is called the *Kolmogorov–Sinai entropy* [KS entropy, Kolmogorov (1958), Sinai (1959)], but in fact all order-q Renyi entropies computed on the joint probabilities are entropies in the spirit of Kolmogorov and Sinai, who were the first to consider correlations in time in information theory.

The limit $q \to 0$ gives the *topological* entropy h_0. As D_0 just counts the number of nonempty boxes in a partition, h_0 counts the number of different unstable periodic orbits embedded in the attractor.

Example 11.10 (Data taken with an analogue/digital converter). Assume that the output of an experiment is digitised by an 8-bit analogue/digital converter. Then the finest reasonable partition consists of 2^8 intervals of size 2^{-8}. If on this level of resolution the transition probability from any interval I_j to any other interval I_k were the same (and thus 2^{-8}), the joint probabilities would factorise and we would find $H_q(m) = m\, 8 \ln 2$ independent of q (this factorisation happens if we observe an uncorrelated noise process), and hence $h_q(m) = 8 \ln 2$. It is easy to verify that every correlation between successive time steps can only reduce the entropy. This example illustrates that the normalisation $\frac{1}{m}$ in Eq. (11.25) is reasonable. Let us stress that the finite resolution of the data in this example is responsible for the absolute value found, and that for a real valued random process this value would diverge to infinity as in Example 11.8. $\square$

Example 11.11 (Markov model of order one: the Bernoulli shift). A *Markov model* of order m is a probabilistic model, where the transition probabilities are functions of the positions during the last m time steps. This means in particular that $H(m'+1) - H(m')$ for $m' > m$ equals h, whereas it is usually larger for $m' < m$. The *Bernoulli shift* $x_{n+1} = 2x_n$ mod 1 is a deterministic map. If we use the binary partition $I_0 = [0, 1/2[$ and $I_1 = [1/2, 1[$ and encode orbits as sequences of the labels 0 and 1, we obtain a Markov model of first order with transition probabilities $p_{0,0} = p_{0,1} = p_{1,0} = p_{1,1} = 1/2$. Inserting this in Eq. (11.24) yields the Kolmogorov–Sinai entropy $h_{\text{KS}} = \ln 2$. $\square$

In particular the last example helps us to understand what the Kolmogorov–Sinai entropy describes. Applying the Bernoulli shift to a real number in the unit interval means that in a binary representation of the number we shift all bits one step to the left and discard the leftmost one. Thus originally invisible information in the "insignificant" rightmost bits after some steps becomes very important, whereas the information which was originally stored in the most significant bits has got lost. This flow of information from the small to the large scales is typical of chaotic systems and is another aspect of their "sensitive dependence of initial conditions". We can even understand the precise value we find for the Bernoulli shift $x_{n+1} = 2x_n \bmod 1$. The graph of the map looks like saw teeth. Due to the stretching by a factor of two the unit interval is mapped onto itself twice, the map is two-to-one. For the attractor to be an invariant set, the average stretching and folding rate has to add up to zero, so that, loosely speaking, the entropy is the logarithm of the average number of folds. In fact, the *topological entropy* h_0 assumes exactly this value. For nonuniform transition probabilities the values for the other entropies are typically of the same order of magnitude but may differ slightly. Glancing back at Eq. (11.24) we can understand that the following relation holds:

$$H_q(m) \geq H_{q'}(m) \quad \text{if} \quad q' > q. \tag{11.27}$$

Let us finally relate the entropy to the predictability of a system. When we know the present state of the system with only finite precision (as in the example of the 8-bit data), this corresponds to a binary representation where we know only a certain number of bits. Since the entropy gives the rate at which bits are shifted to the left, we can estimate after how many time steps all the known bits are used up and have disappeared to the left. This time t_p is given by $\epsilon_0 e^{h t_p} = 1$, where ϵ_0 is the precision of the measurements which are taken to be scaled to the unit interval. Thus on average after a time $t_p = -\frac{\ln \epsilon_0}{h}$ predictability is lost completely, and we have to increase the precision of the measurement exponentially for a linear increase of the prediction time.

We should stress, however, that these arguments apply on average only. The fluctuations can be dramatic. In fact, systems such as the Lorenz model or the corresponding laser data set (Appendix B.1) can easily be predicted during the oscillatory episodes. All the uncertainty is concentrated at those instances when the amplitudes suddenly drop; see Fig. B.1.

11.4.4 The ϵ-entropy per unit time

Before we come to the computation of entropies on real data sets, we want to extend the concept of the Kolmogorov–Sinai entropy to a quantity with an explicit length scale dependence, called the ϵ-*entropy per unit time* or *coarse grained dynamical entropy.* (See Gaspard & Wang (1993) for a detailed introduction.) In Section 11.4.3 we purposefully denoted the chosen partition of the state space by $\mathcal{P}_\epsilon$, where ϵ defined the radius of the largest partition element. Hence, implicitly, the block entropies $H_q(m, \epsilon)$ contain a length scale. Whereas the definition of the KS-entropy implies the limit $\epsilon \to 0$, we will discuss here that the ϵ-dependence of suitably defined coarse grained entropies reveals additional information about the system.

The difficulty in such a concept lies in the fact that there are arbitrarily many different partitions whose largest cell diameter is ϵ – hence, when fixing ϵ, the corresponding partition is not unique. In order to achieve uniqueness, Gaspard & Wang (1993) define

$$h_q(\epsilon) = \inf_{\mathcal{P}_\epsilon} \lim_{m \to \infty} H_q(m + 1, \mathcal{P}_\epsilon) - H_q(m, \mathcal{P}_\epsilon), \qquad (11.28)$$

where the infimum is taken over all ϵ-partitions with fixed ϵ. Typically, in the limit $\epsilon \to 0, h_q(\epsilon) \to h_q$.

When we discuss data analysis with entropies in the next subsection, we will replace the partition by a covering, which is uniquely defined for every given data set and hence avoids the estimation of any infimum, and we will typically compute h_2 instead of h_1 because of its better finite sample properties. Of course, the limit $m \to \infty$ cannot be computed from data, but often the behaviour of (m, ϵ) for large finite m allows us to extrapolate to $m \to \infty$.

We should stress that relaxing the limit $\epsilon \to 0$ destroys the invariance of $h_q(m, \epsilon)$ under smooth coordinate transforms. It is hence a quantity whose single numerical values are not very informative. Instead, its dependence on ϵ and on the block length m contains essential information.

The general behaviour of $h_q(m, \epsilon)$ for non-negative q is depicted in Fig. 11.3 for different situations. At large $\epsilon, h_q(m, \epsilon) = 0$ trivially. For a deterministic dynamical system and when m is larger than the attractor dimension, $h_q(m, \epsilon) \to h_q(m) = $ *const.* for small ϵ. Depending on the system and on how large m actually is, $h_q(m)$ can be used as to estimate the KS-entropy of the system. For a stochastic system (and for a deterministic system when m is smaller than the attractor dimension), instead, $h_q(m, \epsilon) = -\ln \epsilon + h^{(c)}(m)$ on the small scales, where the rôle of the constant $h^{(c)}(m)$ will be discussed later. Consequently, for a deterministic system with measurement noise, on the very small scales the ϵ-entropy also diverges, but often there is an intermediate range of length scales where $h_q(m, \epsilon) \approx h_q$. When a deterministic system is to be characterised by its entropy, then this plateau on intermediate length scales has to be identified. Relevant for the discussion of how far high-dimensional and high entropic data can be identified as deterministic is the large ϵ regime of the dashed curve: it behaves exactly as stochastic data, namely like $-\ln \epsilon$. Hence, if because of lack of data points or because of the

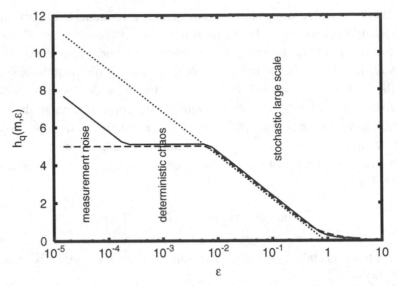

Figure 11.3 The general behaviour of the ϵ-entropy $h_q(m, \epsilon)$ $(q > 0)$ for three different types of systems: deterministic chaos (dashed curve), deterministic chaos with additive measurement noise of small amplitude (continuous curve), and strongly stochastically driven systems such as AR(1)-model (dotted). The crossover from large ϵ where $h_q(m, \epsilon) = 0$ to smaller ϵ where $h(m, \epsilon) \approx$ *const.* for deterministic data is typically indistinguishable from the stochastic behaviour $h(m, \epsilon) \simeq -\ln \epsilon$.

amplitude of measurement noise the plateau is not accessible, deterministic data are indistinguishable from stochastic data. There is no feature on the large length scales which indicates the presence of determinism.

The analysis of the m-dependence additionally reveals information about the memory of the system. In Eq. (11.28) the limit $m \to \infty$ was involved, in analogy with the definition of the KS-entropy in Eq. (11.25). The rate of the convergence to this limit is related to the strength of long range (linear and nonlinear) correlations in the system. When we study a deterministic system, the entropy asymptotically is finite, and we can exchange the limits in m and ϵ here. We call $\lim_{\epsilon \to 0} h_q(m, \epsilon) = h_q(m)$. $h_q(m)$ converges for $m \to \infty$ to h_q from above, i.e., not taking into account all temporal correlations can only enlarge our lack of information about the system.[10]

For stochastic processes with smooth probability densities, on the small scales, $h(m, \epsilon) = -\ln \epsilon + h_q^{(c)}(m)$. The constant $h_q^{(c)}(m)$ is called the conditional *continuous entropy* and given by the continuous entropies of the corresponding block

---

[10] Grassberger (1986) has suggested to use $\sum_{m=1}^{\infty} h_q(m) - h_q = C_{emc}$ as a measure for complexity. We will not consider this further, since time series information is typically too sparse for a sound estimate of C_{emc}.

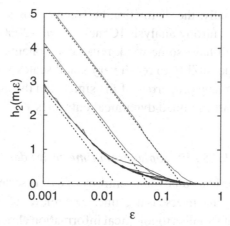

Figure 11.4 The ϵ-entropies of an AR(2)-process from a time series. The dashed lines indicate the $-\ln \epsilon$-behaviour, and the continuous curves are $h_2(m, \epsilon)$ for $m = 0, 1, 2, \ldots$ from right to left. Curves for $m \geq 2$ are superimposed on each other, since the AR(2)-process forms a Markov-chain of order 2.

entropies of block length m by $h_q^{(c)}(m) = H_q^{(c)}(m + 1) - H_q^{(c)}(m)$. Those are defined by

$$H_q^{(c)}(m) = \frac{1}{1 - q} \ln \int \left(\rho^{(m)}(\mathbf{s})\right)^q d\mathbf{s}. \tag{11.29}$$

Here, $\rho^{(m)}(\mathbf{s})$ is the probability density of the process in the m dimensional embedding space, and $d\mathbf{s}$ denotes a volume element in this space. It is easy to verify that if the process has no correlations at all and hence $\rho^{(m)}(\mathbf{s}) = \prod_{k=1}^{m} \rho(s_k)$, the continuous entropies are $H_q^{(c)}(m) = m H_q^{(c)}(1)$ and $h_q^{(c)}(m) = H^{(c)}(1) \; \forall m$. If the data for which this analysis is done represent a Markov chain of order l then $h_q^{(c)}(m) = h_q^{(c)}(l) \; \forall m \geq l$.

Example 11.12 (The ϵ-entropy of an AR(2) process). An AR(2) process $s_{n+1} = a_1 s_n + a_2 s_{n-1} + \xi_n$ forms a Markov chain of order 2. Hence we expect that ϵ-entropies diverge as $-\ln \epsilon$ and that the conditional continuous entropies $h_2^{(c)}(m)$ are identical for $m \geq 2$. Figure 11.4 gives an impression of how well this can be recovered from a time series of length $10\,000$, using the TISEAN routine d2. This routine computes the $q = 2$-entropies. The exact numerical values of $h_1^{(c)}(m)$ in this case could be computed analytically. $\square$

As it is frequently stated in the literature, the entropy is a concept that applies to both deterministic and stochastic systems. For sufficiently large m, the ϵ-dependence for stochastic and for deterministic signals is entirely different, whereas, for sufficiently small ϵ, the m-dependence sheds light on the memory in the system. So formally, the distinction between chaos and noise can be fully performed through the entropy analysis [Cencini *et al.* (2000)]. In practice, every deterministic signal will have some weak noise components, and also the lack of data and hence the impossibility to reach very small scales ϵ will render this issue uninteresting, but still many properties of the signal and the underlying system can be inferred through coarse grained dynamical entropies.

11.4.5   Entropies from time series data

For two reasons a numerical value of the entropy of a time series is interesting for its characterisation. Firstly, its inverse is the time scale relevant for the predictability of the system. Secondly, it supplies topological information about the folding process. Since entropies as also Lyapunov exponents are measured in inverse units of time, the numbers obtained from a time series are given in units of the inverse sampling interval. Sometimes, in particular when the duration of a sampling interval is known, it is reasonable to convert entropies into units of 1/s. Also, entropies of flow data in a Poincaré surface of section are to be divided by the average time in between two intersections in order to be compared to the entropy of the original flow data. Notice also that when you compute entropies with the algorithms suggested below with a time lag τ larger than unity, the results must be divided by τ.

Unfortunately, entropies are difficult to extract from time series, mainly because their computation requires more data points than dimensions and Lyapunov exponents. Apart from the fact that we cannot take the supremum over all possible partitions (i.e, we cannot proceed to arbitrarily small ϵ), the limit $m \to \infty$ in Eq. (11.25) constitutes the crucial problem. We have to provide enough statistics in a possibly high dimensional space.

A straightforward implementation of the above definition of the Kolmogorov–Sinai entropy would require *box counting*. If the data are scaled to the unit interval, phase space is covered by $1/\epsilon$ equal intervals of size ϵ and one determines the joint probabilities $p_{i_1,...,i_m}$ by just counting the number $n_{i_1,...,i_m}$ of segments of the time series falling into the sequence of intervals $I_{i_1}, \ldots, I_{i_m}$. This amounts to the evaluation of an m dimensional histogram. Even when smart data structures are used which allocate memory to only those boxes which really are occupied,[11] the finite size of the boxes introduces severe edge effects and scaling is very poor. We would therefore rather use overlapping boxes as in dimension estimation. In order to sample the measure in the most efficient way, the partition elements are not uniformly distributed in space but are centred around points which are distributed according

---

[11] In fact this reduces the required memory dramatically. Only approximately $\epsilon^{-D_0} e^{mh}$ boxes are occupied out of ϵ^{-m}.

to the measure. Such a technique is called *importance sampling*. Obviously, the points of the time series themselves are a sample of the desired distribution.

Let us first consider the case where $q > 1$ and use partition elements of fixed radius. When we take into account the fact that the partition elements have already been chosen with probability p, we arrive at Eq. (11.11). We compute the time series estimate, Eq. (11.16), for increasing embedding dimensions m. Then

$$C_q(m, \epsilon) \propto \epsilon^{(q-1)D_q} e^{(1-q)H_q(m)} . \tag{11.30}$$

Remember that in principle we should perform a supremum over all possible partitions and, if one does not *a priori* know a finite (so-called *generating partition*) which yields the supremum, one has to perform a limit $\epsilon \to 0$. Of course, due to the finiteness of the data set and due to noise on the data this limit is meaningless for the algorithm. Instead we have to find a scaling range in which the results are independent of ϵ. Such a scale independence can only be expected if we see a plateau in the corresponding plot of the local scaling exponents $D_q(m, \epsilon)$. Otherwise we have to abandon this kind of analysis.

For values of ϵ inside the plateau in the dimension plots, the factor ϵ^{D_q} in Eq. (11.30) is almost constant and we can determine the exponent h_q by plotting

$$h_q(m, \epsilon) = H_q(m + 1, \epsilon) - H_q(m, \epsilon) = \ln \frac{C_q(m, \epsilon)}{C_q(m + 1, \epsilon)}, \tag{11.31}$$

both versus m for various ϵ, and versus ϵ for various m. This gives us a clear feeling for the robustness against the change in both parameters. For sufficiently large m this should converge towards a constant h_q. How large m has to be depends on the (nonlinear) correlations in the system. For order-m Markov processes one finds $h_q(m') = h_q$ for $m' > m$, as mentioned before. Most dynamical systems cannot be mapped to Markov processes with a finite memory, such that we have to choose m as large as possible for a fair estimate of h_q. In this limit, however, we get problems due to the lack of neighbours, since, as Eq. (11.30) shows, we lose about e^{-h} neighbours when passing from m to $m + 1$.

When computing the dimension of an attractor we saw that the most robust dimension is the correlation dimension D_2. The same goes for the entropies: h_2 is the most computable, simply due to the fact that for $q = 2$ the correlation sum is an arithmetic average over the numbers of neighbours and thus yields meaningful results even when almost all reference points have lost their neighbours. Thus when computing h_2 we can reach the largest values of m.

In practice, we use the same correlation sums $C_2(m, \epsilon)$ which we determined to estimate the correlation dimension. The core of this algorithm is the program d2 of the TISEAN package, discussed in Section A.6.

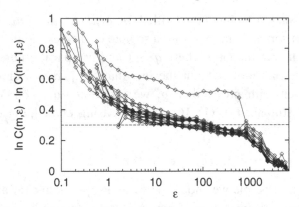

Figure 11.5 Estimate of the correlation entropy of the NMR laser data after noise reduction. The correlation sum $C_2(m, \epsilon)$ has been computed for $m = 2, \ldots, 20$. The curves are $\ln C_2(m, \epsilon)/C_2(m + 1, \epsilon)$. Apart from the curve for $m = 2$, there is a reasonable plateau at $h_2 = 0.3 \pm 0.05$.

Example 11.13 (Order-two entropy of NMR laser data). In Fig. 6.4 we have shown the correlation sums for the NMR laser data, Appendix B.2. The larger m is, the lower is the corresponding curve. We understand this now. So by just plotting the results in a different fashion we can estimate the correlation entropy h_2. In Fig. 11.5 we show the numerical values of $h_2(m, \epsilon)$, also for values of ϵ outside the scaling range. For $\epsilon < 1$ the entropy increases since we feel the noise contaminating the data. For $\epsilon > 1000$ scaling is violated due to the finite size of the attractor and we systematically underestimate the entropy. A comparison of Fig. 11.5 to the schematic drawing in Fig. 11.3 shows that reality poses unexpected problems for a precise estimate of the entropy. □

When computing the proper Kolmogorov–Sinai entropy h_1 from a generalised correlation integral using Eq. (11.31) we again encounter the problem of empty neighbourhoods. We would rather make use of the fixed mass approach that we pursued when we estimated D_1 from data. To this end, let us include the m-dependence in Eq. (11.18) [Badii & Politi (1985), Grassberger (1985)]:

$$\left\langle \epsilon_p(m)^{(1-q)D_q} \right\rangle \propto \left(p e^{H_q(m)} \right)^{1-q}, \qquad p \to 0. \tag{11.32}$$

The limit $q \to 1$ yields:

$$\begin{aligned} h_1(m, p) &= H_1(m + 1, p) - H_1(m, p) \\ &= D_1(\langle \ln \epsilon_p(m) \rangle - \langle \ln \epsilon_p(m + 1) \rangle). \end{aligned} \tag{11.33}$$

If $h_1(m, p)$ approaches a finite limit for increasing m within the scaling regime, this limit is an estimate for h_{KS}. All the necessary quantities can be computed with the routine c1 of the TISEAN package.

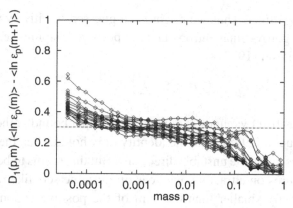

Figure 11.6 Estimate of the Kolmogorov–Sinai entropy of the NMR laser data. Curves $D_1(p, m)(\langle \ln \epsilon_p(m) \rangle - \langle \ln \epsilon_p(m + 1) \rangle)$ versus p are shown for $m = 3, \ldots, 15$ (counting downward on the left end of the range). A plateau is visible for $0.001 < p < 0.1$, leading to the estimate $h_{KS} = 0.3 \pm 0.05$.

Example 11.14 (Kolmogorov–Sinai entropy of NMR laser data). For the cleaned NMR laser data (Appendix B.2) we computed $\langle \ln \epsilon_p \rangle_m$ for $m = 2, \ldots, 15$ and values of p logarithmically spaced in the range $1/N < p < 1$, as in Example 11.7. We form $D_1(p, m) = d \ln p / d \langle \ln \epsilon_p(m) \rangle$ and thus plot $D_1(p, m)(\langle \ln \epsilon_p(m) \rangle - \langle \ln \epsilon_p(m + 1) \rangle)$ versus the mass p in Fig. 11.6. There is a clear scaling region and we estimate $h_{KS} = 0.3 \pm 0.05$. $\square$

11.5 How things are related

Lyapunov exponents, dimensions and entropies are different ways of describing properties of the same invariant measure. All these quantities characterise aspects of the same underlying dynamics. Thus it is natural to seek the relations between them.

11.5.1 Pesin's identity

Positive Lyapunov exponents characterise the exponential divergence of nearby trajectories. The fact that trajectories diverge directly implies a loss of information about their future position. Let us take into consideration the uncertainty about the knowledge of the present state of the system due to measurement errors by representing it as a small ball around the observed value, whose diameter is given by the noise amplitude. The time evolution will deform the ball into an ellipsoid, whose semi-axes are stretched and contracted by factors $e^{\lambda_i t}$. Thus although the total volume of the ellipsoid generally shrinks due to dissipation, the uncertainty about the future position grows with the lengths of the expanding semi-axes. If there are several positive exponents λ_i, each of the corresponding unstable directions contributes to this effect. Thus we might have guessed intuitively the relation between the Kolmogorov–Sinai entropy and the positive Lyapunov exponents called *Pesin's identity* [Pesin (1977)]:

$$h_{KS} \leq \sum_{i:\lambda_i > 0} \lambda_i .$$ (11.34)

The sum of all positive Lyapunov exponents is an upper bound of the Kolmogorov–Sinai entropy [Ruelle (1978)], and the identity only holds when the invariant measure is continuous along the unstable directions. But this seems to be true in general for natural measures on attractors, whereas on unstable sets like repellers the entropy can be strictly smaller than the sum of the positive exponents. However, Pesin's identity can be generalised taking fractality into account [Ledrappier & Young (1985)]:

$$h_{KS} = \sum_{i:\lambda_i > 0} D_i \lambda_i = - \sum_{i:\lambda_i < 0} D_i \lambda_i ,$$ (11.35)

where the dimensions D_i are called *partial dimensions* and are the information dimensions of the sets formed by the intersection of the attractor with the unstable or stable directions related to λ_i.[12]

Very often, Pesin's identity is the only way to obtain a good estimate of the Kolmogorov–Sinai entropy of a time series, since its direct computation often fails due to a lack of data. However, for a consistency check we recommend computing at least the correlation entropy h_2 (which we obtain as a byproduct when we compute the correlation dimension for a range of embedding dimensions m; see Section 11.4.5). As a lower bound of h_{KS}, h_2 has to be smaller than the sum of the positive Lyapunov exponents.

For multi-fractal measures, h_2 can in fact be considerably smaller than h_{KS}. To give an example, for the Hénon map (Exercise 3.1) the values $h_{KS} \approx 0.419$ and $h_2 \approx 0.318$ are reported by Grassberger & Procaccia (1983). Nevertheless, experience shows that many attractors reconstructed from Poincaré sections of flows are more homogeneous, such that signatures of their multi-fractality often disappear in the systematical and statistical errors of the estimation of the order-q entropies and dimensions. Even if you find large differences between h_2 and the sum of positive Lyapunov exponents for your data, you should first check your numerical results and their robustness against changes of parameters of the algorithms. Only if you are really sure that the disagreement is significant should you interpret it as a signature of multi-fractality which also has to be visible when computing D_q for different values of q.

---

[12] The second equality is true only for invertible systems. It does not hold, for example, for one dimensional chaotic maps.

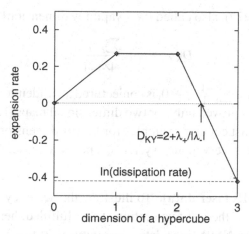

Figure 11.7 Expansion rate (sum of the contributing Lyapunov exponents) of a typical hypercube as a function of its dimensionality for a flow with Lyapunov exponents λ_+, 0, and λ_- with $|\lambda_-| > \lambda_+$.

11.5.2 Kaplan–Yorke conjecture

Less obviously than the entropy, the information dimension is also related to the Lyapunov exponents. The attractor is invariant under the dynamics. This means in particular that when considering it as a D_1 dimensional sub-volume of the state space, it neither shrinks nor stretches over the course of time. Furthermore, it attracts points from its neighbourhood and is thus stable under small perturbations. Knowing the spectrum of Lyapunov exponents, we can seek for the value of D_1 such that a typical volume of this dimensionality has the required properties. Obviously, a finite one dimensional sub-set with probability 1 will be stretched exponentially fast with a rate given by the maximal Lyapunov exponent. Thus, assuming $\lambda_1 > 0$, it is not invariant. If the second Lyapunov exponent is negative and, moreover, $\lambda_2 < -\lambda_1$, a typical two dimensional area is still stretched in one direction with $e^{\lambda_1 t}$, but in the other it shrinks with $e^{\lambda_2 t}$ and thus in total shrinks with a rate $e^{-(|\lambda_2|-\lambda_1)}$. Thus it is quite plausible that a fractal object which occupies the contracting directions only in a Cantor-like way with a *partial dimension* $D_1 - 1 < 1$, is invariant under time evolution, if $\lambda_1 + (D_1 - 1)\lambda_2 = 0$, i.e. $D_1 = 1 + \lambda_1/|\lambda_2|$. In Fig. 11.7 we sketch this for a typical flow. These considerations suggest that the integer part of the attractor dimension is the maximal number of exponents that one can add (in descending order) such that their sum remains positive. The fractional part is found by a simple linear interpolation.[13] Consequently, the Kaplan–Yorke dimension

---

[13] Now we can understand how a linear filter with feedback, $\hat{x}_n = x_n + a\hat{x}_{n-1}$, for example, can increase the attractor dimension: the passive, linear degree of freedom added to the system introduces one more Lyapunov exponent which is typically negative and of small modulus. This can alter the fractional part of the dimension.

[Kaplan & Yorke (1979), also called the Lyapunov dimension] given by

$$D_{KY} = k + \frac{\sum_{i=1}^{k} \lambda_i}{|\lambda_{k+1}|}, \qquad (11.36)$$

where $\sum_{i=1}^{k} \lambda_i \geq 0$ and $\sum_{i=1}^{k+1} \lambda_i < 0$, is conjectured to be identical to the information dimension D_1. This is proved only for two dimensional maps [Ledrappier & Young (1985)], and there exist counter-examples for higher dimensions. Still, $D_{KY} = D_1$ seems to be valid for very many systems, and is a reasonable approximation otherwise.

Example 11.15 (NMR laser data). To illustrate the accuracy up to which we can expect the identities of the last two sections to be fulfilled, here we collect all the information about the NMR laser data (Appendix B.2) that we have obtained so far. It is summarised in Table 11.2. Although there is a good agreement between the Kaplan–Yorke dimension and D_1, we have large error bars. The uncertainties of the Lyapunov exponents computed by the tangent space method were discussed before. $\square$

Further reading

For a general introduction to information theory see Kullback (1959). More specific material concerning the information theory of dynamical systems can be found in Eckmann & Ruelle (1985). A comparison of different algorithms for Lyapunov exponents is carried out in Geist *et al.* (1990). Generalised dimensions are discussed, for example, in Grassberger (1983) and (1985), Badii & Politi (1985), Paladin & Vulpiani (1987), Grassberger *et al.* (1988) and Badii (1989).

Exercises

11.1   In Section 11.2.3 we argued that the repeated multiplication of Jacobians onto a tangent space vector and its successive projection onto a subspace E_k yields the k-th Lyapunov exponent through the average stretching rate of this vector. This and the following exercise can shed more light on this procedure.

   The Jacobians are not in general symmetric and can have nonorthogonal eigenvectors. Let $\mathbf{M}$ be such a non-normal matrix (i.e., $M^\dagger M \neq M M^\dagger$). Assume you know that $\mathbf{e}_1$ is an eigenvector with eigenvalue Λ_1. Take a normalised vector $\mathbf{u}$ with $\mathbf{u} \cdot \mathbf{e}_1 = 0$ and assume that $\mathbf{u}$ is a linear combination of $\mathbf{e}_1$ and a second eigenvector $\mathbf{e}_2$ (not orthogonal to $\mathbf{e}_1$) with the eigenvalue Λ_2. Show that the component of $\mathbf{Mu}$ orthogonal to $\mathbf{e}_1$ equals $\Lambda_2 \mathbf{u}$.

11.2   Show that the above remains true if the matrix $\mathbf{M}$ can be written as $\mathbf{M} = \mathbf{B} \cdot \mathbf{A}$ (where neither $\mathbf{e}_1$ nor $\mathbf{e}_2$ are eigenvectors of $\mathbf{A}$ or $\mathbf{B}$ but of $\mathbf{M}$) and the projection step is applied twice. First you reduce $\mathbf{Au}$ to its part that is orthogonal to $\mathbf{Ae}_1$, multiply this by $\mathbf{B}$ and project the result onto a subspace orthogonal to $\mathbf{BAe}_1$.

11.3 Show that $\tilde{H}_q(\epsilon) = -\ln \epsilon$ is indeed the maximal value for all ϵ-partitions of the uniform distribution, Example 11.8.

11.4 Revisit Exercise 6.3. Compute the relevant invariants and verify numerically that the Kaplan–Yorke conjecture and Pesin's identity hold. Which embedding dimension do you need for a faithful representation of the dynamics? Can you identify the spurious Lyapunov exponents?

Chapter 12

Modelling and forecasting

When we try to build a model of a system, we usually have the ultimate goal of establishing the equations of motion which describe the underlying system in terms of meaningful quantities. Writing down the behaviour of the relevant components of the system in a mathematical language, we try to combine all we know about their actions and interactions. This approach may allow one to construct a simplified but useful image of what happens in nature. Most of the knowledge we have about the inherent mechanisms has been previously derived from measurements. We call such models *phenomenological models*. In some cases, as in classical mechanics, one is able to derive a dynamical model from first principles, but even the so-called first principles have to be consistent with the empirical observations.

In order to establish a useful phenomenological model, one needs specialised knowledge about the system under study. Therefore, the right place to explain how to make a model for the time variations of some animal population is a book on population biology. On the other hand, there are techniques for constructing models that are based almost purely on time series data. These techniques are generally applicable and thus we consider their treatment appropriate for this book.

The problem treated in this chapter lies at the heart of time series analysis. What can we infer about the dynamical laws governing a system, given a sequence of observations of one or a few time variable characteristics of the system? We suppose that the external knowledge about the system is limited to some assumptions that we may make about the general structure of these laws. Since our observations of the system are most likely incomplete, the solution of this inverse problem will not be unique. The ambiguity will be partially resolved by some natural restrictions that we can impose on the analysis. In particular, within a class of models we will prefer the *simplest model consistent with the observations*. Sometimes we have some external criterion with which to distinguish "good" and "bad" models (e.g. good: computable, stable, reproducible; bad: unstable, hard to manage, etc.).

Models based entirely on time series data have the drawback that the terms they contain do not usually have a meaningful physical interpretation. This lack of an interpretation is obviously not a failure of the individual methods, but is fundamental to this approach. Nevertheless, a successful model can reproduce the data in a statistical sense, i.e. under iteration it yields a time series with the same properties as the data. In this sense such models are often superior to the phenomenological models. The first step of the time series approach consists in choosing an appropriate model equation containing free parameters, which in a second step are adapted to the observed dynamics by some fitting procedure. If these parameters do have a phenomenological meaning, such a model is called *parametric*. More often, the model equation will have no phenomenological meaning but is just a suitable functional form, whence its parameters cannot be interpreted in terms of known phenomena and the model is called *nonparametric*.[1]

A faithful time series model can be used for many purposes, such as prediction, noise reduction and control (see Chapter 15), or we can use it to create an arbitrary amount of artificial data which shares, for example, the power spectrum, distribution of values, attractor dimension, entropy, and Lyapunov exponents of the data. Let us stress, however, that the construction of a faithful model in the above sense and the construction of the optimal predictor may result in different mathematical expressions. Optimised predictions are not designed for a good reproduction of long term dynamics but instead focus on short time horizons into the future.

Eventually, it would be desirable to combine the insight of the phenomenological approach with the statistical accuracy of a time series model. This is to a large extent an open problem. Nevertheless, prior information can be included in the modelling procedure by the choice of an appropriate model class, as we will see below. Symmetries, for instance, can be exploited quite easily. Let us finally mention that some of the predictive methods we will develop below (the locally constant and linear approaches) do not yield a globally valid model in the strict sense.

In our understanding, modelling is a creative task, different from the computation of a single number such as a dimension. Therefore, there is no single general algorithm but, instead, a collection of different approaches. While we could give a recipe for the calculation of dimensions, we can only provide the cooking equipment for modelling.

As we have said earlier in this book, there can be different sources of predictability in a time series. One is the existence of linear correlations in time which are reflected by a nonzero value of the autocorrelation function. If this is all the information we

---

[1] Do not be confused: a nonparametric model will typically contain many more fitting parameters than a parametric model.

want to include in the model (or if this is all the structure there is), then the best choice is a linear stochastic model, a *moving average* (MA), an *autoregressive* (AR), or a combination of MA and AR (ARMA). Typical examples are coloured noise processes or Brownian motion, where the simple fact that the resulting path is continuous allows us to predict the future position inside a radius growing with the square root of the time difference.

An alternative origin of predictability is (nonlinear) determinism. In its strongest form, it implies that equal states have an equal future, and similar states will evolve similarly, at least over short times. In this case one will try to construct a deterministic model. If the observed time series appears to be very irregular, this can be explained by a deterministic model if it is able to produce chaos – for which it must be nonlinear. In systems with linear (harmonic) dynamics, determinism and linear correlations are synonymous. This is not the case for chaotic dynamics. Autocorrelation functions of chaotic signals generally decay exponentially fast, and the spectra have a broad band component. Therefore, the construction of a (predominantly) deterministic model for chaotic data cannot rely on autocorrelations.

The successful modelling of chaotic data obtained in laboratory experiments together with the willingness to accept that there is almost no pure low dimensional chaos outside the laboratory has stimulated some successful attempts to combine nonlinearity and stochasticity in modelling and predictions. These recent results on the construction of Fokker–Planck equations and Markov chains from data can be, in some sense, considered as a synthesis of linear stochastic modelling and nonlinear deterministic modelling. Their discussion will hence be postponed to Section 12.7, since the methods rely conceptually and algorithmically on deterministic modelling techniques to be introduced below. Let us, however, start with a brief review of linear stochastic models.

12.1 Linear stochastic models and filters

In this section we recall the most relevant facts about linear stochastic models, which for decades have been the prevailing paradigm in time series analysis. We will introduce some details about the Gaussian linear processes, the autoregressive (AR) and moving average (MA) models. Beyond the educational aspect, this knowledge is of relevance since these models can serve as a reference when testing the null hypothesis of linear random data as in Chapter 7. Also, *local linear models* to be introduced later are a very useful concept for the modelling and prediction of nonlinear deterministic processes. Since the construction of global and local linear models from data relies on the same ideas, part of Section 12.3.2 is based on the discussion in the next subsection. Nonlinear filters will only be mentioned for completeness. We postpone the discussion of two other very important classes of stochastic processes, namely Markov models and Fokker–Planck equations, to the end of this chapter.

12.1.1 Linear filters

The most popular class of stochastic models for time series analysis consists of linear filters acting on a series of independent noise inputs and on past values of the signal itself.[2] A problem which is immediately obvious is that we do not observe this series of inputs. When modelling such a system we have to estimate all the parameters from the outputs only, which will lead to some ambiguity at least.

If the estimated spectrum of a time series is of the form of coloured noise, i.e. there are no prominent peaks, the *MA model* (*moving average*) is a good candidate. It is a filter on a series of Gaussian white noise inputs η_n,

$$x_n = \sum_{j=0}^{M_{\mathrm{MA}}} b_j \eta_{n-j},$$
(12.1)

where $\langle \eta_n \eta_m \rangle = \delta_{nm}$ and $\langle \eta \rangle = 0$. Rescaling all parameters b_j by the same constant hence adjusts the variance of the output signal to the variance of the observed data, whereas the relative values of the b_j with respect to each other determine the shape of the power spectrum and of the autocorrelation function. The parameter b_0 is nonzero, and M_{MA} is called the order of the process. Note that x_n is also a Gaussian random variable with zero mean. If the time series to be modelled has a non-zero mean, the MA model just describes the fluctuations around the mean. Since the input noises have a constant (white) power spectrum and a δ-like autocorrelation function, the power spectrum of the output time series of length N is given by $S_k^{(x)} = |\sum_{j=0}^{M_{\mathrm{MA}}} b_j e^{i2\pi kj/N}|^2$. Here, $k \in \{0, 1, \ldots, N/2\}$ are the wave numbers, yielding the discrete frequencies supported by the finite time series as $f = k/N$. Correspondingly, the autocorrelation function of the signal is $c_\tau = \sum_{j=0}^{M_{\mathrm{MA}}} b_j b_{j-\tau}$, where it is understood that $b_j = 0$ for $j < 0$ and $j > M_{\mathrm{MA}}$. Sometimes, such a model is called a *finite impulse response* filter, since the signal dies after M_{MA} steps, if the input is a single pulse, $\eta_0 = 1$ and $\eta_n = 0$ for $n \neq 0$.

Alternatively, in an *AR* (*autoregressive*) *model*, the present outcome is a linear combination of the signal in the past (with a finite memory), plus additive noise:

$$x_n = \sum_{j=1}^{M_{\mathrm{AR}}} a_j x_{n-j} + \eta_n.$$
(12.2)

---

[2] Section 2.4 had the same caption but the setting was different there: we wanted to use a filter on the data in order to reduce noise. Here we want to interpret the data as the result of filtering white noise.

This describes an AR model of order M_{AR}, where η_n is white Gaussian noise as above. Thus, again, x_n is a Gaussian random variable. If an AR model fulfils the condition that all roots of the polynomial $P(z) = -\sum_{j=0}^{M_{AR}} a_j z^{M_{AR}-j}$ with $a_0 = -1$ lie exactly on the unit circle, the model without noise inputs possesses marginally stable harmonic solutions. Due to the noise the amplitude of these oscillations changes in a non-stationary way. If there are zeros outside the unit circle, oscillation amplitudes are even unstable and x_n diverges to infinity. Only if all roots are inside the unit circle, the process is stationary and resembles noisy damped harmonic oscillations with varying amplitudes.[3] The spectrum of the process obtained from a sequence of N data points is uniquely determined by the coefficients a_j, where k is again the wave number as above: $S_k^{(x)} = 1/|1 - \sum_{j=1}^{M_{AR}} a_j e^{i2\pi kj/N}|^2$. You recognise in the denominator the polynomial involved in the stability condition of the model, multiplied by $z^{-M_{AR}}$, and z being replaced by the complex variable $e^{i2\pi kj/N}$ which is confined to the unit circle. Hence, the zeros of the denominator are outside the unit circle for a stable models, and exactly on the unit circle for a marginally stable model. The latter corresponds to the fact that the excitation of an undamped harmonic corresponds to a pole in the spectrum, whereas damped harmonics create maxima with a finite height and a width related to the damping. Thus, an AR model is particularly appropriate if the spectrum of a time series is dominated by a few peaks at distinct frequencies.

In principle, all Gaussian linear stochastic processes can be modelled with arbitrary accuracy by either of the two approaches. But the number of parameters (i.e. the *order* of the model) might be extremely large (in fact it can be infinite) if we try, for example, to model a noisy harmonic oscillation by an MA process. An obvious generalisation is a combination of the AR and the MA models, the so-called ARMA (autoregressive moving average) process. This yields a power spectrum with both: poles and a polynomial background.

The properties of all three models are only well understood if the increments, or inputs η, are Gaussian white noise. Real world data are often not Gaussian distributed. If one wants to model them by one of these processes, one usually could assume that a nonlinear transformation distorts the output of the Gaussian random process and changes the distribution to the one observed. We call such nonlinearities *static*, in contrast to nonlinearities in the dynamics of a system. They conserve the property of time reversal invariance, which is also characteristic for linear stochastic processes. Before fitting an ARMA model to such data, one should render the distribution Gaussian by empirically inverting the nonlinear transformation.

---

[3] In the language of dynamics you find this condition by a linear stability analysis for the fixed point $x_k = 0 \ \forall k$ of the deterministic counterpart of Eq. (12.2), dropping η_n.

Still, the optimal order of the model remains unknown. Since it is obvious that one can reproduce the data better the more parameters the model contains, it is necessary to employ a criterion which limits the number of coefficients in order to prevent over-fitting. See Section 12.5 for details. ARMA models have already been mentioned in the introduction to Chapter 7, where we also presented an example of a typical time series.

The distinction of AR, MA and ARMA models and the determination of the corresponding coefficients from a time series are only relevant if we are interested in the model *per se*. If our purpose is to make forecasts of future values, let us recall that the noise inputs η_n are not known to us and must be averaged over. This leaves only the AR part as a predictive model. In order to determine the coefficients a_j, $j = 1, \ldots, M_{AR}$, we can minimise the one-step prediction error, as described in Section 2.5.

Example 12.1 (Linear prediction of NMR laser data). In Chapter 1 we compared a simple nonlinear prediction scheme to something which we vaguely called the "best" linear predictor of the NMR laser data (Fig. 1.2). These predictions were made with an AR model of order $M_{AR} = 6$, where it turned out that the improvements resulting from increasing M_{AR} were negligible. This is in agreement with the fact that the autocorrelation function of these data (Appendix B.2) decays very fast. The one-step prediction error there was found to be $e = 962$ units. □

12.1.2 Nonlinear filters

Autoregressive models can be generalised by introducing nonlinearity. One important class consists of *threshold autoregressive (TAR) models* [Tong (1983)]. A single TAR model consists of a collection of the usual AR models where each single one is valid only in a certain domain of the delay vector space (separated by the "thresholds"). For the construction of the model one divides the reconstructed phase space into patches, and determines the coefficients of each single AR model as usual, using only data points in the corresponding patch. TAR models are thus piecewise linear models and can be regarded as coarse-grained versions of what we call locally linear models below (Section 12.3.2).

Alternatively, AR models can be extended by nonlinear terms in Eq. (12.2). They then assume the form of global nonlinear models and will be discussed in Section 12.4. Nonlinearity makes it impossible to analyse the properties of the model as in the linear case. We cannot usually obtain the power spectrum of the process analytically, or its autocorrelations, or the distribution of the output. As long as the nonlinearity is weak, one can still regard these models as filters on the noise input. This is no longer appropriate for strong nonlinearities, where we would rather consider the deterministic part as the dominant feature and the noise only as a perturbation.

12.2 Deterministic dynamics

To model deterministic dynamics, or a dominant deterministic part of some mixed system, we first have to embed the scalar observables as usual. In the following we want to assume that the attractor can be reconstructed reasonably well by an m dimensional delay embedding with unit delay, but it will be obvious how to apply the methods for different delays and for other embedding spaces. In particular, one can profit from vector valued (multivariate) time series.

Since the time series data are discretely sampled over time, a deterministic model is always a map, and in a delay embedding space it reads

$$\hat{\mathbf{s}}_{n+1} = (s_{n-m+2}, \ldots, s_n, F(s_{n-m+1}, s_{n-m+2}, \ldots, s_n)). \tag{12.3}$$

Thus all we need for a forecast is a prediction of s_{n+1}, that is, we want to find a scalar valued function $F(\mathbf{s})$ such that:

$$\hat{s}_{n+1} = F(\mathbf{s}_n). \tag{12.4}$$

If the data are flow data,[4] successive measurements are strongly correlated, and it might be advantageous to consider this explicitly by modifying Eq. (12.4):

$$\hat{s}_{n+1} = s_n + \Delta t \ f(\mathbf{s}_n). \tag{12.5}$$

This is motivated by Euler integration of the differential equation, Eq. (3.2), with a time step Δt.

After these more general considerations, we have to find a criterion for the quality of the model, which can be turned into a prescription of how to construct it. The usual way is to write down a cost function which is then to be minimised. The choice of the proper cost function depends on the purpose of the modelling. If we need predictions for some practical use, we can take the term "cost" quite literally and weight bad predictions by the harm they cause. A different approach is to maximise the *likelihood* of the observations with respect to the model. If the prediction errors can be assumed to follow a Gaussian distribution and to be independent of each other, maximising the likelihood amounts to minimising the mean squared *prediction error*,

$$e = \sqrt{\frac{1}{N} \sum_{n=1}^{N} \left(s_{n+k} - F^{(k)}(\mathbf{s}_n)\right)^2}, \tag{12.6}$$

[4] See Section 11.2.2 for a definition.

were we have written the expression for k-step ahead predictions. Most often we will use $k = 1$. The question of the appropriate cost function will be taken up in Section 12.5.

Now we have to select a general form for the function F (respectively, f) containing enough freedom so that it is capable of representing the data. The free parameters in F are optimised by minimising the cost function. In the following we shall discuss different model classes. In order to understand their strengths and drawbacks it is sometimes helpful to give a geometrical interpretation of our task. Equations such as (12.4) describe a scalar field over $\mathbb{R}^m$, or, in other words, an m dimensional hypersurface in $\mathbb{R}^{m+1}$. If F is nonlinear, this surface is bent; for linear F it is just a hyperplane. Thus we have to find a surface which best describes a cloud of points $(s_{n+1}, \mathbf{s}_n)$ scattered in $\mathbb{R}^{m+1}$.

12.3 Local methods in phase space

If the data base is large and the noise level small, local methods can be very powerful. They derive neighbourhood relations from the data and map them forward in time. They are conceptually simpler than global models, but, depending on our purposes, they can require a large numerical effort.

12.3.1 Almost model free methods

Whereas one generally constructs a model of the dynamics as in Eq. (12.4), one can renounce this when working locally. In Section 4.2 we mentioned Lorenz's "method of analogues" as the simplest algorithm in this class. To predict the future of a point in delay embedding space, one searches for its closest neighbour and uses the image of the neighbour as the prediction. This scheme obviously does not involve any modelling, and the only underlying assumption is that the dynamics are deterministic on the m dimensional space and that it is continuous. Unfortunately, when we use it for more step predictions, it starts to reproduce a past part of the time series and thus in the end creates a periodic orbit. Therefore, the simple prediction algorithm does some averaging over several points: we search for all close neighbours within some distance. The prediction is the average of the images of all these points. This can be very successful and might even be superior to advanced methods when the deterministic part of the data is too weak or the state space too high dimensional.

Sugihara & May (1990) use a weighted average of the images of exactly $m + 1$ neighbours which form a simplex containing the current point, a method which we cannot recommend, since the weighting is very *ad hoc* and the demand for a simplex can introduce unnecessarily large distances. In the worst case, due to folding processes, the true image of the point may be outside the image of the simplex. There are other similar suggestions to exploit the idea of continuity.

12.3.2 Local linear fits

From Section. 12.1.1 we know already what a linear prediction model in an m dimensional delay embedding looks like. For predictions, only an *autoregressive model* of order m, AR(m) (or the AR part of an ARMA model) can be used, such that the deterministic forecast is a linear superposition of the m last observations, Eq. (12.2). The noise term responsible for the *stochastic* nature of the AR model will be set to zero for predictions, although also the future measurement to be predicted might be uncertain due to measurement noise, even though we assume otherwise *deterministic dynamics*. The coefficients of the AR model can be determined as described in Section 2.5.

We know that the dynamics generated by the deterministic part of an AR model consists of damped harmonic oscillations which settle down on a stable fixed point, and that it is unable to produce chaos. Instead of using a single model for the global dynamics, one can use a new local model for every single data item to be predicted, so that globally arbitrarily nonlinear dynamics is generated:

$$\hat{\mathbf{s}}_{n+1} = \mathbf{A}^{(n)}\mathbf{s}_n + \mathbf{b}^{(n)} \tag{12.7}$$

or, using delay coordinates,

$$\hat{s}_{n+1} = \mathbf{a}^{(n)} \cdot \mathbf{s}_n + b^{(n)}. \tag{12.8}$$

We use the superscript (n) to remind us that these equations are expected to be valid only within some neighbourhood $\mathcal{U}_n$ of $\mathbf{s}_n$. Hence, when fitting the parameters by the minimisation of the cost function Eq. (12.6) we extend the sum only over those indices $k \in \{1, \ldots, N\}$ whose corresponding delay vectors are close to the current vector $\mathbf{s}_n$. This introduces the neighbourhood size as an additional parameter. The collection of all these local linear models constitutes a nonlinear dynamics, and the single local models are interpreted as the linearisation of the global nonlinear dynamics at the considered points in phase space. In addition to continuity we now demand implicitly that the first derivative of F also exists.

Obviously, here we are constructing an approximation to the tangent plane to the surface $F(\mathbf{s})$ mentioned above at a given point $\mathbf{s}_n$. If the size of the neighbourhood $\mathcal{U}_n$ of the current point is large compared to the inverse curvature of the true surface, this approximation may be very poor. The neighbourhood size is constrained from below by two requirements. First, in order to determine the coefficients of the local model with a reasonable statistical error, we need enough neighbours. Thus short time series (short on a scale depending on the number of points needed to

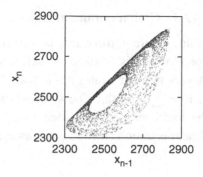

Figure 12.1 The attractor formed by iterated local linear fits of the feedback laser dynamics (Example 12.2). Compare with Fig. B.7.

resolve the deterministic structure of the attractor) lead to large neighbourhood diameters. Second, the diameter of the neighbourhood has to be larger than the measurement noise amplitude, since we otherwise fit the properties of the noise and not the deterministic structure. If due to these constraints we cannot avoid large neighbourhoods, we have to use a global, nonlinear model.

Locally, clean attractors can be embedded in fewer dimensions than are required for a global reconstruction. Covariance matrices of local neighbourhoods are thus sometimes close to singular and the fits are unstable. Experimental data usually contain enough noise to avoid this effect. When computing local linear fits of the NMR laser data, we find on average better results on the original data than on those after noise reduction. Local linear models have been used with reasonable success to determine Lyapunov exponents from time series, see Example 11.5.

Example 12.2 (Iterated predictions of the feedback laser data). To demonstrate the potential of local linear fits by a successful example we could impress you again with the NMR laser data. Instead, we use the data from the autonomous feedback laser (Appendix B.10). Beginning with an arbitrary five dimensional delay vector of the experimental data, we construct a local neighbourhood of at least 20 neighbours and at least 4 units in diameter. On these points we determine the local linear model to compute the image of the current vector. At its position, we again collect neighbours and so on. Thus we create a new trajectory by iteration, which is a much more powerful test than just computing the average forecast error (which is about 0.7% on the training set). See Fig. 12.1 for the resulting attractor. The maximal Lyapunov exponent is estimated to be $\lambda_+ \approx 0.006 \pm 0.001$, in reasonable agreement with the estimate from the Poincaré map (the error is a rough estimate obtained from fits with different minimal neighbourhood sizes). The second exponent is $\lambda_0 \approx -0.0008$, which is consistent with the expected value of zero for a flow. $\square$

12.4 Global nonlinear models

The idea of global modelling is straightforward: we just have to choose an appropriate functional form for F which is flexible enough to model the true function on the whole attractor. A very popular strategy is to take F to be a linear superposition of basis functions, $F = \sum_{i=1}^{k} \alpha_i \Phi_i$. The k basis functions Φ_i are kept fixed during the fit and only the coefficients α_i are varied. Inserted into the cost function Eq. (12.6), this leads to linear equations for the fit parameters, which simplifies the determination.

12.4.1 Polynomials

Global linear models can be regarded as the first approximation in a Taylor series expansion of $F(\mathbf{s})$ around the current observations $\mathbf{s}_n$. The obvious generalisation is to use a polynomial. Since F acts on an m dimensional space, it has to be a multivariate polynomial, which for order l has $k = (m + l)!/m!l!$ independent coefficients. Despite their large number they can usually be readily determined as long as the data do not contain too much noise. Polynomials have the advantage that we are very familiar with them, which gives us some hope that we can understand the result. The k coefficients can be determined by the inversion of a single $(k \times k)$ matrix. Sometimes, however, prediction functions obtained by polynomial fits give rise to unstable dynamics under iteration since the polynomial basis functions diverge for large arguments and trajectories may escape to infinity.

Example 12.3 (Polynomial fit for NMR laser data). The dynamics of the NMR laser data (Appendix B.2) can be described extremely accurately by a polynomial of degree seven in a three dimensional delay embedding space. Since the time series has length 38 000 and the noise level is only around 1.1%, all 120 coefficients are statistically well defined. Of course they are not all equally relevant, but without careful investigation one cannot eliminate the irrelevant terms. The data possess a rms amplitude of about 2290 in their natural units, and the one-step prediction error is about $e = 41$ units. (For the data after noise reduction, we find an improved prediction error of $e = 12$ units. Of cause, a realistic prediction task can only involve data as they come from the experiment.) □

Example 12.4 (Differential equation model for a nonlinear resonance circuit). The precise functional form of the nonlinear electric resonance circuit described in Appendix B.11 is unknown. However, since the Landau theory of phase transitions of second order predicts as simplest model a quartic double well potential, it is reasonable to expand the unknown nonlinearity into a Taylor series, so that the following model is fitted to the data q ($\dot{q}$ and $\ddot{q}$ are obtained from the data by

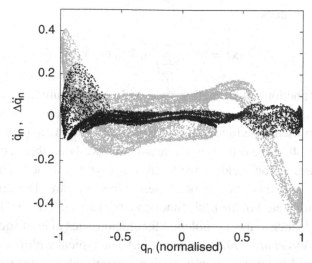

Figure 12.2 The signal $\ddot{q}$ to be predicted (grey) and the prediction errors as a function of the coordinate q. The pronounced spatial structure of the errors suggests that the fitted model does not describe all deterministic dependencies. Enlarging the model class (see text) does not just reduce the error but also renders its distribution almost q-independent.

suitable differences, employing Savitsky–Golay filters [Press *et al.* (1992)]):

$$\ddot{q} = R\dot{q} + \sum_{j=1}^{7} a_j q^j + A\cos\omega t + B\sin\omega t$$

However, prediction errors of the fitted model with a 7th order polynomial still contain spatial structureSee (see Fig. 12.2), indicating that they are not only due to noise but due to non-resolved determinism. A more general ansatz is a bivariate polynomial $\sum_{j=0}^{J} \sum_{k=0}^{K} a_{jk} q^j \dot{q}^k$, admitting nonlinear friction and forces that depend on $\dot{q}$. With $K \geq 7$ and $J \geq 3$ this model yields prediction errors without spatial structure. By the help of symmetry arguments, the resulting terms can be sorted into generalised friction terms and generalised repelling forces, and coefficients of terms whose total power $j + k$ is even are almost zero, in agreement with the overall symmetry of the dynamics. For details see Hegger *et al.* (1998). □

12.4.2 *Radial basis functions*

Another very flexible model class was introduced into the field of time series analysis by Broomhead & Lowe (1988) and is called *radial basis functions*. One defines a scalar function $\Phi(r)$ with only positive arguments r. Additionally, one has to select k centres $\mathbf{y}_i$ on the attractor. Including the constant function as the zeroth basis function, F reads

$$F(\mathbf{x}) = \alpha_0 + \sum_{i=1}^{k} \alpha_i \, \Phi(\|\mathbf{x} - \mathbf{y}_i\|) \,. \tag{12.9}$$

Typical basis functions Φ are bell-shaped with a maximum at $r = 0$ and a rapid decay towards zero with increasing r. But also increasing functions or even singular ones have been used. The function F is modelled by adjusting the coefficients α_i of the functions Φ. If the centres $\mathbf{y}_i$ are reasonably well distributed on the attractor, the linear superposition yields a well-behaved (hyper-)surface. The number and positions of the centres $\mathbf{y}_i$ have to be chosen with some care. The centres and other parameters (width, etc.) of the basis functions $\Phi(r)$ are kept fixed. Determining the coefficients α_i is then a linear problem. (Technical issues in least squares fitting are discussed in Press *et al.* (1992), for example.) The typical width of the function Φ can be optimised by systematically testing several values, since for every fixed value the numerical fit procedure is very fast. Fitting becomes nonlinear and thus much more difficult when the centres are also to be optimised.

Example 12.5 (Radial basis function fit for NMR laser data). We cover the NMR laser attractor (Appendix B.2) in three dimensions by a regular grid and place a centre $\mathbf{y}_i$ in every cube which contains attractor points. For a $13 \times 13 \times 13$ grid, for example, this yields 140 centres. The function Φ was chosen to be a Lorentzian, $\Phi = 1/(1 + r^2/a^2)$, where the optimal value for a turned out to be $a = 1770$. After minimisation, the average one-step prediction error was found to be $e = 42$ units. (After noise reduction: $e = 10$ units.) $\square$

12.4.3 Neural networks

Neural networks are supposed to be capable of learning complex tasks such as recognition, decision making or predictions. Therefore, it is only natural to employ them for our present purpose. In fact, the networks to be used here are again nothing but nonlinear global models of the dynamics, the structure of which can be understood quite easily.

One particular class which has been used for time series modelling comprise the so-called *feed-forward networks* with one *hidden layer*. This means that we have one layer of input units, one layer of *neurons*, and one layer of output units. The input layer consists of m units if we want to work in an m dimensional embedding space. Each unit provides one component of a delay vector $\mathbf{s}$. The input to the i-th neuron in the hidden layer is a weighted sum of these components which can be written as a scalar product $\mathbf{b}^{(i)}\mathbf{s}$. Inspired by the nervous system, the function of a neuron is usually a smoothed step function, a *sigmoid function* such as $\Phi^{(i)}(\mathbf{s}) = 1/(1 + e^{\mathbf{b}^{(i)}\mathbf{s} - c^{(i)}})$. In the case of scalar predictions, the output layer consists of a single unit which generates a weighted sum of the output of the neurons. The whole net with k neurons can thus be expressed as the function

$$F(\mathbf{s}) = \sum_{i=1}^{k} \frac{a^{(i)}}{1 + \exp\left(\mathbf{b}^{(i)}\mathbf{s} - c^{(i)}\right)}. \tag{12.10}$$

It takes a delay vector as its argument and the weight parameters $a^{(i)}$, $\mathbf{b}^{(i)}$, and $c^{(i)}$ have to be determined by a fit. Since the minimisation problem is nonlinear in the parameters, and in order to avoid over-fitting, we should not choose the net too large. The standard way to determine the parameters is, in this context, called *training* and the most popular algorithm is *error back-propagation*. This is nothing other than a gradient descent method where a cost function is minimised by presenting all *learning pairs* $(s_{n+1}, \mathbf{s}_n)$ of the one-step prediction error individually to the function F. The parameters are iteratively modified. See e.g. Weigend *et al.* (1990) for more details.

Apart from being inspired by the neural structure that is presumed to perform cognitive functions in the brain, the neural net is just one particularly flexible model class for the approximation of functions with limited data. Thus, instead of training by back-propagation, we prefer to insert Eq. (12.10) into the one-step prediction error of Eq. (12.6) and apply a routine for nonlinear minimisation such as a conjugate gradient or a simplex algorithm [see e.g. Press *et al.* (1992)]. With some experience in nonlinear minimisation this yields satisfactory results. Since the minimisation procedure is nonlinear in the parameters now, we have to deal with a new problem. The solution is not usually unique since several local minima can exist. Therefore, one has to restart the minimisation with different initial conditions (trial parameters $a^{(i)}$, $\mathbf{b}^{(i)}$, $c^{(i)}$), and/or use more sophisticated algorithms such as *simulated annealing* in order to escape from shallow relative minima. Learning algorithms do not seem to be equally good at finding the *right*, that is the global, minimum. Unfortunately, the literature is full of contradictory claims and it is hard to make a general statement. Therefore, the successful training of neural nets often requires a considerable amount of experience.

Example 12.6 (Neural network fit for NMR laser data). To stick to our favourite example, we apply a net with nine neurons in the hidden layer and three input units to the NMR laser data (Appendix B.2) and find a forecast error of $e = 75$ units. Using more neurons, we did not succeed to arrive at lower minima, but we had to stop the numerical computation before reaching clear convergence. The result is not even as good as the locally constant predictor of Section 4.2 ($e = 52$ units) and we suspect that the neural net is not a very effective basis for this data set. See the discussion below. □

12.4.4 *What to do in practice*

Nonlinear global models can be used quite successfully if the underlying dynamics are sufficiently smooth and the data are not too noisy. The examples presented here are nontrivial with respect to the dimensionality of the embedding space, the shape of the hypersurface, and the degree of instability (Lyapunov exponent about 0.4 bit/iteration). Below we will confirm that the resulting fits are indeed excellent. What the examples also show is that there is some variability in the results depending on details of the fits, in particular with respect to convergence of nonlinear minimisation schemes (affecting neural network results) and sensitivity to noise (mainly in radial basis function fits).

You may ask why we have introduced three different classes of nonlinear models when they all yield reasonable results. Unfortunately, this is not always the case, and for the NMR laser data we encountered problems with the neural net. We can only repeat that the determination of the dynamics is not as straightforward as other analysis tasks. A successful fit requires an appropriate general expression for F. If you want to construct a good model of your data, you have to perform and compare different fits. If the data are sufficiently noise-free to compute the noise level using the method described in Section 6.7, you know the lower bound of the one-step prediction error and can thus estimate the quality of your fit. Moreover, iterating the resulting dynamics yields an impression of how efficiently it can reproduce the attractor, but this is more a qualitative than a quantitative means (we shall say more about model verification below). It is worth trying to understand why a certain functional form for F performs unsatisfactorily. For instance, in a one dimensional space, two sigmoid functions are enough to create a one-humped function. If one wants a bell-shaped structure in two dimensions, one already needs at least three sigmoids, and in higher dimensions even more. Since the NMR laser data require a polynomial of degree seven (all lower-order polynomials were unsuccessful) we have to expect a very bumpy surface, such that the neural net was perhaps too small. We have not been able to use larger nets due to the numerical effort. Even in the polynomial fit it might be worth reducing the number of coefficients systematically by eliminating those whose contributions are small. From the practical point of view, there is no definite means of determining the optimal size of the model (compare next section, however). The degree of the polynomial, the number of radial basis functions or hidden units in the neural net have to be found by exploration and experience. Sometimes, prior information such as symmetries or the behaviour of the function F for large arguments can be respected by the choice of the form of the model. Summarising, there is much room for optimisation.

From the theoretical point of view, the lack of structural stability of almost all systems is a serious problem. It means that the slightest deviation of a model system from the "true" dynamics might create a completely different attractor. Fortunately, stable periodic orbits in the vicinity of chaotic attractors often are of very small measure in parameter space. Many structurally unstable systems seem to be stochastically stable, which means that under the influence of weak dynamical noise the properties of the closest chaotic attractor become visible. Finally, let us mention that there are cases of data sets where only one out of several model classes yield functions F which are able to reproduce the attractor under iteration, and where we cannot understand why one class of functions works and another does not. Therefore, we are sceptical about claims that a certain method of fitting gives optimal results in general.

In view of these problems, it can be advantageous in certain situations to consider several alternative model classes for F. One interesting suggestion which we want to mention is the use of functions [e.g. polynomials, Brown *et al.* (1994), Giona *et al.* (1991)] which are orthogonal on the attractor, i.e. $\int \Phi_i(\mathbf{x})\Phi_j(\mathbf{x})d\mu(\mathbf{x}) = \delta_{ij}$, where μ is the invariant measure represented by the distribution of the data, and the Φ_i are the basis functions. The function F is then a linear superposition of the functions Φ_j. The advantage is that the coefficients can be determined by a simple projection, but the nontrivial step is the construction of the orthogonal functions.

12.5 Improved cost functions

The different approaches of the last section have one aspect in common that is usually not questioned. They all rely on the minimisation of the one-step mean squared prediction error. Unfortunately, there are different reasons why this is not always the optimal cost function.

12.5.1 *Overfitting and model costs*

Overfitting can occur since the more free parameters a model has the better it can be adapted to the data. Our goal is, however, to find a model which to some extent generalises the information contained in the data. The data are just a finite output sequence of some process and we are looking for a model of that process and not just of the finite data set. When *overfitting* the data by providing too many adjustable parameters, peculiarities of the actual trajectory, including the noise and intrinsic fluctuations, are erroneously interpreted as global features of the system.

As we will discuss in Section 12.6, overfitting becomes evident if the in-sample prediction error is significantly smaller than the out-of-sample error.

In order to avoid this problem, cost functions must not be straightforwardly compared for models with different numbers of coefficients. In order to make such a comparison possible it is necessary to add a term for the *model-costs* to the minimisation problem. In a general setting, this is done by adding an appropriate function of the number of adjustable parameters to the *likelihood function*. Remember that the usual cost function, the mean squared prediction error, is derived from the *maximum likelihood* principle in the special case of Gaussian errors. Suppose we start from a general function for the dynamics, $F_{a_1,...,a_k}(\mathbf{s})$, depending on k adjustable parameters a_j, $j = 1, \ldots, k$. We want to find the particular set of parameters $\{\hat{a}_j\}$ which maximises the probability $p(\{s_n\}; \{a_j\})$ to observe the data sequence $\{s_n\}$ when the underlying model is $F_{a_1,...,a_k}(\mathbf{s})$. That is, we maximise the *log-likelihood* function

$$L = -\ln p(\{s_n\}; \{a_j\}) \qquad (12.11)$$

with respect to the parameters $\{a_j\}$. If the errors $s_{n+1} - F_{\{a_j\}}(\mathbf{s}_n)$ are Gaussian distributed with variance e^2, the probability of the data is

$$p(\{s_n\}; \{a_j\}) = \frac{1}{(e\sqrt{2\pi})^N} \exp\left[-\frac{\sum_{n=1}^{N}(s_{n+1} - F_{\{a_j\}}(\mathbf{s}_n))^2}{2e^2}\right]. \qquad (12.12)$$

Estimating the variance by $\hat{e}^2 = \sum_{n=1}^{N}(s_{n+1} - F(\mathbf{s}_n))^2/N$ we obtain

$$L_{\text{Gauss}} = -\frac{N}{2}\ln\hat{e}^2 - \frac{N}{2}\ln 2\pi - \frac{N}{2}, \qquad (12.13)$$

which is maximal when the mean squared error $\hat{e}^2$ is minimal.

The complexity of a model F is now taken into account by adding a term $-\alpha k$ to the log-likelihood function. The maximum of L_k is then determined for models with different numbers k of adjustable parameters and the model which yields the maximal L_k is selected.

For the constant α, two main suggestions have been made. If the purpose of the analysis is to minimise the expectation value of the out-of-sample prediction error, information theoretic arguments lead to the choice $\alpha = 1$. This is the *Akaike information criterion* [AIC, Akaike (1974)]. It has, however, been pointed out that if the data have been generated by a process with k coefficients, a model with additional, redundant parameters may lead to better predictions. In such a case, Akaike's criterion prefers the redundant model to the one which generated the data.

If the main interest is the model itself, it is preferable to use $\alpha = \frac{1}{2}\ln N$, following Rissanen (1980). This result is obtained in a very intuitive way if it is assumed that the *description length* of an optimal encoding of the data should be minimised. We can encode the data in two parts: first we encode a description of the model (including the parameters) and then we encode the residual errors when the model is run. For a given resolution, we need fewer bits to encode the errors if they are smaller than the original data. The price of this reduction is the encoding of the model. The residual errors can be encoded with an average length $-L = \ln p(\{s_n\}; \{a_j\})$, plus a resolution dependent constant. The parameters need not be encoded with infinite precision either. If we have N data points, the number of relevant bits in each parameter will typically scale as $\ln \sqrt{N}$, which gives a total description length of $-L + \frac{k}{2} \ln N$.

Thus the modified log-likelihood functions are

$$L' = -\ln p(\{s_n\}; \{a_j\}) - \begin{cases} k & (Akaike) \\ \frac{k}{2} \ln N & (Rissanen). \end{cases} \tag{12.14}$$

Note that the above expressions are asymptotically valid for large N. The corresponding changes to the mean squares cost function for Gaussian errors are

$$e'^2 = \frac{1}{N} \sum_{n=1}^{N} (s_{n+1} - F(s_n))^2 \times \begin{cases} e^{\frac{k}{N}} & (Akaike) \\ e^{\frac{k}{2N} \ln N} & (Rissanen). \end{cases} \tag{12.15}$$

Although these criteria are derived from quite general principles, some ambiguity still remains. One concern is that coefficients are only penalised by number, no matter how complicated the specification of the model is otherwise. There is no unique way to quantify the cost of choosing a complicated model class.[5] Finally, the very fact that there are at least two different suggestions for a penalty of additional parameters leads to the suspicion that there is no unique "optimal criterion" for all situations. In any case, we can conclude that a smaller model should be preferred, even if it yields slightly larger in-sample prediction errors.

12.5.2 The errors-in-variables problem

The *errors-in-variables* problem arises when an ordinary least squares procedure is applied when not only the dependent variables are noisy, but also the *independent* variables. Eq. (12.6) implicitly divides the variables into *independent* and *dependent* ones. The least squares fit only leads to an unbiased result if the independent variables s_n are noise free. Since all variables are elements of the same noisy time series, this assumption is clearly wrong. For noise levels of more than about 10% the consequence becomes obvious in numerical simulations and one obtains

---

[5] We could, for instance, label possible model classes by numbers 1 to ∞. If we label our favourite, most intricate model by a small number, it can be specified with the few bits necessary to specify that number.

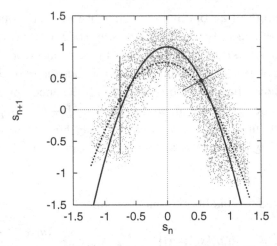

Figure 12.3 Illustration of the errors-in-variables problem. The points are noisy in all their components. The standard one-step prediction error is the average of the vertical distances, whose minimisation leads to the dashed curve and a value for the error of 26% of the signal's variance. The continuous graph is the function underlying the data (logistic map) and which should be found by the optimal fit, but which, however, creates a one-step error of 34%.

systematically wrong results; see Schreiber & Kantz (1996) and Jaeger & Kantz (1996) for discussion and additional references. One can avoid this problem by treating all variables in a symmetric way. In ordinary least squares fitting, the cost function is the sum of the vertical distances between the noisy point and the surface. Here we consider instead the sum of the orthogonal distances (see Fig. 12.3). Such a minimisation problem is called *total least squares* in the literature of statistics.

Although this is an important achievement, one can still do better. As we can clearly see from Fig. 12.3, our procedure to find the dynamics consists of the reduction to a merely *geometrical* problem. Once we have converted the time series into the vectors in the reconstructed state space, we fit a surface to a collection of points, ignoring the fact that the points are not mutually independent. Since they represent delay vectors, they have common components, and shifting one point also induces changes in other points. For noise levels larger than 15–20% this again leads to systematic errors. The result of the total least squares problem does not only consist of the optimal graph of $F(\mathbf{s})$, but also in hypothetical noise-free points on the graph when constructing the orthogonal distances between noisy points and graph. These noise free points in the delay embedding space cannot be composed to a single noise free time series, since without additional constraints they are mutually inconsistent (every noisy observation s_n appears as a coordinate in m different delay vectors and hence is associated to m different noise free points). The effect can be avoided by simultaneous minimisation of multi-step errors. One can quite successfully minimise the sum of the k-step errors for $k = 1, 2, 3, \ldots$ (in practice we were able to proceed up to $k = 12$) and successfully reconstruct dynamical equations without bias up to noise levels of the order of the signal. More details can be found in Jaeger & Kantz (1996) and Kantz & Jaeger (1997).

Let us finish the discussion of possible cost functions by mentioning that optimisation of the prediction quality is not the only approach for finding good model parameters. Parlitz (1996) uses the concept of *auto-synchronisation* for this purpose. Brown *et al.* (1994) argue that a model can be considered appropriate if it synchronises with the observed data if coupled appropriately. Synchronisation is a sufficient condition but not necessary: it can fail even for identical signals if the wrong type of coupling is chosen. Parlitz optimises model parameters by minimising the deviation from synchronisation. This is done dynamically while the model is coupled to the data and thus constitutes a very attractive concept for real time applications, in particular when slow parameter changes have to be tracked.

12.5.3 Modelling versus prediction

In the last subsection we have assumed that we were interested in modelling. These considerations yield the surprising implication that in the presence of strong noise, finding the model closest to some assumed underlying system and finding the best predictor are different tasks. By definition, the best predictor is always found be the minimisation of the prediction error, Eq. (12.6). The reason for this is simple. If we model the system, the equations should be applied to an initial condition and should then be iterated. Therefore, they act on clean values, and in numerical simulations we aim at identifying model equations which are as close as possible to those which create the original data. In contrast to this, a predictor obtains the noisy values as input and should take this into account. One can easily verify using simple examples that the exact deterministic dynamics does not do the best job with this respect. Instead we have to find an effective dynamics. Therefore, if we want to perform predictions, it is by definition optimal to minimise the mean squared vertical distances in Fig. 12.3 given by Eq. (12.6), or, for Δn-step predictions, the Δn-step equivalent of Eq. (12.6).

12.6 Model verification

Up to now we have been vague about when a fit of the dynamics is "good". The actual value of the cost function at its minimum is only a rough indicator, since for noisy data it can be quite large anyway and thus rather insensitive to small changes of F. More importantly, the remaining individual errors $s_{n+1} - F(s_n)$ can be small but systematic, and they can even be spuriously small on the data due to over-fitting.

The simplest step which should be always performed is to sub-divide the data set into a *training set* and a *test set*. The fit of the parameters is performed on the data of the training set, and the resulting F is inserted into the cost function and its value is computed for the test set. If the error is significantly larger on the second set, something has gone wrong. Either the data are not stationary, or one has *over-fitted* the training set data. Changing the model function by eliminating free parameters can reveal the difference.

Example 12.7 (Determination of the period of a driving force). In Example 12.4 we described the fit of a non-autonomous ODE model to data from an electric resonance circuit. This system is driven by a sinusoidally oscillating voltage. How can we determine the period of this voltage in the most precise way? We perform a one-step prediction with a rather simple polynomial map in a two dimensional delay embedding space, including explicitly a sinusoidal time dependence:

$$\hat{q}_{n+1} = F_p(q_n, q_{n-1}) + A \cos \frac{2\pi n}{\tau} + B \sin \frac{2\pi n}{\tau} .$$

We compute the average in-sample and out-of-sample prediction errors as functions of τ. Notice that by splitting the driving term $C \cos(\phi_n + \Psi) = A \cos \phi_n + B \sin \phi_n$ the nonlinear fit for the unknown phase shift Ψ is replaced by a linear fit for two amplitudes A and B. Since F_p is a polynomial and hence also linear in its parameters, the fits can be performed so easily that we can test a huge number of values τ. In Fig. 12.4 one observes that around $\tau = 25.45$ sampling intervals, the error has a minimum. The out-of-sample error has an interesting signature: the test set is a segment of the time series immediately following the training set. If τ is slightly detuned, the phase Ψ is adjusted by the fit such that the phase-mismatch is smallest on average inside the training set. If the time interval covered by the training set times the detuning of the frequency is exactly half a period, then, with the optimal Ψ, the phase mismatch is maximal in the test set, hence leading to the enhanced out-of-sample error. Both errors together yield a very precise estimate of the period τ.   □

We mentioned above that the errors $s_{n+1} - F(\mathbf{s}_n)$ can be systematic. A glance back at Fig. 12.2 shows that remaining forecast errors (the *residuals*) in fact can have rich structure, indicating that part of the determinism has not been identified by the model. Here, however, we are speaking about more subtle effects, which might be so weak that a statistical analysis of the residuals does not always reveal them. This will, e.g., be the case when the data represent a chaotic attractor, while the model favours a stable limit cycle. In particular when the difference between the embedding dimension and the attractor dimension is large (say, > 2), there is much freedom to construct dynamical equations to which the observed trajectory

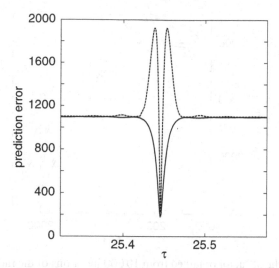

Figure 12.4 The average in-sample (continuous line) and out-of-sample (broken line) one-step error of the resonance circuit data as a function of the period τ of the sinusoidal driving term. The out-of-sample error has a very sharp minimum. τ is measured in multiples of the sampling interval.

is an approximate solution, but an unstable one. Thus, under iteration points would escape from the observed attractor. In such a case the Lyapunov exponents that one would compute with the tangent space method (cf. Section 11.2) would all be wrong.

The most severe test that you can perform is to iterate your equations. Select an arbitrary point of your observed data as initial condition, and create your own trajectory by your fitted model. When you embed it in the same way as your observed data, the synthetic data should form a very similar attractor. A positive result is highly nontrivial, since with an approach which is local in time (and thus in space) one has reconstructed global properties of an attractor. In an ideal case, this attractor should look like a skeleton in the body of your noisy observations. If this is the case, you may say that you possess a real model of the dynamics, and you can use your synthetic data for further investigations of your system. The sampling of the output of a model for the estimation of parameters and errors is called *bootstrapping* in the literature. Since your new time series is noise free and can be made arbitrarily long, you can compute all the invariants from it.

In particular, all the *unstable periodic orbits* embedded in the chaotic attractor should be present and should be located almost at the same points. The ability of locating periodic orbits of a short period accurately is of high practical relevance, e.g. for *chaos control*, the stabilisation of periodic orbits by small perturbations; see Chapter 15.

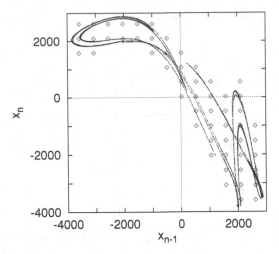

Figure 12.5 The attractor obtained from 10 000 iterations of the radial basis function fit of the NMR laser data, to be compared with Fig. 1.3. The symbols are the locations of the centres of the radial basis functions fit projected into the plane

Example 12.8 (NMR laser dynamics). Since the noise level of the NMR laser data (Appendix B.2) is rather low, we find a very good description of their dynamics by minimisation of the mean squared one-step prediction error. The results for the (out-of-sample) one-step prediction errors for the different methods are summarised in Table 12.1. Note that the locally constant and locally linear models are usually counted as parameter free since no parameters are adjusted by minimising the overall cost function.

No matter if we employ locally linear fits, a global polynomial or a set of radial basis functions, we can construct synthetic time series which reproduce the original attractor extremely well. We show the result of the fit with radial basis functions in Fig. 12.5. In all cases we find reasonable Lyapunov spectra. The numerical values are presented in Table 11.1 on page 209. □

12.7 Nonlinear stochastic processes from data

We should always be aware of the fact that a deterministic differential equation with a few degrees of freedom is an idealisation as a description of a real experiment. It is almost impossible to isolate a real system from the rest of the universe. Even with great care, experimentalists can only minimise but never exclude perturbations of their system under study by external influences. In the laboratory, these perturbations often are small enough to be ignored in our data analysis, so that we consider only measurement noise. Good examples for this are the NMR laser data (Appendix B.2) or the electric resonance circuit (Appendix B.11). More generally, however,

Table 12.1. *Root mean squared, out-of-sample prediction errors e for various prediction methods (NMR laser data). k is the number of adjustable parameters in the model*

Model	k	e
constant	–	2288
AR(6)	6	962
locally constant	–	52
locally linear	–	38
polynomial	120	41
radial basis functions	140	42
neural network	45	75

one can not always neglect dynamical noise. A model including dynamical noise can be formulated as a (nonlinear) stochastic process. If the number of deterministic degrees of freedom is much higher than we can resolve in our data analysis, a non-linear stochastic process might be a more appropriate description than the attempt of some deterministic modelling. In this section we therefore discuss two classes of nonlinear stochastic processes which can be identified from data and which can be used in data analysis.

12.7.1 Fokker–Planck equations from data

The stochastic counterpart of the equation of motion of a deterministic dynamical system in continuous time is the Langevin equation:

$$\dot{\mathbf{x}} = \mathbf{f}(\mathbf{x}) + G(\mathbf{x})\xi. \tag{12.16}$$

Apart from the phase space vectors $\mathbf{x}$ and a deterministic vector field $\mathbf{f}$, both elements of the $\mathbb{R}^d$, a noise term composed of a (typically diagonal) $\mathbf{x}$-dependent $d \times d$ dimensional tensor G multiplied with a vector valued white noise process ξ enters the right hand side of the equation. Excellent discussions of the theory of stochastic processes and their description by Langevin and Fokker–Planck equations can be found in, e.g., Risken (1989), van Kampen (1992) or Gardiner (1997).

   As in a deterministic chaotic system, the individual solution of Eq. (12.16) as initial value problem $\mathbf{x}(t = 0) = \mathbf{x}_0$ together with a given realisation of the noise process $\xi(t)$ for all t is highly unstable, so that alternatively the time evolution of phase space densities $\rho(\mathbf{x}, t)$ are studied. Here, $\rho(\mathbf{x}, t)d\mathbf{x}$ is the probability to find an individual solution inside a small volume element $d\mathbf{x}$ around $\mathbf{x}$, if the initial condition of this solution was randomly chosen according to the initial density $\rho(\mathbf{x}, t = 0)$. Due to the noise, these densities exist and are smooth, i.e., fractal measures do not occur. The equation of motion for the phase space density $\rho(\mathbf{x}, t)$ is called the Fokker–Planck equation:

$$\frac{\partial \rho(\mathbf{x}, t)}{\partial t} = -\sum_i \frac{\partial}{\partial x_i} D_i^{(1)}(\mathbf{x})\rho(\mathbf{x}, t) + \sum_{i,j} \frac{\partial}{\partial x_i} \frac{\partial}{\partial x_j} D_{ij}^{(2)}(\mathbf{x})\rho(\mathbf{x}, t). \qquad (12.17)$$

The transition from the Langevin equation to the Fokker–Planck equation and back is uniquely given by the way the drift term $D_i^{(1)}(\mathbf{x})$ and the diffusion tensor $D_{ij}^{(2)}(\mathbf{x})$ correspond to the deterministic and stochastic part of the Langevin equation:[6]

$$D_i^{(1)}(\mathbf{x}) = f_i(\mathbf{x}) + \sum_k G_{kj}(\mathbf{x})\frac{\partial}{\partial x_k} G_{ij}(\mathbf{x}), \qquad D_{ij}^{(2)}(\mathbf{x}) = \sum_k G_{ik}(\mathbf{x})G_{jk}(\mathbf{x}).$$

$$(12.18)$$

If we can assume that a description by Eq. (12.16) or Eq. (12.17) is suitable for a given data set, then one has two choices: either one tries to fit the Langevin equation to the observed data, or one tries to fit the Fokker–Planck equation. Some deeper thinking reveals that the discrete sampling of data leads to systematic errors when directly fitting Eq. (12.16). The vector field that one finds with the conditional average $\hat{\mathbf{x}} = \frac{1}{2\Delta}(\mathbf{x}(t + \Delta) - \mathbf{x}(t - \Delta))|_{x_t = \mathbf{x}}$ is not, as expected, $\mathbf{f}(\mathbf{x})$, although an ensemble average of Eq. (12.16) for fixed $\mathbf{x}$ yields $\langle \dot{\mathbf{x}} \rangle = \mathbf{f}(\mathbf{x})$ [Just *et al.* (2003)].[7] Moreover, such an approach cannot easily extract the potentially state dependent noise amplitude $G(\mathbf{x})$.

Instead, as was shown in the pioneering work of Friedrich & Peinke (1997), one can directly exploit the time series data to find estimates of the drift- and the diffusion terms of the Fokker–Planck equation. Under the assumption of time independence of the parameters, drift and diffusion can be determined by the following conditional averages:

$$D_i^{(1)}(\mathbf{x}) = \lim_{\Delta \to 0} \frac{1}{\Delta} \langle x_i(t + \Delta) - x_i(t) \rangle_{\mathbf{x}(t) = \mathbf{x}} \qquad (12.19)$$

and

$$D_{ij}^{(2)}(\mathbf{x}) = \lim_{\Delta \to 0} \frac{1}{2\Delta} \langle (x_i(t + \Delta) - x_i(t))(x_j(t + \Delta) - x_j(t)) \rangle_{\mathbf{x}(t) = \mathbf{x}}. \qquad (12.20)$$

In practice, since $\mathbf{x}(t)$ equals $\mathbf{x}$ only with zero probability, one will introduce some tolerance, i.e., exploit the averages on a neighbourhood of $\mathbf{x}$ with a suitable neighbourhood diameter ϵ. Also, the time interval Δ is limited from below by the sampling interval of the data. Hence, finite sample corrections can be necessary, in particular for the diffusion term. As derived in Ragwitz & Kantz (2001) a useful

---

[6] We use here the Stratonovich interpretation of stochastic integrals.
[7] If the tensor G is constant, Eq. (12.19) of cause yields f_i.

(albeit incomplete) first order correction in Δ is given by the following estimate of the diffusion term:

$$D_{ij}^{(2)} \approx \frac{\langle (x_i(t+\Delta) - x_i(t))(x_j(t+\Delta) - x_j(t)) \rangle_{\mathbf{x}(t)=\mathbf{x}} - \Delta^2 D_i^{(1)}(\mathbf{x}) D_j^{(1)}(\mathbf{x})}{2\Delta \left(1 + \Delta \frac{\partial D_i^{(1)}(\mathbf{x})}{\partial x_j}\right)}$$

$$(12.21)$$

In these expressions, the knowledge of Δ in physical units is not required for consistent estimates, both $D^{(2)}$ and $D^{(1)}$ can be measured in arbitrary temporal units.

A very convincing application of this method is the analysis of highway traffic flow in terms of the two variables density of cars $q(t)$ and velocity $v(t)$ measured at a certain point on a highway. Kriso *et al.* (2002) identify the drift and the diffusion terms in a two dimensional Langevin equation of the variables q and v. The analysis reveals fixed points and metastable states of the deterministic part of the motion, hence yielding the dynamical understanding of what car drivers like and what they hate, namely free flow and congestion.

The chief problem of this approach is that it only works if the full state vectors $\mathbf{x}$ can be measured. Anything like a time-continuous version of a stochastic embedding theorem (if it existed) could only lead to a differential equation of higher temporal order and hence would lead outside the realm of the Fokker–Planck equation, which is almost unexplored. Therefore, despite the elegance and theoretical appeal of this concept, we discuss in the following subsection an alternative concept which is more *ad hoc* but potentially of larger practical relevance.

12.7.2 *Markov chains in embedding space*

Both autoregressive models and deterministic dynamical systems can be regarded as special classes of *Markov models*, which have been mentioned already in Chapter 3. Markov models rely on the notion of a state space. As with a deterministic system, the current state uniquely determines the future evolution. But rather than providing a unique future state, here only a probability density for the occurrence of the future states is obtained. In continuous time, a Langevin equation can be shown to generate a Markov model. In discrete time and space, a Markov model is described by a simple transition matrix. A univariate Markov model in continuous space but discrete time is often called a *Markov chain*, where the order of the Markov chain denotes, how many past time steps are needed in order to define the current state vector.

In the framework of nonlinear deterministic systems, time discrete Markov models with a discrete state space occur naturally when one studies symbolic dynamics. The big issue there is typically to find a partitioning of the state space (which establishes the correspondence between symbols and real valued state vectors) which creates a symbolic dynamics with finite memory. The probabilistic nature of the dynamics is here introduced through a coarse graining, by which precise knowledge of the present state is lost. This is typically not our concern in time series analysis, so that we will not discuss this model class any farther.

We concentrate here on scalar real valued Markov chains of order l. Such a model produces a sequence of random variables x_k. Its dynamics is fully characterised by all transition probability densities from the last l random variables onto the next one, $p(x_{n+1}|x_n, \ldots, x_{n-l+1})$ with $\int dx_{n+1} p(x_{n+1}|x_n, \ldots, x_{n-l+1}) = 1$ and $p(.|.) \geq 0$. As we will show below, the natural way to analyse data produced by such a model will be the time delay embedding space of dimension $m = l$. However, before we can discuss this, we have to think about how natural it is to expect a time series to represent a Markov chain.

Example 12.9 (AR(1) model as a Markov chain). The AR(1) model $x_{n+1} = ax_n + \xi_n$ with Gaussian white noise ξ_n of variance σ^2 creates a first order Markov chain. The transition probabilities are easily derived through their definition $p(x_{n+1}|x_n) = p(x_{n+1}, x_n)/p(x_n)$, where $p(x_{n+1}, x_n)$ is obtained as $p(x_{n+1}, x_n) = \int d\xi \, p(x_n) \mathcal{N}(\xi, \sigma) \delta(x_{n+1} - ax_n - \xi)$. Evidently, $p(x_{n+1}|x_n) = \mathcal{N}(x_{n+1} - ax_n, \sigma)$, where we denote here by $\mathcal{N}(., \sigma)$ the Gaussian distribution with variance σ and zero mean. Values of x farther in the past have no influence on the probability of the values for x_{n+1}.

12.7.3 No embedding theorem for Markov chains

Now let us assume again that a reasonable description of a noisy dynamical system is a vector valued Langevin equation Eq. (12.16). However, as usual, we assume that the measurement apparatus provides us with a time series of a single observable only. It is well known that the Langevin dynamics generates a Markov process in continuous time in the original state space (see e.g. Risken (1989)). It is hence tempting to believe that our scalar, real valued time series values represent a Markov chain of some finite order. If something like an embedding theorem existed, we would expect that the order of the Markov chain were related to the dimensionality of the original state space.

Unfortunately, this conjecture is wrong! Generally, a scalar time series obtained from a time continuous vector valued Markov process is not Markovian, i.e., it does not have a finite memory (see, e.g., van Kampen (1992)). Nonetheless, in most cases, the memory is decaying fast, so that a finite order Markov chain may be a reasonable approximation, where the errors made by ignoring some temporal correlations with the past are then smaller than other modelling errors resulting from the finiteness of the data base.

In the following we will show how the Markov chain model of some underlying Langevin dynamics can be used for predictions and for modelling. The performance of our predictor will give some hints on the validity of the Markov approximation, and we will recall that Kolmogorov–Sinai-type entropies will yield additional insight.

12.7.4 Predictions for Markov chain data

In the following paragraphs we will assume that our time series $\{s_n\}$, $n = 1, \ldots, N$, represents a Markov chain of order l. As mentioned before, the dynamics of a Markov chain of order l is fully determined by the set of all transition probabilities $p(s_{k+1}|s_k, \ldots, s_{k-l+1})$, where stationarity ensures the invariance of these transition probabilities under time shift. Delay vectors of dimension $m = l$ and unit time lag evidently form the states onto which these transition probabilities are conditioned. Under the assumption of their smoothness in this argument we can now estimate these distributions from the time series data. Let us consider a small neighbourhood $\mathcal{U}_\epsilon(\mathbf{s}_n)$ of the actual delay vector $\mathbf{s}_n = (s_n, s_{n-1}, \ldots, s_{n-l+1})$. Because of the smoothness of the transition probability, the future observations s_{k+1} of all delay vectors $\mathbf{s}_k \in \mathcal{U}_\epsilon(\mathbf{s}_n)$ are subject to (almost) the same transition probability and hence form a finite sample of $p(s_{n+1}|\mathbf{s}_n)$.

We recall here that the best predictor in the root mean squared sense is the mean value of a distribution, if nothing more than the marginal distribution of the data is known. Hence, the best prediction for the Markov chain is

$$\hat{s}_{n+1} = \int s' p(s'|\mathbf{s}_n)\,ds' \approx \frac{1}{\|\mathcal{U}_\epsilon(\mathbf{s}_n)\|} \sum_{k:\mathbf{s}_k \in \mathcal{U}_\epsilon(\mathbf{s}_n)} s_{k+1}\,. \tag{12.22}$$

This is exactly the zeroth order predictor Eq. (4.7), which, as we argued in Section 12.3.2, is the lowest order local approximation of a globally nonlinear model in the deterministic case. In the deterministic case, the transition probability is a δ-distribution, $p(s_{n+1}|\mathbf{s}_n) = \delta(s_{n+1} - F(\mathbf{s}_n))$, if $F(\mathbf{s})$ is the unique deterministic mapping in the delay embedding space. For the deterministic case, it was sometimes reasonable to pass over to higher order approximations, whereas in the stochastic Markov case, the "true" $p(s_{n+1}|\mathbf{s}_n)$ might be a broad distribution, such that its $\mathbf{s}_n$-dependence cannot easily be approximated beyond the assumption of constancy on some neighbourhood.

12.7.5 Modelling Markov chain data

The fact that modelling and prediction are sometimes different tasks becomes most evident for stochastic systems. Whereas the goal of prediction is well defined as soon as the cost function to be minimised is given, modelling is a vague concept. At least, a good model should be able to generate typical realisations of the observed process.

Example 12.10 (Modelling and prediction for an AR(1) process). Given an AR(1) process $x_{n+1} = ax_n + \xi_n$, the best prediction following a current observation s_n is $\hat{s}_{n+1} = as_n$, where, in a real time series task, a had to be extracted from the data. Iterating this prediction yields $as_n, a^2 s_n, \ldots, a^k s_n$, with $\hat{s}_{n+k} \to 0$ for large k. In fact, since the correlations between the actual observation and future values decay exponentially in time, the best prediction far into the future would be the mean value of the distribution, which is 0. However, such an exponentially relaxing segment of a time series is an extremely improbable and atypical realisation of the AR process, so that iterated predictions in this case will not be useful for modelling the data.   □

If data are taken from a stochastic system, a reasonable model must be able to produce fluctuations. Therefore, it is natural to include randomness into the model. In fact, if we are given the transition probabilities of a Markov chain, the way to produce a sample path is to successively draw values at random, according to $p(s_{n+1}|s_n)$, where the actual s_n is constructed from the l last random numbers produced. Exactly the same can be done if the transition probabilities are extracted from the time series: instead of using the mean $\hat{s}_{n+1} = \frac{1}{||\mathcal{U}_\epsilon(s_n)||} \sum_{k:s_k \in \mathcal{U}_\epsilon(s_n)} s_{k+1}$ as the next value of our model trajectory, we just choose randomly one of the s_{k+1} from the neighbourhood. The only drawback is that the model trajectory we thus generate contains only values which have been inside the "training set", but apart from that it is a new realisation of the stochastic process with the correct temporal correlations and the correct distributions. Due to the randomness it will also not become periodic. Such a scheme has been called *local random analogue prediction* in Paparella *et al.* (1997), although in our interpretation the term *modelling* would be more appropriate.

Example 12.11 (Modelling an AR(1) process from time series data). Assume that we have a time series representing an AR(1) process like in Example 12.10. As argued in Example 12.9, the transition probability from s_n to s_{n+1} is a Gaussian $\propto$ $\exp((s_{n+1} - as_n)^2/2\sigma^2)$. The future observations s_{k+1} of s_k inside a neighbourhood of the current state s_n (one dimensional embedding space since the order of the Markov chain is unity) form a finite random sample according to this distribution. Hence, picking one of them at random is equivalent to producing a "new" random number according to the Gaussian, with the advantage of being model free in the sense that we neither need to know that the transition probability is a Gaussian nor do we have to fit any of its parameters. □

12.7.6 Choosing embedding parameters for Markov chains

Until now we showed that the concept of a time delay embedding space and of neighbourhoods in this space is the natural framework for data analysis of Markov chains of known order l. We also mentioned already that a typical experimental data set (one observable, discrete time) of Langevin dynamics does not represent a Markov chain. Nonetheless, due to a fast decay of temporal correlations, in most cases one can expect that the approximation by a finite order Markov chain will capture most aspects of the dynamics. Here we will discuss the choice of the order l (i.e., of the embedding dimension m), and of the time lag τ. In particular, it is reasonable to expect a smooth convergence of the optimal embedding parameters to those values which are suitable for a purely deterministic system, when we reduce the noise amplitude in a model such as Eq. (12.16) to zero.

In order to determine the best value for m, in the deterministic case one often uses the method of false nearest neighbours discussed in Section 3.3.1. However, it will not work in the stochastic case, since stochasticity leads to false neighbours no matter how large one chooses m. The most simple and direct way to find an approximate embedding is to compute the forecast error with the predictor Eq. (12.22) as a function of the embedding dimension and to select the model order which minimises this error. This does not yield an estimate of the order of a reasonable approximation of the underlying process itself, but instead takes into account the finite sample effects due to the finiteness of the time series. Let us recall that the number of data points inside an ϵ-neighbourhood of some delay vector decreases dramatically with increasing embedding dimension, such that the finite sample thus obtained may represent the "true" transition probability density in an extremely poor way. (There might be just a single neighbour or none at all.) Increasing the neighbourhood diameter increases the number of neighbours found, but comes into conflict with the locality requirement expressed by Eq. (12.22). Hence, even if a 10 dimensional embedding represented the temporal correlations in a wonderful way, the finite sample effects could render the predictions in this space almost worthless. When we search for the minimal prediction error, we identify exactly this tradeoff between taking into account temporal correlations and still working with sound statistical averages in terms of sample size.

A theoretically more appealing concept is the computation of dynamical entropies. As we discussed in Section 11.4.4, the numerical estimates of the ϵ-entropy of a stochastic process for various embedding dimensions, $h(m, \epsilon)$, asymptotically follow the law $h(m, \epsilon) = -\ln \epsilon + h^{(c)}(m)$, where the constant $h^{(c)}(m)$ is the continuous entropy in m dimensions. If the data represent a Markov chain of order l, then all $h^{(c)}(m) = h^{(c)}(l)$ for $m > l$. Hence, the length of the memory can be detected by an entropy analysis (see also Example 11.12). If data are not exactly Markovian, then still the $h^{(c)}(m)$ typically converge to a finite value, so that one can determine a reasonable value l for a Markovian approximation.

The optimal time lag is another issue, which, as in the deterministic case, does not have any unique formal solution. As investigated by Ragwitz & Kantz (2002), the optimal time lag will be closer to unity when the the relative contribution of the stochastic driving to the total power of the signal is increased. The explanation is intricate and somewhat against intuition. The interested reader is referred to the article by Ragwitz & Kantz (2002). Unfortunately, the optimal order of the approximate Markov chain may depend on the time lag chosen. Hence, in the end one has to optimise the two parameters m and τ simultaneously.

These considerations show that a time delay embedding space and information extracted from local neighbourhoods in such spaces can yield reasonable results even if data are stochastic.

12.7.7 *Application: prediction of surface wind velocities*

Let us illustrate the possibilities of Markov modelling by a quite ambitious goal: the prediction of wind speeds. It will turn out that the answer to the time series issue related to these data requires a different evaluation of the transition probabilities than just the computation of their means.

Wind speeds measured at the surface of the earth represent an immense reduction of information about the turbulent wind field. Since turbulent gusts are a major danger for the safe operation of wind energy turbines, their prediction is of high technological and economical relevance. A predicted gust could be made innocent by a simple adjustment of the pitch angle of the rotor blades, which reduces their mechanical load from the wind field. For this purpose, a prediction horizon of up to a few seconds is sufficient. Evidently, the longer into the future predictions are to be made, the less accurate they are.

Several benchmarks for predictions exist in this case. Persistence means that on average, the future wind speed can be supposed to be identical to the actual one. Using the actual value as prediction for the future supplies a trivial but rather good predictor for the absolute speed, but it is evidently worthless for the prediction of gusts, which are characterised by a sudden increase of the wind speed. The trend extrapolation from the last two measurements $\hat{s}_{n+1} = 2s_n - s_{n-1}$ yields good short time predictions for smooth data but is worse than persistence for our turbulent data. Beyond that, linear auto-regressive models can exploit linear correlations in the data. The quality of the predictions by the different prediction schemes is compared by their root mean squared prediction errors Eq. (12.6) as a function of the prediction horizon.

We use data sets which stem from the research centre in Lammefjord in Denmark. The absolute wind speeds measured by cup anemometers are recorded with a sampling rate of 8 Hz. For details see Appendix B.13. The comparison of prediction errors shows an about 10% superiority of the locally constant scheme Eq. (12.22) on prediction horizons larger than half a second. This alone would not be an astonishing result.

Since within our treatment turbulent gusts appear randomly according to a rather broad distribution, where our prediction represents its mean value, it is not surprising that the predicted wind speed typically does not deviate dramatically from the present one. A much more appropriate procedure is to predict how probable it is on average that during the next Δn samples the wind speed increases by more than g m/s with respect to its present value. This probability can be computed from the conditional probability by

$$prob(\text{gust} > g) = \frac{1}{\Delta n} \sum_{k=1}^{\Delta n} \int_{s_n+g}^{\infty} ds_{n+k} \, p(s_{n+k}|\vec{s}_n) \,, \tag{12.23}$$

where the conditional probability density is again obtained from the data as described above. Since by construction for every prediction we have a single realisation only, one might ask how to check the correctness of the predicted probabilities. This is done by the *reliability plot* known from weather forecasting (see also Example 12.13). One gathers all those events among the predictions where the predicted probability *prob*(gust) in Eq. (12.23) is inside $[r, r + dr]$ and counts in how many of these cases actually a gust followed. If our predicted probabilities were perfect and the sample were large enough, we would expect that a fraction of exactly r of these events were gust events. As we show in Fig. 12.6, we are pretty close to such a situation, where, however, the large predicted probabilities occur too rarely for a meaningful statistical test. Finally, we show in Fig. 12.7 two transition probabilities, where in one situation a gust to follow is very likely, in the other very unlikely. The actual observation s_n has almost the same value close to 9 m/s in both cases, but the $m - 1$ past values are very different, so that the conditioning onto a delay vector (i.e., assuming a Markov chain of order $m > 1$) is essential for a meaningful prediction.

A more general investigation of two and three dimensional turbulent data are described in detail in Ragwitz & Kantz (2000). A few words on the supposed origin of predictability in a turbulent wind field are in order. Our measurements

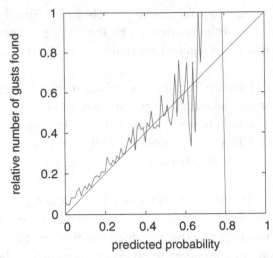

Figure 12.6 The reliability plot for 20 000 predictions during 24 h of Lammefjord data "tag191.dat" containing wind speeds between 2.8 and 18.8 m/s. The curve is truncated where for large predicted probabilities no events occurred. If the predictions were perfect, the empirical line should coincide with the diagonal (8 Hz data, prediction 1.5 s ahead, $m = 6$ with a time lag 2).

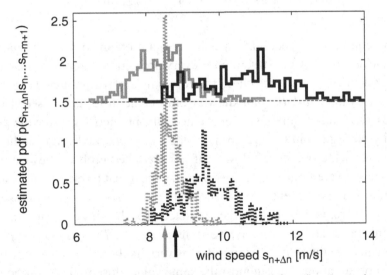

Figure 12.7 Estimated conditional probability density functions: grey in a situation where $prob(\text{gust} > 1) = 0.09$, black for $prob(\text{gust} > 1) = 0.7$. The arrows indicate the wind speeds s_n at the moment when the prediction was performed. The lower histograms show the pdfs after 3/8 s ($\Delta n = 3$), the ones shifted by 1.5 are for 1.5 s ahead ($\Delta n = 12$). Evidently, the conditioning on the $m - 1$ past components which are not shown introduces the very difference of the two situations (data and parameters as in Fig. 12.6).

are local in space. For nonzero mean wind speeds, the wind field drifts across our measurement locations. Hence, what we want to predict is not a result of the local dynamics, but it is spatially remote in the moment when the prediction is to be made. Hence, predictability can only be possible due to spatio-temporal correlations. Such correlations are known as eddies or coherent structures in turbulence research. It seems that a typical source of a sudden increase of the wind speed is the backward border of a coherent structure drifting across the measurement device, and that the approach of this border can be partly predicted from the data.

12.8 Predicting prediction errors

In the last two sections of this chapter we will address general issues of forecasting which are not directly related to a specific prediction method. In particular, it will be irrelevant whether we interpret our data as a deterministic or a stochastic process.

A typical experience from weather forecasting which everyone of us has made is that there are weather situations where we can do a very precise prediction without any work (e.g., if there is a stable anti-cyclone), and that there are other weather conditions where even the most sophisticated weather models produce forecasts which fail. Hence, it is very reasonable to ask whether we can predict, how precise our actual prediction really is.

12.8.1 Predictability map

Following the idea of the state space, we may assume that the predictability should depend on the state of the system and be similar throughout small neighbourhoods. If so, then all predictions starting from the same small volume in state space can be assumed to be equally accurate or inaccurate. Hence, the most straightforward way of predicting the prediction error is again to learn from the past. With the chosen prediction scheme (e.g., the locally constant predictor), one performs predictions on the given data (training set). From these predictions, one can compute all individual prediction errors. The prediction errors can be accumulated in a histogram on the state space, which can either be represented graphically (predictability map) or can serve as a look-up table: the expected error of a new prediction on the test set is given by the mean of the errors made on the training set inside a small volume in state space around the actual delay vector. As an equation, this reads

$$\hat{e}_{n+1} = \frac{1}{||\mathcal{U}_\epsilon(\mathbf{s}_n)||} \sum_{k:s_k \in \mathcal{U}_\epsilon(\mathbf{s}_n)} e_{k+1} \,. \tag{12.24}$$

Here $e_{k+1} = |\hat{s}_{k+1} - s_{k+1}|$ is the error of the prediction $\hat{s}_{k+1}$ made for s_{k+1} with the chosen prediction scheme, and $\mathcal{U}_\epsilon(\mathbf{s}_n)$ is the ϵ-neighbourhood of $\mathbf{s}_n$ as usual.

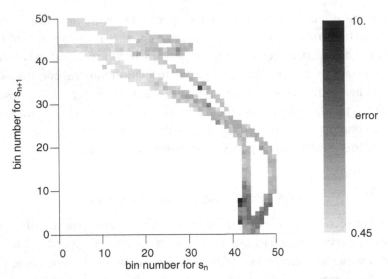

Figure 12.8 The mean prediction errors for predictions starting from the corresponding bin in the histogram of the NMR laser data are encoded in grey-scale. Evidently, predictability is worse on the lower right part of the attractor and best in the upper left.

Example 12.12 (Predictability map of NMR laser data). We cover a two dimensional delay embedding of the NMR laser data by a grid of boxes and accumulate the individual prediction errors (obtained from predictions in a three dimensional embedding) in a histogram built on these boxes. The mean value inside each box gives an estimate for the prediction errors, as shown in Fig. 12.8. Since these data represent an almost deterministic chaotic system, the mean prediction errors are closely related to the magnitude of local divergence rates found in a linear stability analysis. □

12.8.2 Individual error prediction

If the dimension of the phase space is too large to build a histogram on it, or if the length of the time series is insufficient for this task, we might try to predict each prediction error individually. Before we can propose a method for the prediction of prediction errors, we have to define more properly what we intend to predict. A prediction error is a random variable with some distribution. Once the next observation is made, the current realisation of this random variable is given. But a prediction of the prediction error can only be a prediction of the mean of this distribution, from which the actual realisation will be drawn. On the other hand, a single observation is unsuitable to verify or falsify a statement about a distribution.

Hence, the correct way to compare predictions and realisations of objects wich occur only once is what in weather forecasting is called the reliability function: assume that we have an algorithm which tells us the mean value $\hat{e}_n$ of the distribution from which the prediction error e_n will be drawn. Then gather all errors e_k from a long sequence of predictions for which the predicted mean $\hat{e}_k$ was inside the interval $[\hat{e}_n - \Delta, \hat{e}_n + \Delta]$ with some small Δ. If all these e_k really were produced by distributions whose means were $\hat{e}_k$, then this empirical sample should itself represent a distribution with mean $\hat{e}_n$.

Example 12.13 (Prediction of probability for rain). Let us assume that the weather forecast predicts a probability of $r\%$ that it will rain tomorrow. The next day one can observe whether or not it does rain. How can we check whether the predictions are good? The simplest is to collect all days for which the prediction of rain was $r\%$, and count whether in fact $r\%$ of them were rainy. More sophisticated tests are possible. □

Now, how can we actually predict the prediction error? In (almost) deterministic systems, prediction errors are essentially caused by errors in the estimation of the initial condition (measurement noise) and by modelling errors (too rough approximation of the local dynamics by our model equations), amplified by the exponential divergence due to chaos. In stochastically driven systems, additional unpredictability enters through the unknown stochastic inputs coupled into the dynamics. In both cases, an additional inaccuracy comes from the fact that also the observation to be predicted is subject to measurement noise. Our claim is that all these inaccuracies are, to a good amount, contained in the spread of the images of the neighbours of the current delay vectors. Without noise and without chaos, the images of the neighbours are as close as the neighbours themselves. With either chaos or noise, the images of the neighbours are on average farther apart than the neighbours. Hence, if noise amplitudes or exponential stretching rates are state-dependent, then the distance of the images of the neighbours can vary, and they can be used as a measure for the prediction error to be expected. One could argue that the ratio of the mean spacing of the images and the mean spacing of the neighbours is the proper estimate of how much the local instability deteriorates the predictions. Since this ratio defines an error amplification factor, an estimate of the expected error can only be derived from it if one has an additional estimate of the error in the initial condition. Also the neighbourhood diameter contributes to the error, since in regions of the phase space where the density of points is low, this diameter is large (in order to guarantee a minimal number of neighbouring points) and hence modelling errors are large. On the other hand, a strongly stochastically driven system, the spread of the images is even roughly independent of the size of

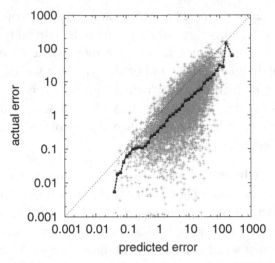

Figure 12.9 Predicted versus actual forecast errors of the NMR laser data (grey) and the reliability curve (black boxes), i.e., the mean value of the actual errors conditioned on values of the predicted error. A perfect error prediction would yield the diagonal.

the initial neighbourhood. Therefore, in practice the following quantity turns out to be a good prediction $\hat{e}_{n+1}$ of the prediction error $e_{n+1} = |s_{n+1} - \hat{s}_{n+1}|$:

$$\hat{e}_{n+1} = \sqrt{\frac{1}{\|\mathcal{U}_\epsilon(\mathbf{s}_n)\|} \sum_{k:\mathbf{s}_k \in \mathcal{U}_\epsilon(\mathbf{s}_n)} (s_{k+1} - \hat{s}_{n+1})^2} \,, \qquad (12.25)$$

where $\hat{s}_{n+1}$ is the predicted time series value, which in the case of the simple predictor from Eq. (4.7) is also the mean of s_{k+1} in the neighbourhood.

Example 12.14 (Local prediction errors in NMR laser data). We make forecasts using the TISEAN routine `zeroth` (which is the implementation of the simple predictor Eq. (4.7)), compute the individual forecast errors e_n, and the predicted errors $\hat{e}_n$ according to Eq. (12.25) for the Poincaré map data of the NMR laser (data set Appendix B.2). Predicted error and true error are plotted versus each other in Fig. 12.9 as light grey symbols. We introduce a binning in the predicted error (on the log-scale) and compute the mean of the actual errors for each bin (reliability plot, see Example 12.13 and before), i.e., we perform a conditional averaging of the grey symbols on vertical lines. These averages are plotted as black symbols. Evidently, albeit the correlation between the predicted errors and the actual errors is far from perfect, they form a sample whose mean value is close to the predicted error. The error prediction scheme over-estimates the errors by about a factor of 3 in this example. □

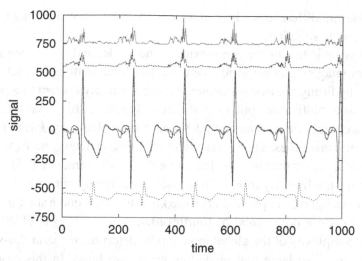

Figure 12.10 Predictions and prediction errors of human ECG data. From top to bottom: actual error, predicted error, predicted and actual signal, signal from which the prediction starts (the latter one rescaled, all others only with different off-sets).

Example 12.15 (Predicting errors in ECG prediction). We try to predict a human ECG about a quarter of a second ahead, which is about one quarter of the single heart beat wave form. It turns out that the wave form of the ECG is pretty much reproducible, so that low prediction errors occur. The main instability in an ECG recording is given by fluctuations of the onset of a new heart beat. Hence, the P-wave and the QRS complex are hard to predict correctly. Exactly this is visible in Fig. 12.10, where the top trace shows these errors. The second trace shows the predicted errors according to Eq. (12.25). For comparison, we show also the true and the predicted signal, and at the bottom (with reduced amplitude) the signal from which the prediction was started: naturally, it is lagging behind the predicted signal. Evidently, predicted errors are a good prediction of the actual errors. (Routine `zeroth`, $m = 8$, $\tau = 5$, at least 10 neighbours required, sampling rate 250 Hz.) □

12.9 Multi-step predictions versus iterated one-step predictions

If the temporal horizon for which one wants to predict the observable is larger than unity (in terms of the sampling interval), then in principle two alternatives exist: predicting the full horizon, Δn, into the future in a single step, or iterating a one-unit-ahead prediction Δn times. Both methods cannot avoid the exponential divergence between true future and predicted future, if a system is chaotic, but there are relevant differences which make the one method or the other favourable, depending on the situation.

The first issue is that of the complexity of the model and the related modelling error. Let us imagine that we were able to find a global nonlinear function $s_{n+1} = F(\mathbf{s}_n)$, e.g., by fitting the free parameters in some suitably chosen functional form for F such as a multivariate polynomial. Let us further assume that for reasons of statistical accuracy, we are unable to determine more than, say, k free parameters with our finite time series. Then it is evident that we can determine $F(\mathbf{s})$ itself with much higher accuracy than we can determine the Δn-th iterate of F for the simple reason that the nonlinearity in $F^{(\Delta n)}$ increases dramatically with Δn. For instance, in a one dimensional example, if $F(s)$ were known to be a quadratic function, then $F^{(2)}(s)$ is already a polynomial of fourth order, and $F^{(\Delta n)}(s)$ is of $2^{\Delta n}$th order! Hence, the complexity of the global model to be determined for an Δn-step ahead prediction can be so large that modelling errors are huge. In this case, iterated one-step predictions are preferable.

If we restrict ourselves to the locally constant model, the situation is very different: if no dynamical noise is present, we possess very accurate values for $F^{(\Delta n)}(\mathbf{s}_k)$ from the time series. So one can assume that the mean of the neighbour images $s_{k+\Delta n}$ is a very good prediction of $s_{n+\Delta n}$ if the initial neighbours are close. Nonetheless, if we start from a neighbourhood of diameter ϵ, the diameter of the image of this neighbourhood reaches unity after $\Delta n_{\max} : \epsilon e^{\lambda \Delta n_{\max}} = 1$ time steps, and this $\Delta n_{\max}$ is the prediction horizon, after which the predictions saturate and become constant at the mean value.[8] One might again think that iterated one-step predictions are better, since this limitation through the neighbourhood size does not exist. However, iterated one step errors suffer from an accumulation and enhancement of not only the uncertainty in the initial condition, but also from each small prediction error after every step of the iteration. This additional inaccuracy does not exist in the multi-step prediction, so that we have a trade-off between two effects. Moreover, for large noise levels, models suffer from misestimation, which also will be iterated. Numerical experience shows that all effects are essentially of comparable magnitude, hence one has to test both methods for the given data set.

Iterated predictions differ in one essential property from multi-step predictions: the latter become constant after a sufficiently large time horizon, whereas the first often continue to show a dynamics typical of the time series. So, iterated errors perform a smooth transition from prediction to modelling: after a time interval identical to the maximal prediction horizon of the multi-step error, the iterated

---

[8] If a chaotic attractor is split into subsets such as in the case of two band chaos in the logistic equation, or if a system is periodically driven, this statement has to be modified in an evident way.

prediction and the true observation lose their correlation completely, although the iterated prediction can in favourable situations be used to produce a new time series which has all the ergodic properties of the original one.

Due to these differences, the limit of predictability is very nicely visible for the multi-step error even without computing the prediction error, simply through the constancy of the prediction. Moreover, the mean value is the best prediction when correlations are lacking, whereas the false time dependence of the iterated prediction many steps ahead produces an average prediction error of $\sqrt{2}$ the size of the multi-step error.

As a final remark we should say that in some data sets we found that the error of a Δn-step ahead prediction can be reduced by increasing the time lag τ of the embedding space with respect to what is optimal for a one-step error.

In summary, iterated predictions are essentially superior when we use more sophisticated modelling techniques, which have a limitation in the complexity of the dynamics to be modelled. In all other cases, both prediction schemes have comparable prediction errors but very different dynamical properties for Δn beyond the horizon of predictability.

Further reading

With the material presented in this chapter you should be able to obtain useful results in modelling and forecasting. However, we had to omit many technical aspects which are as diverse as the possible time series you might want to predict. A good impression of the different facets and many good ideas can be found in the proceedings volumes by Casdagli & Eubank (1992) or Weigend & Gershenfeld (1993). Some classical references are Farmer & Sidorowich (1987), Crutchfield & McNamara (1987), Casdagli (1989), and Smith (1992). Some relevant new modelling techniques have been discussed in the recent papers by Voss *et al.* (1999) and by Timmer *et al.* (2000).

Exercises

12.1 In Section 12.5 we claimed that the exact deterministic dynamics is not the best one-step predictor in the presence of noise. Convince yourself that this is true even for a simple linear relation. Assume a large set of "noisy" pairs (X_i, Y_i) formed according to the following rule: $X_i = x_i + \xi_i$, $Y_i = ax_i + \eta_i$, where for simplicity the x_i have zero mean and variance σ_x but are arbitrary otherwise. The noises ξ_i and η_i are uncorrelated Gaussians with zero mean and variance σ. Without noise we would have $Y_i - aX_i = 0$. Show that the minimisation of $e^2 = \langle (Y_i - \hat{a}X_i)^2 \rangle$ leads to $\hat{a} \neq a$ and that the resulting minimal error is smaller than the one when you insert a for $\hat{a}$.

12.2   Write a program to perform local linear fits by minimisation of the one-step prediction error (for hints see Section 2.5). Apply it to data from the Hénon map. Since you know the dynamics underlying the data, you can compute the exact linearised map at every point. Plot all points in the two dimensional delay embedding space where the fitted map and the exact map deviate considerably. Convince yourself that these are points on the attractor where the size of the neighbourhood is large compared to the inverse curvature (i.e. the approximate local radius of the attractor).

12.3   Use the TISEAN routines `zeroth` and `nstep` to produce multi-step and iterated one-step errors for some chaotic data set, e.g., from the Hénon map.

12.4   Plot a time series from the Hénon map in the following three dimensional representation: $(s_n, s_{n-1}, s_{n+\Delta n})$ for different $\Delta n = 2, 3, \ldots$. Convince yourself that for $\Delta n = 2$ the data would nicely fit in a parabolic surface, and that for larger Δn the attractor gets more and more wrinkled, such that a global model would be much more complex.

Chapter 13

Non-stationary signals

Like in almost every other work on time series, throughout this book stationarity must be assumed almost everywhere. However, in reality non-stationarity is the rule and stationarity the exception, for two reasons. First, virtually every process we find in nature (i.e., outside the numerical universe on a digital computer) is not stationary, since its parameters depend on time, even if these dependences may be very weak and even if a skillful experimentalist has tried his best to exclude them. Second, since the opposite, stationarity, cannot be constructively proven for a given finite time series, the natural assumption would be that a given sequence of data is not stationary.

Unfortunately, statistical methods typically require stationarity. In the theory of statistics, expectation values are defined through ensemble averages. For stationary processes, the ergodic theorem enables one to replace an ensemble average by a time average. Exactly this is typically done in time series analysis, where most often we have a single time series only and not an ensemble of time series. Hence, the estimators of statistical entities such as the mean value are evaluated on windows in time. If the underlying process is non-stationary, two time scales compete with each other and time averages lose their precise meaning.

Example 13.1 (Estimated mean of a time series from a process with a drift). Let us consider the non-stationary process of univariate random numbers with the probability density $p(x_n, n) = \frac{1}{\sqrt{2\pi}} \exp\{-(x_n - e(n))^2/2\}$ with $e(n) = \sqrt{n}$. Our process thus is a sequence of independent random numbers from a Gaussian distribution with unit width and a time dependent centre $e(n)$. If we had N different realisations of the process, $X^{(k)} = \{x_1^{(k)}, x_2^{(k)}, \ldots, x_n^{(k)}, \ldots\}$, $k = 1, \ldots, N$, we could define and estimate a time dependent mean by averaging over the different realisations at equal time, $\hat{e}(n) = \frac{1}{N} \sum_{k=1}^{N} x_n^{(k)}$ (ensemble average). In the limit $N \to \infty$ this converges evidently to $e(n)$, and is also unbiased.

If a single realisation only is given, we can average over adjacent observations, $\tilde{e}_l(n) = \frac{1}{2l+1} \sum_{k=n-l}^{n+l} x_k$ with a suitable choice of l. For $e(n) = const.$ this would be again a valid estimate of the mean value. For the n-dependent process, due to the assumed square root dependence, $\tilde{e}_l(n)$ can resolve the value $e(n) = \sqrt{n}$ only if l is very small. If l is a small number, then the statistical fluctuations of $\tilde{e}_l(n)$ are huge and might be so huge that the systematic n-dependence is hardly visible. More discouragingly, the only unbiased estimate of $e(n)$ is x_n itself, which is a worthless estimate, and every quantity with smaller variance will introduce a bias. □

Example 13.2 (Multi-trial experiments in neurology and NMR spectroscopy).
In many sciences, the response of a system to an external stimulus is studied. The recorded time series then are explicitly non-stationary: before the stimulus, a kind of rest state is observed, after the stimulus some excitation and a return to rest will be visible.

In neurology, the response of individual nerve cells or groups of cells in the brain to sensoric stimulation of the living being is investigated in order to learn more about the functional areas in the brain and the mechanism of information processing. It is standard to repeat such experiments many times and to average over the different trials at fixed times lags following the stimulus.

In NMR spectroscopy, sometimes very weak resonances are to be identified, which are often hidden in the background noise. By an energy pulse, a molecular system is excited, and the exponentially decaying energy emission is recorded. Again, very many trials (up to 10^6) are performed and superimposed, in order to enhance the weak signal components. □

Hence, the problem is not that non-stationary processes could not be properly characterised, but that typical time series analysis tasks require an analysis without having access to an ensemble of realisations of the process. In such a case our information about the process is very restricted, such that we cannot determine its characteristic properties by standard statistical tools.

In this chapter we want to discuss some attempts to cope with this problem. We will first recall some tools to detect non-stationarity, and then suggest two different approaches for the analysis of non-stationary data for certain restricted settings. We want to stress that although these concepts have already proven their usefulness, there is yet no sound theory available.

13.1   Detecting non-stationarity

Stationarity requires that joint probabilities of a process, $p(s_i, s_{i+1}, \ldots, s_{i+m})$, to observe a sequence of values s_k for a quantity s for the specified time instances, for all m are invariant under time shift $i \mapsto i + k$ for arbitrary k. An evident consequence of this is that if there exists a mathematical model of the process it has to be autonomous (i.e., no explicit time dependence in the equations), and that in a physical realisation all system parameters including the influences from the environment must be strictly fixed. Statisticians speak of *weak stationarity*, if the marginal probabilities and the two-point probabilities are time independent. If only linear Gaussian processes are considered, this is sufficient, since all higher moments are completely determined by mean values and covariances. However, if we assume nonlinearities to be involved, weak stationarity is not enough.

There are many suggestions in the literature of how to detect non-stationarity. Since severe non-stationarity is typically very evident and most often can be identified by visual inspection of the data, most of these methods either test for feeble non-stationarity or look for non-stationarities which are not so obvious.

We want to recall here that even slight non-stationarity can sometimes lead to severe mis-interpretations. One such concept which we made use of is the method of surrogate data. Those types of surrogates which can be easily generated are created under the hypothesis of stationarity. Hence, all surrogate time series represent a stationary process. If a test statistics reveals differences between an ensemble of surrogates and the original time series, then this may be due to effects of some non-stationarity in the original data where this is the only reason that the null hypothesis is violated.

Also intermittent behaviour can be due to small parameter fluctuations around some critical value. In order to distinguish dynamical intermittency for fixed parameters as discussed in Section 8.4 from the alternation between non-intermittent dynamics of different types due to non-stationarity, one has to study the statistics of the intermittent phases and chaotic bursts in detail and to compare it with the predictions for dynamical intermittency.

As first tests for non-stationarity we suggested the computation of moving window means and variances in Example 2.2. In Witt *et al.* (1998) it is thoroughly discussed how to use these and other statistics for a sound test against stationarity. In order to detect slight non-stationarity, refined test statistics are necessary, since quantities computed on moving windows show statistical fluctuations even for stationary processes. These "normal" statistical fluctuations are estimated in order to be compared to those which one finds on the data. In practice, the method of surrogate data can often replace these intricate tests, since the iteratively amplitude adjusted Fourier transform surrogates of Section 7.1 represent a stationary process, and hence one can use these in order to extract the merely statistical fluctuations of moving window estimates for one's data set. As the following example shows, it is essential to preserve the supposed temporal correlations of the stationary data inside one's surrogates, otherwise wrong interpretations are possible.

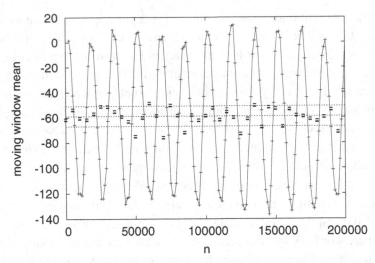

Figure 13.1 Mean value of data from the electric resonance circuit (Appendix B.11) computed on time windows of 5000 (+) (covering about 200 periods of oscillation of the data) with three quarters overlap, and on randomly shuffled data (filled boxes). The time ordering of the latter is arbitrary. The dashed lines represent their mean plus/minus the standard deviation.

Example 13.3 (Windowed mean of data from an electric resonance circuit).
The data of the resonance circuit described in Appendix B.11 shows a clear periodic fluctuation of its windowed mean value, as shown in Fig. 13.1. As a comparison, windowed means obtained from the same data after random shuffling (notice: for a stationary time series, the mean value is invariant under random shuffling) are shown, together with their mean value and standard deviation. Since the time series values vary inside an interval [−12000 : 12000], a fluctuation of the mean value of ±60 is not dramatic, but it is evidently significant in a statistical sense. One could conclude that here one of the experimental parameters oscillates slowly. However, this conclusion is wrong! The oscillations are simply aliasing effects of the window, which is not a perfect multiple of the period of the data. Hence, the magnitude of these fluctuations has to be compared to fluctuations of windowed means of stationary *correlated* data. In fact, the given data are stationary within the precision of all statistical tests we performed.   □

Moreover, we suggested to study the power in the low-frequency range of the power spectrum in Section 2.3.1, and the computation of cross-prediction errors in Section 4.4. Also, we showed that recurrence plots are useful tools also to detect non-stationarity in Section 3.6. At the end of Section 13.2.3 we refer to a quantitative analysis of recurrence plots.

13.1.1 Making non-stationary data stationary

Often, passing from an observable to its temporal derivative renders a non-stationary data set stationary. Of course, once sampled in discrete time, derivatives are replaced by *increments*, $r_n = s_{n+1} - s_n$.

Example 13.4 (Non-stationarity of a diffusion process). A mathematical model of a diffusion process in discrete time is a marginally stable AR(1) model, i.e., $x_{n+1} = x_n + \xi_n$, where ξ_n are Gaussian independent random numbers with zero mean and unit variance (which is the discrete time version of *white noise*) and with $x_0 = 0$. Despite the fact that all parameters of this process are time independent, its ensemble marginal probability depends on time in the following well known way: $p(x_n) = \frac{1}{\sqrt{2\pi n}} e^{-(x^2/2n)}$, and the mean squared displacement $\langle (x_0 - x_n)^2 \rangle \propto n$. Hence, all joint probabilities of this process are n-dependent. What is time independent, however, are all transition probabilities, e.g., $p(x_i|x_{i-1}) := p(x_i, x_{i-1})/p(x_{i-1})$. Increments $r_n = x_{n+1} - x_n$ are here evidently white noise and hence stationary.

Passing over to increments (or *returns*) is a very popular method in the analysis of financial data. If one is interested in deterministic structure, however, this procedure is delicate, since it strongly enhances noise (compare Section 9.3.1 on derivative coordinates).

Other methods of rendering data stationary are widely used in some part of the literature of data analysis of natural phenomena, often called *detrending*. If the underlying process is nonlinear (why would we else use nonlinear time series analysis tools?), detrending can only be employed if it is guaranteed that the trend has nothing to do with the dynamics, but is, e.g., due to the calibration of the measurement devices. In such cases, subtraction of a baseline drift or a time dependent rescaling is possible. Else, due to the non-validity of a superposition principle, trends should not be removed from nonlinear processes since they afterwards just appear more stationary without being so. In the mathematical language of dynamical systems this means that due to the lack of structural stability of almost all realistic dynamical systems, a change in a parameter which seems to introduce a simple drift in the data cannot be compensated by a change of coordinates (rescaling).[1] Even if the data appear more stationary afterwards (e.g., constant mean and constant variance), their dynamical properties such as temporal correlations are still as time dependent as before.

Only in fluctuation analysis in the spirit of Hurst exponents, additive detrending is well established, even if the mechanisms creating power-law long range correlations

---

[1] Two dynamical systems which can be made identical by a change of variables are called *conjugate* to each other.

are most surely nonlinear, since the scaling of the empirical variance of data in the length of the data segment is robust against detrending. For more details see Section 6.8.

13.2 Over-embedding

13.2.1 *Deterministic systems with parameter drift*

In this section we discuss a particular source of non-stationarity. We assume that a small number of parameters influencing the otherwise deterministic system dynamics is slowly time dependent, potentially making jumps from time to time. It will become evident what we precisely mean by "slow". Intuitively it means that during the typical coherence time of the deterministic dynamics the parameters should be constant to a good approximation. In continuous time, we hence start from the non-autonomous ordinary differential equation

$$\dot{\mathbf{x}} = \mathbf{f}(\mathbf{x}, \mathbf{a}(t)) \quad \text{with } \mathbf{a}(t) \text{ given,} \tag{13.1}$$

where $\mathbf{x} \in \mathbb{R}^d$ is the state vector and $\mathbf{a}(t) \in \mathbb{R}^k$ is a predefined vector valued function with the slowness assumption mentioned above, driving the non-stationarity. As always, we assume to have recorded a time series $\{s_n\}$ obtained by a scalar measurement function $s_n = s(\mathbf{x}(t = n\Delta t))$ and a sampling interval Δt, without precise knowledge about the dynamics and the parameters.

Our goal here is to reconstruct the extended phase space formed by the state space of the system plus the space spanned by the time dependent parameters. Knowing a point in this space simultaneously fixes the state variable $\mathbf{x}(t)$ and the actual values of the system parameters $\mathbf{a}(t)$. Hence, a small neighbourhood in this extended phase space is mapped onto a small neighbourhood some time later, since all points represent similar state vectors which are subject to identical equations with similar parameter settings. Since we required the time dependence of the parameters to be slow, a typical trajectory will explore the subspaces of fixed parameters much faster than it will move transversally to it. Before we show why this extended phase space is useful, we have to discuss how to reconstruct it from data.

Example 13.5 (Hénon map with drifting parameter). Let us consider a time series $\{x_n\}$ of the Hénon map $x_{n+1} = 1 - a_n x_n^2 + b x_{n-1}$. For every triple of measurements one can solve for $a_{n-1} = \frac{x_n - b x_{n-2}}{x_{n-1}^2}$. Under the assumption of slow variation of a, $a_n \approx a_{n-1}$, we hence can predict x_{n+1} from the triple (x_n, x_{n-1}, x_{n-2}). Due to smoothness, the unknown map $(x_n, x_{n-1}, x_{n-2}) \mapsto x_{n+1}$ can be reconstructed from neighbours of (x_n, x_{n-1}, x_{n-2}) in a three dimensional space without knowing the actual value of a_n. □

The example shows that an additional past variable, x_{n-2}, allows us to identify simultaneously the state vector of the system and the actual value a_n. With the particular measurement function $s_n = x_n$ for this particular system, a three dimensional time delay embedding is sufficient for the reconstruction of the extended phase space. More generally, following Whitney's embedding theorem, we conjecture that if we want to reconstruct the space of k parameters together with a d dimensional state space of the system, we should use a delay embedding with $m > 2(d + k)$. This is called over-embedding [Hegger *et al.* (2000)].

A sufficiently small neighbourhood of a point in this embedding space is populated by points which represent not only very similar states of the system but also very similar settings of the system parameters. Hence all data analysis and data manipulation tasks for which such an identification is necessary can be performed in the over-embedding space even for non-stationary data.

Example 13.6 (Non-stationary electric resonance circuit). We illustrate the concept of over-embedding using (almost) stationary experimental data and simulate a baseline drift of the measurement apparatus by adding a sinusoidal modulation, $\tilde{s}_n = s_n + \frac{\sigma}{8} \cos(n/1000)$. The data stem from the periodically driven nonlinear resonance circuit Appendix B.11, and σ is their standard deviation. The stationary data can be embedded in 4 dimensions. With the time dependent rescaling, we do not only lose neighbours, but we also gain false neighbours, since the attractors obtained for different offsets intersect each other. Due to the attractor dimension of slightly more than 2, already in 4 dimensions these intersections have measure zero, but they lead to a statistically significant contribution of false nearest neighbours. The fraction of false nearest neighbours as a function of the embedding dimension is shown in Fig. 13.2 for the stationary and the non-stationary data, which confirms that an increase of the embedding dimension solves the problem. □

Since very often a good embedding of the stationary data is in fact twice as large as the attractor dimension, over-embedding sometimes is not really required, but it can be replaced by an enlargement of the embedding window through a larger time lag τ.

13.2.2 *Markov chain with parameter drift*

We argued in Section 12.7.2 that the natural representation of time series data which form an lth order Markov chain is the time delay embedding space of dimension $m = l$. In analogy with the last subsection, we want now to generalise the concept of over-embedding to Markov chains [Ragwitz & Kantz (2002)]. So we assume that the transition probabilities of order l, $p(x_i|x_{i-1}, \ldots, x_{i-l})$, which define the

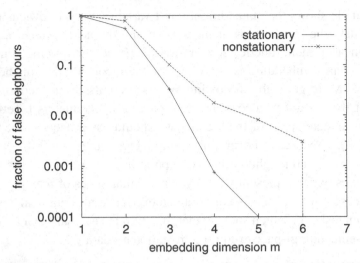

Figure 13.2 The fraction of false nearest neighbours as a function of the embedding dimension for the stationary and the non-stationary data from an electric resonance circuit.

Markov process, are functions of a set of k slowly time dependent parameters $\mathbf{a}(t)$. One natural form of such a process is a stochastic differential equation with drifting parameters, which reads:

$$\dot{\mathbf{x}} = \mathbf{f}(\mathbf{x}, \mathbf{a}(t), \xi) \,. \tag{13.2}$$

We recall that the process that is represented by a scalar time series $s_n = s(\mathbf{x}(t_n))$ is generally not Markovian, even for constant $\mathbf{a}$. Typically it can be approximated reasonably well by a Markov chain, the order of which here is assumed to be l. Hence, the non-stationary process is characterised by conditional probability densities $p(s_{n+1}|\mathbf{s}_n, t) = p(s_{n+1}|\mathbf{s}_n, \mathbf{a}(t))$. In the deterministic case above, Eq. (13.1), we argued that additional $2k$ delay coordinates allow us to identify the manifold representing the equation of motion for a particular setting of $\mathbf{a}$. Due to the stochastic nature of the Langevin equation, such a manifold does not exist in the noisy case. Instead, the more knowledge about the past we use, the better we can identify the parameter vector $\mathbf{a}$. The main message is that in the stochastic case, there is little hope for a rigorous statement, but the recipe is the same as in the deterministic case. The instantaneous properties of a non-stationary process can be extracted from data by increasing the embedding dimension with respect to what would be reasonable if the process was stationary.

Example 13.7 (Prediction of random numbers with non-constant mean). Let us study a sequence of independent Gaussian random numbers with unit variance but slowly time dependent mean $e(n)$, $p(s_n, n) = \frac{1}{\sqrt{2\pi}} \exp\left(-(s_n - e(n))^2/2\right)$, where $|e(n) - e(n+1)| \ll 1$. Ignoring the non-stationarity, the optimal prediction would be the mean value of the time series with a forecast error identical to the standard deviation of the data. Due to the slowness of the variation $e(n)$, a much better prediction here is the time dependent mean value $e(n+1) \approx e(n)$ of the distribution, leading to an rms forecast error of unity, i.e, the standard deviation of our distribution for constant e.

Let us now study the performance of the locally constant predictor Eq. (4.7) as a function of the embedding dimension and of the neighbourhood size. In Fig. 13.3 we show in panel (a) a sequence of random numbers according to $p(s_n)$ where $e(n)$ itself is generated by a slow AR(2) model. As shown in panel (b), the rms errors decrease with increasing m if ϵ is small, whereas they are identical to the standard deviation of the whole data set for large ϵ. In some interval of small ϵ, the prediction error indicates convergence to the theoretical limit of unity for large m. The good performance of the locally constant predictor relies on the correct identification of past states evolving under the same stochastic rules as the present state. $\square$

In the last example, one could in principle find an estimate of $e(n+1)$ through a windowed average, $\hat{s}_{n+1} = \frac{1}{l+1} \sum_{k=n-l}^{n} s_k$ with some suitable time window l. However, such an approach requires a matching of time scales between l and the temporal dependence of $e(n)$. It cannot make use of the recurrence of e in the past, and it cannot deal with sudden jumps in the parameters. Hence, time windowed models are very often inferior to over-embedding.

13.2.3 Data analysis in over-embedding spaces

In the reconstructed extended space of system states and parameters, we could do our standard analysis inside the subspaces of constant parameters, i.e., we could perform predictions, we could use the implicit knowledge of the equations of motion for noise reduction, but we could also compute Lyapunov exponents or attractor dimensions. The problem lies in the fact that the subspaces of constant parameters, which are trivial in the original extended space, are unknown in its reconstructed version. Hence, all tasks which require to work *exactly* inside these subspaces are impossible, and unfortunately, dimension estimation and everything which depends on the fractality of the invariant measure are among these tasks. If we assume that our deterministic system has a fractal attractor, then most surely its structure sensitively depends on the parameters **a**. Dimensions will be estimated incorrectly in an uncontrollable way, since fractality will be wiped out when we mix different subspaces, even if they correspond to very similar system parameters. Think of the Hénon map as an example.

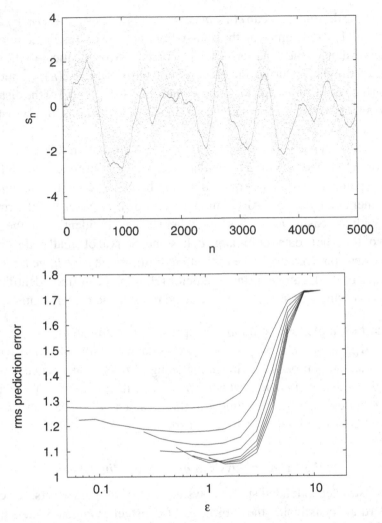

Figure 13.3 Panel (a): the non-stationary Gaussian random numbers with unit variance and varying mean $e(n)$ (continuous line). Panel (b): rms prediction errors obtained with the predictor Eq. (4.7) for embedding dimensions $m = 1$ to 8 (top to bottom) as a function of the neighbourhood diameters ϵ.

The situation is more favourable if we want to exploit the equations of motion themselves. We already presented examples related to forecasting, and we will discuss noise reduction for human voice in detail in the next subsection. Let us only point out here that the identification of neighbours in the over-embedding space identifies the useful redundancy in the time series in the past. The alternative of using only a sufficiently short past time window where the parameters can be assumed to be close enough to the actual values would be very inefficient in this respect. Moreover, such a conventional approach (called segmentation) leads to large errors when parameters make sudden jumps. If parameters jump, one could think of a manual segmentation of the data set, but apart from the practicability issue, also here the re-occurrence of a parameter vector remains undetected and the past data do not enrich the data base. Despite the advantages of the over-embedding space, it is not the universal remedy of everything: predictions are made under the assumption that parameters inside the prediction horizon are unchanged. Hence, iterated predictions yield stationary dynamics. Evidently also the nature of the time dependence of the parameters themselves is not resolved, and their temporal evolution cannot be predicted. We will hence discuss in Section 13.3 a complementary approach.

Although the over-embedding space does not offer a direct access onto the parameters it offers access to temporal correlations of the parameters. Two delay vectors in the over-embedding space are only close if the dynamical states are similar and if the parameters are similar. Hence, all regions in a recurrence plot of such delay vectors which are non-empty refer to a pair of temporal episodes where the parameter values are very similar. This observation was first published by Casdagli (1997).

Example 13.8 (Human ECG data). The most evident non-stationarity of ECG data is introduced through the variations of the heart rate. There are more subtle effects introduced by the relative position of the heart with respect to the ECG-electrodes on the skin, which may change due to posture changes of the human, and there may be physiological effects on the actual shape of the wave form of a single heart beat (see Fig. B.4). Assuming stationarity, a two dimensional embedding appears to be sufficient, since the heart dynamics is essentially limit-cycle dynamics. The recurrence plot of a 5 dimensional embedding shown in Fig. 13.4 contains white patches, which reflect the mutual mismatch of the heart rate or other parameters during the two time episodes represented by the time index pairs (i, j). $\square$

We recall that as far as non-stationarity is concerned, a meta-recurrence plot shows essentially the same information. A meta-recurrence plot typically contains less data than a recurrence plot, since it contains already a compression of the time axes. One does lose, however, the information about state vectors, e.g., the signature of almost periodicity in Fig. 13.4. See Fig. 3.8 on page 46 for a meta-recurrence plot for ECG data.

In non-stationary processes, time translational invariance is explicitly broken, such that the time of the occurrence of a certain dynamical feature can enter the focus of interest. Hence, additional statistical quantities based on time indices can be introduced. An example is presented in Ragwitz & Kantz (2002), where the temporal alignment of two non-stationary processes is achieved by the computation of an average cross-neighbour index. The cross-neighbour index is the mean value of the time indices of all neighbours a delay embedding vector from one time series

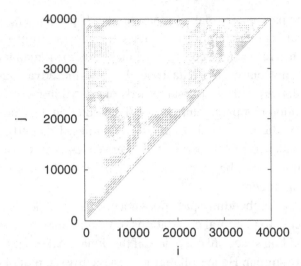

Figure 13.4 A recurrence plot in 5 dimensions with a delay of 80 ms of the human ECG data (Appendix B.7). The almost parallel lines reflect the almost periodicity of the heart dynamics ("limit cycle"), whereas the white patches are due to non-stationarity, which becomes evident in an over-embedding space.

finds among the delay vectors of the second series. It therefore tells one, at which time the dynamics of the second time series was closest to the dynamics of the first one at the time the delay vector was selected.

Exploiting the same idea of broken time translational symmetry in a recurrence plot, Rieke *et al.* (2002) have suggested a quantitative analysis in over-embedding spaces as a measure for non-stationarity. They study different statistical estimates based on the distribution of the indices of neighbours of a given point in the time series.

Finally, we want to recall work by Parlitz *et al.* (1994). They proposed a particular way for identifying unobservable system parameters which implicitly also rely of the concept of over-embedding. They assume that a system parameter of interest can be measured directly only with high effort. Hence they create a data base which is the joint recording of this particular parameter together with scalar time series data of the dynamics of the system. For monitoring, they exclusively observe the dynamics. In a suitably high dimensional embedding space, they search neighbours of the current delay vector inside the data base. Since the vectors from the data base were recorded together with the values of the now unknown parameter, an estimate of the latter is the average of the values assigned to the neighbours found.

13.2.4 Application: noise reduction for human voice

The local projective noise reduction scheme of Section. 10.3.2 can also be applied to human voice. Of course, we do not claim that human voice represents a deterministic

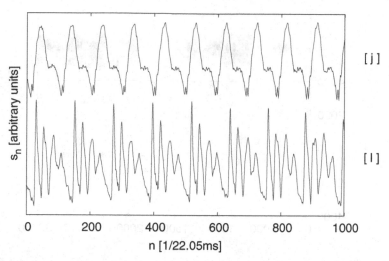

Figure 13.5 45 ms of recordings of two different phonemes resembling [l] like in "bell" and [j] like in "yeast" (arbitrary off-set for better visibility). Notice the almost periodicity. A single wave covers about 5 ms.

chaotic system, but there is a lot of structure which resembles very much limit cycle dynamics with time dependent parameters. In Fig. 13.5 we show the wave forms of two different phonemes (phonemes are the units of spoken language comparable to syllables of written language). Notice that the wave form is speaker dependent, since the human auditory system exploits essentially only the power spectrum of the sound signal, ignoring the relative phases. Articulated voice is a concatenation of phonemes. For noise reduction, it is relevant to identify the actual dynamics (corresponding to the manifold onto which we project) and the actual state vector. Both are sufficiently determined using an embedding window of about 5 ms (covering one oscillation inside a phoneme) and about 20 dimensions. In a recurrence plot one therefore observes that pairs of neighbouring points in this over-embedding space stem from the same phonemes and hence contain the redundancy which is needed for the noise reduction (see Fig. 13.6). Since in voice filtering noise magnitudes of the order of the signal itself are to be treated, it is nontrivial to identify the correct neighbours also in the presence of noise (the embedding parameters from above are reasonable for low noise amplitudes). The way out is to use a time lag of unity and hence to increase the embedding dimension to about 100 for data points with a sampling rate of 22 050 kHz. When then distances are computed through the Euclidean norm, it is easy to verify that the squared distance of two noisy vectors is in good approximation the squared distance of the noise-free vectors plus the variance of the noise times the embedding dimension [Kantz *et al.* (2001)]. As we emphasised, voice dynamics is non-chaotic, and hence no exponential divergence in time has to be compensated. Hence, for noise levels of

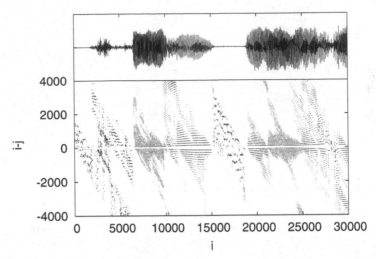

Figure 13.6 Recurrence plot (Section. 3.6) of a segment of human voice for delay vectors with $m = 80$, $\tau = 1$, in a rotated coordinate frame. The white tongues reaching towards the horizontal reflect the ability of the delay embedding to distinguish between different phonemes, the horizontal line structure reflects the periodicity inside the phonemes with time dependent period length, which corresponds to raising and lowering the pitch of the voice.

100% and more (i.e. a signal to noise ratio of 0 dB) on typical words we find a noise suppression by factors of better than 1/5 in the rms sense, which can be translated to a gain of better than 14 dB. The procedure appears to be of practical interest, for more details see Kantz *et al.* (2001).

13.3 Parameter spaces from data

We will very briefly introduce a second concept for the treatment of non-stationarity. The essential drawback of over-embedding is that we cannot use it to extract the properties of the parameters. We might have, e.g., indirect evidence when parameters vary periodically through a meta-recurrence plot, but we cannot "see" the parameters themselves. In this section we propose a concept to extract quantities which are equivalent to the unobserved time dependent system parameters.

We continue to assume a slow parameter variation in an either deterministic or stochastic system. In order to make these parameter variations visible, we hence need a compression of the time scale. Also, in order to focus on parameters, we want to get rid of the fast time dependence of the state vectors. Both goals are achieved when we compress the original time series into a much shorter time series of *feature vectors*. The procedure is straightforward: we choose suitable statistical quantities such as mean, variance, the value of the autocorrelation function at selected time lags, but also (if reasonable) nonlinear quantities such as Lyapunov exponents, dimension, entropy, or better, average Poincaré recurrence times, forecast errors, coefficients of particular model equations fitted to the data. These quantities are computed on (typically overlapping) segments of a time series. Hence, the data inside each time window are condensed into a set of numbers, which are interpreted as components of a vector, the feature vector. The key idea is again to create a kind of embedding space, namely here for the unobserved system parameters, eliminating the state space of the system by a kind of projection. This projection is achieved by time averaging, which replaces the single state space vectors by an ergodic average. If by assumption all joint probabilities of our process depend on k parameters only, then every single feature is a functional on these joint probabilities for almost fixed values of the parameters (slowness assumption). If features are chosen suitably, a feature vector represents the values of the k parameters, and the observed variation of the feature vectors in time is equivalent to the variation of the unobserved parameters.

So, what are the main claims? We propose that feature vectors of sufficiently many features have the following two properties:

- If a set of feature vectors obtained from segmenting a time series spans a set in feature space with dimension $d > d_p$, then at least d_p parameters have been varied independently during the time spanned by the series.
- The path in feature space created by the time ordered set of feature vectors represents the path in parameter space of the unknown parameters. In particular, a periodic path in feature space reflects that those parameters which can be resolved in the chosen feature space vary periodically with the observed period.

Several remarks have to be made. In particular, this concept contains a number of practical problems which at first sight are technicalities. In Güttler *et al.* (2001) different examples are discussed in order to present some evidence for the above claims.

Apart from the slowness assumption for the parameters a smoothness assumption is essential for the argumentation above: all joint probabilities must be smooth functions of the system parameters. This is clearly violated by deterministic systems in the vicinity of bifurcations. However, as it was argued in Güttler *et al.* (2001), a small amount of dynamical noise, which can always be assumed to be present in dynamical systems, formally ensures this property.

The clear difficulty in the application of this method lies in the choice of suitable features: evidently, certain features might be blind to certain parameter variations (see example below), and it is evident that we might be able to construct a process where almost nobody would guess the right feature to make the parameter variation visible.

The second difficulty becomes evident when we compare the reconstruction of the parameter space through features to phase space reconstruction. The charming property of the time delay embedding is that all coordinates of the time delay vectors create the same marginal distribution, i.e., when computing the distance between two delay vectors evidently all coordinates should have equal weight. If instead, we had a truly multivariate time series, where one observable is measured in units such that it covers an interval on the real numbers which is orders of magnitude larger than some other observable, a meaningful measure for the distance of vectors has to introduce either weights for the different coordinates, or one should use rescaled observables. But both ways contain the same problem: if one coordinate is constant plus pure low amplitude noise, then rescaling guarantees that we see a contribution of unity to our dimension estimation from this coordinate, instead of zero as expected for a fixed point. We may encounter exactly this problem when trying to extract from the set of feature vectors a dimensionality of the manifold in feature space where they are confined to. A possible, yet not fully explored remedy would be to identify for each single feature, by the construction of stationary surrogates, how large its statistical fluctuation for stationary data would be, and to normalise the observed variance to the surrogates' variance. The dimensionality of the manifold should then be determined on scales larger than unity, whereas around unity the dimension estimates should reach the dimension of the feature space.

Example 13.9 (Saw-tooth maps). We consider a family of saw-tooth maps, $x_{n+1} = a x_n$ mod 1. For every integer $a > 1$, the invariant measure is uniform, i.e., mean and variance of the distribution are identical. A smooth and slow variation of a through an interval, say, from 1 to 5, hence would pretend a periodic parameter variation, if our feature vector only contained mean and variance. Adding more features such as a forecast error, the Lyapunov exponent, or simply a three-point correlation, would solve the problem: regardless of how a fluctuates inside its interval, all feature vectors would lie on a single line in feature space.

As a final remark, the feature space analysis is also suited for data classification problems. If instead of a single non-stationary time series a set of time series from different experiments of similar physical systems is given, one might like to know how many parameters are varying from experiment to experiment.

Exercises

13.1   Create two time series of the logistic equation of length 1000, one for $a = 1.9$, one for $a = 2$ (TISEAN routine `henon` with `-B0`). Shift/rescale one of them such that means and variances of the two are identical. Plot both data sets in a two dimensional time delay embedding. You should be able to observe two parabolae intersecting each other.

13.2 Concatenate these two time series and make a recurrence plot of them. For clarity, if you use 1000 data for each parameter, choose a neighbourhood size of $r = 0.008$. In the upper triangle of the plot, use a one dimensional embedding, in the lower one a three dimensional one. Is the result an impressive confirmation of the usefulness of over-embedding?

13.3 As discussed in Example 13.5, you should find equivalent results for a time series of the x-coordinate of the Hénon map (if you choose $b = 0.3$ then $a = 1.31$ and $a = 1.4$ are suitable parameters). Now use $s = x^2$ as observable and study the recurrence plots in up to five dimensional embedding spaces (you will have to increase the neighbourhood size for larger m).

Chapter 14

Coupling and synchronisation of nonlinear systems

The reason for the predominance of scalar observation lies partly in experimental limitations. Also the tradition of spectral time series analysis may have biased experimentalists to concentrate on analysing single measurement channels at a time. One example of a multivariate measurement is the vibrating string data [Tufillaro *et al.* (1995)] that we use in this book; see Appendix B.3. In this case, the observables represent variables of a physical model so perfectly that they can be used as state vectors without any complication. In distributed systems, however, the mutual relationship of different simultaneously recorded variables is much less clear. Examples of this type are manifold in physiology, economics or climatology, where multivariate time series occur very frequently. Such systems are generally quite complicated and a systematic investigation of the interrelation between the observables from a different than a time series point of view is difficult. The different aspects which we will discuss in this chapter are relevant in exactly such situations.

14.1 Measures for interdependence

As pointed out before, a first question in the analysis of simultaneously recorded observables is whether they are independent.

Example 14.1 (Surface wind velocities). Let our bivariate time series be a recording of the x-component and the y-component of the wind speed measured at some point on the earth's surface. In principle, these could represent two independent processes. Of course, a more reasonable hypothesis would be that the modulus of the wind speed and the angle of the velocity vector are the independent processes, and hence both x and y share some information of both. If this was the case, and additionally the modulus of the wind speed is varying smoothly, the magnitude of the fluctuations of the wind speed in each of the two components should be related to some extent.

This example shows that the possibility to define a statistical tool which is able to identify interdependence will depend on the presence of redundancy. We can only see that y shares some information with x if there exists any nontrivial information about the x-process. In the following we will always speak about bivariate data, but most tools can be either directly generalised to more than two data sets, or one can study higher dimensional situations by considering all possible pairs of bivariate data.

If the time lagged cross-correlation function of the two signals,

$$c_{xy}(\tau) = \frac{\langle (x(t) - \bar{x})(y(t - \tau) - \bar{y}) \rangle}{\sqrt{\langle (x(t) - \bar{x})^2 \rangle \langle (y(t) - \bar{y})^2 \rangle}} \qquad (14.1)$$

shows a statistically significant deviation from $c_{xy} = 0$, then the two processes are called (linearly) correlated. The time lags at which $|c_{xy}(\tau)|$ assumes its maxima represent time delays of the coupling between x and y. Due to the symmetry of c_{xy} under exchange of x and y, the cross correlations do not tell us anything about the direction of coupling, although it is often supposed[1] that a shift of the maximum towards positive lags indicates that x couples (with some delay) into y. Since c_{xy} just exploits second order statistics, the cross correlations can be identical to zero, even if x and y depend on each other in a nonlinear way.

Example 14.2 (Correlated surface wind speeds). In Fig. 14.1 we display the numerically estimated cross-correlation function based on 20 minutes recordings of wind speed measurements by cup anemometers (Appendix B.13). The three different anemometers are installed on the same mast at 10, 20, and 30 m above ground. Data are recorded with 8 Hz sampling rate. Since the data are strongly non-stationary, neither the auto- nor the cross-correlation function decays to zero. Instead, we find a level of about 0.1 at large time lags. As expected, we find that the correlations get weaker with increasing vertical distance. We also observe a very slight lagging behind of the lower recordings with respect to the upper ones. This is a frequently observed phenomenon: the wind field close to the ground always lags somewhat behind the field in higher layers of air. □

When we want to go beyond second order correlations, we have seen before that this can be achieved by entropy-like quantities which depend on all higher moments of a probability distribution. The time lagged mutual information is a

---

[1] But only partly justified by causality: there could be a time lag in one signal path *after* the coupling.

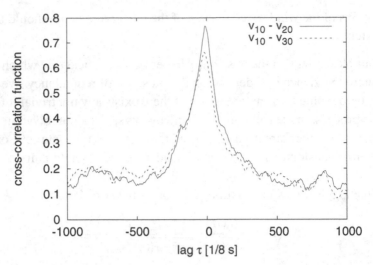

Figure 14.1 The cross-correlation functions Eq. (14.1) for pairs of wind speed measurements, where v_x denotes that the recording is done x meters above ground.

natural extension of the two-point correlation function; compare Green & Savit (1991) and Prichard & Theiler (1995) for similar statistics. Hence, also here we can compute

$$M_{xy}(\tau) = \sum_{i_x, j_y} p_{i_x, j_y} \ln \frac{p_{i_x, j_y}}{p_{i_x} p_{j_y}}, \qquad (14.2)$$

where after the introduction of a partition on the $\mathbb{R}^2$, p_{i_x, j_y} defines the probability to find $(x(t), y(t + \tau))$ in the box labelled (i_x, j_y). The replacement of the Shannon entropies by order-two Renyi entropies suffers from the same difficulties as we encountered them for the time delayed mutual information due to the lack of additivity for all $q \neq 1$-entropies. Despite these conceptual drawbacks, it is most often useful to estimate the mutual information through correlation sums (see also Prichard & Theiler (1995)).

Finally, we want to review two concepts introduced by Arnhold *et al.* (1999). The computation of the mutual information requires stationarity. Hence, its application is questionable when we think of the wind speed example presented above, or of heart–breath coupling etc. The concept of *interdependence* is rather intuitive and very powerful. If x and y have some relationship, then a recurrence of a state vector representing x should imply also a recurrence of a state vector representing y, at least with a larger than zero probability. More precisely, let us use a time delay embedding of dimension m for x and of m' for y, calling the delay vectors $\mathbf{x}_n$ and $\mathbf{y}_n$, respectively. Define by $\mathcal{U}(\mathbf{x}_n)$ a neighbourhood of $\mathbf{x}_n$ containing the l closest neighbours of $\mathbf{x}_n$. Let us assume that their time indices are $k \in \{k_1, \ldots, k_l\}$. Now check whether the delay vectors $\mathbf{y}_k$ with the same indices $k \in \{k_1, \ldots, k_l\}$ are closer to $\mathbf{y}_n$ than an average randomly chosen $\mathbf{y}$-vector. If so, then there is evidently some relation between x and y. In order to define a quantity for interdependence, one has to normalise this average conditional distance

$$d_n(\mathbf{y}|\mathbf{x}) = \frac{1}{l} \sum_{k \in \{k_1, \dots, k_l\}} (\mathbf{y}_n - \mathbf{y}_k)^2 , \qquad (14.3)$$

where $\{k_1, \dots, k_l\}$ are the indices of the l closest neighbours $\mathbf{x}_k$ of $\mathbf{x}_n$.

The normalisation should render $d_n(\mathbf{y}|\mathbf{x})$ independent of a rescaling of y and independent of the time series length. Arnhold *et al.* (1999) make two different suggestions for this purpose.

One can either compare $d_n(\mathbf{y}|\mathbf{x})$ to the average distance of the l closest neighbours of $\mathbf{y}_n$ among the set of all y. Let us define the mean neighbour distance of $\mathbf{y}_n$ as

$$d_n(\mathbf{y}) = \frac{1}{l} \sum_{l:\mathbf{y}_k \in \mathcal{V}(\mathbf{y}_n)} (\mathbf{y}_k - \mathbf{y}_n)^2 , \qquad (14.4)$$

where $\mathcal{V}(\mathbf{y}_n)$ is the neighbourhood of $\mathbf{y}_n$ in m' dimensional delay coordinates containing exactly l neighbours. Then, the statistics to be computed reads

$$I_{xy} = \left\langle \frac{d_n(y)}{d_n(y|x)} \right\rangle_n , \qquad (14.5)$$

where $\langle \cdot \rangle_n$ indicates the ergodic average over the time series. If I_{xy} is small, then there is no interdependence, since in this case the true neighbours of $\mathbf{y}_n$ are much closer to $\mathbf{y}_n$ than those vectors for which the x-process has a recurrence. However, a weak interdependence can hardly be detected, since the value of I_{xy} for independent processes is not *a priori* evident and depends on the properties of the data. Only when I_{xy} comes close to unity, is this a clear confirmation of dependence.

The alternative normalisation is to compare the conditional distance to the mean distance of randomly chosen[2] vectors $\mathbf{y}_k$ to $\mathbf{y}_n$. So we define

$$r_n(y) = \langle (\mathbf{y}_n - \mathbf{y}_k)^2 \rangle_k . \qquad (14.6)$$

Then one can study

$$\mathcal{I}_{xy} = \left\langle \frac{d_n(y|x)}{r_n(y)} \right\rangle_n \qquad (14.7)$$

The way we have defined $\mathcal{I}_{xy}$ it is a measure for independence of x and y, since it is unity if the processes have no relationship at all. In the original article, the same idea was exploited by a different statistics.

---

[2] Note that this is not the variance of y, since $\mathbf{y}_n$ is different from the mean of $\mathbf{y}$. In Arnhold *et al.* (1999), exactly l randomly chosen vectors were used, but this seems to introduce systematic statistical errors for small l.

We have deviated from the literature by our definition of $\mathcal{I}_{xy}$, since this immediately suggests a unification of both concepts. Inspired by how the auto-correlation function usually is normalised we can easily define a quantity that yields values in between zero and unity:

$$\hat{I}_{xy} = \left\langle \frac{r_n(y) - d_n(y|x)}{r_n(y) - d_n(y)} \right\rangle_n . \tag{14.8}$$

If $\hat{I}_{xy} \approx 0$ then there is no relationship between x and y, whereas $I_{xy} \approx 1$ denotes a full deterministic coupling. The relevant property of the interdependence $I_{xy}(\tau)$ is that it can be positive and hence indicate some dependence of the two processes, even if no simple relationship $x = f(y)$ exists.

Of course, in all the above quantities, it is not really necessary to consider delay vectors also for y, i.e., m' may be unity. Additionally, it is often useful to introduce a time lag τ in all the indices of the vectors $\mathbf{y}$, in order to allow for relative time shifts Arnhold *et al.* (1999), which corresponds to a delayed coupling between x and y or to a delayed response. If the time series are non-stationary, the average $\langle \cdot \rangle_n$ can be replaced by averages on time windows, since it might be that also the strength of the interdependence is time dependent.

Example 14.3 (Interdependence of wind speed measurements). Again we use the wind speed data, but this time we consider simultaneous recordings at the same height (10 m above ground) but at different locations in the x-y-plane. Three sites are on a line in northwest–southeast direction. Masts one–two, two–three, and one–three are 20 m, 6 m, and 26 m apart, respectively. A fourth site is 20 m westward of mast one. As we see in Fig. 14.2, during the time spanned by the data, the wind comes from west with a speed of about 8 m/s, since the dependence between mast one and four is strongest, with a time lag of 20 samples (8 Hz data). The interdependence gets weaker with increasing south–north distance of the masts, and the time lag gets larger, indicating that the corresponding component of the wind speed which transports the information is slower. We used embedding dimensions $m = m' = 3$ with time lag $\tau = 2$. We should say that for this example, the cross-correlation function yields the same information at much lower computational effort (see Fig. 14.1).

Due to our lack of suitable data, you might be left with the impression that such a nonlinear quantity is of no use. A nontrivial example where interdependence reveals structure which cannot be seen by the cross-correlation function is given as an exercise, since it involves a simple chaotic map instead of real data.

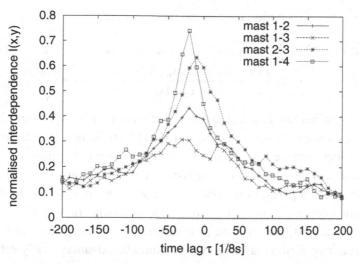

Figure 14.2 Normalised interdependencies of pairs of wind speed time series. See Example 14.3 for an explanation of the labelling.

14.2 Transfer entropy

Most of the measures for dependence discussed in the last section have in common that they are symmetric under the exchange of x and y. As a particular consequence of this, their outcome does not tell us anything about the direction of coupling, if we found coupling at all. Also, it does not yield a direct answer to whether the bivariate time series really does contain more information than one of the univariate time series alone. A concept which was called *transfer entropy* in Schreiber (2000) can help. In order to break the symmetry, it involves at least three arguments in an entropy-like expression.

The *transfer entropy* is a specific version of a Kullback–Laibler entropy or of the mutual information for conditional probabilities. In words, it represents the information about a future observation of x obtained from the simultaneous observation of some past of both x and y, minus the information about the future of x obtained from the past of x alone, $-h_{xy} + h_x$ (entropies represent uncertainties and hence negative information). The transition probability conditioned to m time instances for the variable x and to l time instances of y, $p(x_{n+1}|x_n, x_{n-1}, \ldots, x_{n-m+1}, y_{n-\tau}, y_{n-\tau-1}, \ldots, y_{n-\tau-l+1})$ is compared to the transition probability $p(x_{n+1}|x_n, x_{n-1}, \ldots, x_{n-m+1})$ conditioned on x alone. If the x-process and the lagged y-process are independent, then the first transition probability should not depend on y and hence the two expressions would be identical. If they are not identical, the Kullback–Laibler entropy quantifies the excess of information in the distribution containing the y. Hence,

$$
T_{y \to x}(m, l, \tau) = \sum p(x_{n+1}, x_n, \ldots, x_{n-m+1}, y_{n-\tau}, \ldots, y_{n-\tau-l+1}) \times
$$
$$
\ln \frac{p(x_{n+1} | x_n, \ldots, x_{n-m+1}, y_{n-\tau}, \ldots, y_{n-\tau-l+1})}{p(x_{n+1} | x_n, \ldots, x_{n-m+1})} \tag{14.9}
$$

is the quantity of interest and called the *transfer entropy* in Schreiber (2000). It is evidently not symmetric under exchange of x and y and hence can be used to detect into which direction the coupling in between two observables goes.

The definition of the transfer entropy is done through Shannon's entropy formula and hence requires a partition in space. Since we work in $m + l + 1$ dimensions, such a partition cannot be made too fine, since otherwise no points are left inside the boxes for a safe estimation of the probabilities. A version of the transfer entropy based on order two Renyi entropies, which could be advantageously estimated by the correlation sum, is not always useful since it is not positive semi-definite. In Marschinski & Kantz (2002) therefore an empirical finite sample correction was suggested.

In the previous edition of the book we discussed how to decide, for a bivariate time series with variables $x(t)$ and $y(t)$, whether to use a time delay embedding of x alone or to use a mixed embedding involving also y. The natural question to ask is the following. Given $x(t)$ and trying to predict $x(t + t_p)$, which observable yields more additional information: $x(t - \tau)$ or $y(t - \tau)$ (where the information is to be optimised in τ)? To find an answer, we proposed to compute a combination of two-point and three-point mutual information expressions. The difference of these expressions yields exactly the transfer entropy Eq. (14.9) for $m = 1$ and $l = 1$, where the second time series is either y or, when we want to discuss the time delay embedding, again x itself. The only additional modification was that instead of the conditional probabilities $p(x_{n+1} | \ldots)$, we studied $p(x_{n+t_p} | \ldots)$ with $t_p > 1$, in order to avoid spurious results due to the strong linear correlations in highly sampled flow data.

Example 14.4 (Vibrating string). The amplitude envelopes of the vibrating string (Appendix B.3) were recorded simultaneously in two spatial directions, yielding a vector valued time series (x, y). As we have said above, these two variables span the true configuration space. A comparison of the modified transfer entropies $T_{y \to x}(1, 1, \tau)$ and $T_{x \to x}(1, 1, \tau)$ with $t_p = 40$ shows the clear superiority of the bivariate embedding over a delay embedding in two dimensions (Fig. 14.3). Under violation of causality (negative τ!), it would, however, be optimal to use $y(t + 10)$ instead of $y(t + t_p)$, as one could naively guess.

Example 14.5 (Multichannel physiological data). We apply the same investigation to the simultaneous observables of the sleep data, Appendix B.5. Our finding

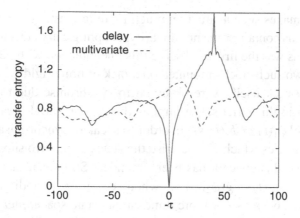

Figure 14.3 Transfer entropies $T_{y \to x}(1, 1, \tau)$ and $T_{x \to x}(1, 1, \tau)$ for the vibrating string data (Appendix B.3). It can be interpreted as the gain of information about the future of x 40 time steps ahead when knowing x at the present time and additionally either x (continuous) or y (broken) at any other time $t - \tau$. When we do not allow the use of future information (causality), the simultaneous knowledge of y is optimal. The best lag for a time delay embedding would be about $\tau = 30$. Note that the first zero of the autocorrelation function appears at $\tau = 40$.

is that the heart rate channel becomes slightly more predictable over short times when the information about the air flow through the nose is incorporated. When we use instead the blood oxygen saturation as additional input, the information equals that of delay coordinates. In all other possible combinations of the three signals (heart rate, air flow, and oxygen concentration in the blood) the delay coordinates are clearly superior (although the total amount of information is small compared to deterministic models). In some cases the additional observable does not yield any information at all: the signature is the same when replacing it by a random reference signal. ☐

14.3 Synchronisation

When two undamped harmonic oscillators are coupled in a linear way, the resulting motion is the linear composition of two harmonic modes. Even if the frequencies of the two oscillators are very similar, they do not synchronise, but instead show a phenomenon called beating: the energy slowly oscillates between the two oscillators, i.e., the amplitudes of their oscillations vary sinusoidally, where one oscillator has a large amplitude when the other has a small one. From a formal point of view, this is the consequence of the existence of a superposition principle. From a more intuitive point of view, this is a consequence of the particular property of harmonic oscillations that the oscillation period is independent of the amplitude of oscillation. Hence, one oscillator has no freedom to adopt its period to the second one.

Nonlinearity makes such a situation much more interesting. Since most natural oscillatory systems contain some nonlinearity, coupling can in fact lead to synchronisation. Huygens was the first to observe this phenomenon with pendulum clocks back in 1665. Two such clocks mounted on a rack of finite rigidity, so that the rack starts to slightly rock itself, were observed to synchronise due to this coupling. Synchronisation in this strict sense implies that the two angles ϕ_1 and ϕ_2 of the two pendula fulfil $\phi_1(t) = \phi_2(t)$ $\forall t$. In order to speak of synchronisation, we hence need two phase angles which characterise the states of the two subsystems.

In recent years, this concept has been extended. So one speaks of *generalised synchronisation* if $\phi_1(t) = g(\phi_2(t))$ $\forall t$, where g is a 2π-periodic function. If the subsystems each have a two or more dimensional state space, *phase synchronisation* denotes a situation where only phases but not the corresponding amplitudes synchronise. This can, e.g., be found in chaotic systems. Even the notion of chaotic synchronisation has been coined [Pecora & Carroll (1990) and (1991)]. A highly recommendable book on synchronisation is Pikovsky *et al.* (2001).

An interesting question in this context is how to establish from multivariate time series data that two or more systems interact with each other and whether they are synchronised. In a nonlinear setting, this is not properly accounted for by linear cross-correlations. In particular, if only the phases of the systems are locked to each other but not the amplitudes, one needs special methods to detect the interrelation.

There are essentially two entry points for time series analysis. The first is via the spectrum of Lyapunov exponents, but is an indirect one and hence needs observations for different parameter values, namely for both the-synchronised and the not-synchronised state. At the transition to synchronisation, one degree of freedom becomes inactive. This should be reflected both by dimension estimates and in the Lyapunov spectrum. For details see Rosenblum *et al.* (1996). We are, however, not aware of any analysis of time series data where synchronisation was observed in this way.

The second more direct approach consists of a comparison of the phase angles of the two subsystems. Here, the difficulty lies in the reconstruction of the phase angles from some arbitrary observables. Often this is not just a technical but a matter of principle. Many systems for which one would like to study (generalised) synchronisation are not usually represented by phase angles as state space variables. Think, e.g., of the Lorenz or Rössler system. Motion on these chaotic attractors does exhibit oscillations, but how precisely should a phase be defined? Very often the possibility to define the presence or absence of synchronisation does not depend on the details of how a phase angle is defined. It can be done in a rather *ad hoc* geometric way, or it can be done by a sophisticated transformation such as the Hilbert transform, or it can be done by a suitable interpolation in between maxima of some observable. In the latter case, one would define the phase to increase by 2π

during the time in between two maxima of the signal, and the phase angle at every instance in between two maxima could, e.g., be $\phi(t) = 2\pi t/(t_{i+1} - t_i)$, where t_i and t_{i+1} are the times of the two successive maxima, and $t_i < t < t_{i+1}$.

It makes sense to require that the phase angle of a subsystem is monotonically increasing in time. Unfortunately, the following rather general way to define phase angles does not always satisfy this. The *Hilbert transform* is a way to continue a given function into the complex plane. Specifically, the Hilbert transform assigns imaginary values to the given real values on the real axis. When it is applied to a time series, this means that we can construct the imaginary part corresponding to the observed real data, where the uniqueness is achieved by the fact that the resulting complex valued function is analytic. Let the signal $s(t)$ be a differentiable real valued function from L^2 (square integrable), then the Hilbert transform reads

$$s_H(t) = \frac{1}{\pi}\text{P.V.} \int_{-\infty}^{\infty} \frac{s(t')}{t - t'}dt',\qquad (14.10)$$

where P.V. denotes the Cauchy principal value of the integral. Then the complex signal reads $z(t) = s(t) + i s_H(t)$, and the phase angle is simply its argument $\phi(t) = \arctan\frac{s_H(t)}{s(t)}$. An instantaneous amplitude can be defined by $a(t) = \sqrt{s^2(t)+s_H^2(t)}$. Unfortunately, the phase thus defined is monotonically increasing only if the input signal has a single dominant peak in the power spectrum or is otherwise close to periodic. Nonetheless, the Hilbert transform has been employed in several of the studies of synchronisation. A numerical estimate of $s_H(t)$ is most easily obtained by a double Fourier transform. It can be verified that the Hilbert transform of a purely harmonic signal $s(t) = a\cos(\omega t + \phi_0)$ is given by a simple phase shift of $-\pi/2$, i.e., $s_H(t) = a\cos(\omega t + \phi_0 - \pi/2)$. Therefore, one can obtain $s_H(t)$ of a time series by computing its discrete time Fourier transform, shifting all phases by $-\pi/2$, and applying the inverse Fourier transform.[3]

In time series analysis, we now need simultaneously recorded observables which are each dominated by the dynamics inside one of the subsystems we wish to study. Hence, in order to study synchronisation between two subsystems, we need at least bivariate time series data. From each of these signals one has to obtain a time dependent phase angle. A phase space reconstruction could, e.g., enable one to use a geometrical definition.

Example 14.6 (Human air flow data). The air flow through the nose of a human as discussed in Example 10.7 on page 192 after noise reduction represents a rather clear limit cycle in a suitable time delay embedding. Defining a point inside the

---

[3] A phase shift of $-\pi/2$ in every Fourier component can easily be achieved by a replacement of the real part by the imaginary part and of the imaginary part by the negative real part.

limit cycle as a centre and the horizontal line through this point as a reference line, we could join every delay vector s_n with the centre point and measure the angle between this line and the horizontal line. Hence, we have a very intuitive definition of ϕ_n, which, through arranging the centre point suitably compared to the shape of the limit cycle, can be tuned to yield monotonically increasing ϕ_n. □

Once we have converted the original data into a sequence of two phase angles, we can try to find a unique relationship between these angles, as it would be necessary for the presence of generalised synchronisation. If there are episodes where these two phase angles evolve with a fixed phase difference, then the detection is even simpler.

Example 14.7 (Synchronisation between breathing and heart beats). Let us report on a very nice analysis of Schäfer *et al.* (1998). In the medical literature there is some controversy about whether the beating of the heart and breathing could be synchronised. Schäfer *et al.* investigated simultaneous recordings of the human ECG and of the breathing similar to the data of Example 14.6. The oscillatory nature of the air flow enabled them to assign to these data a phase angle $\phi_{air}(t)$ by the Hilbert transform. Instead of converting also the ECG data into a sequence of phase angles (which one could do by a linear interpolation from 0 to 2π in between two heart beats), they use the spikes of the ECG as a kind of stroboscope for the continuous phase angle of the air flow data. Hence, they create a sequence $\{\phi_{air}(t_n)\}$, where t_n is the time at which the nth heart beat occurred. Making a histogram of these angles on moving windows of time, they found several episodes of phase locking. In other words, during these episodes the heart rate was an integer multiple of the breath rate, so that the several heart beats during one breathing cycle re-occurred at the same phase angles. □

The latter example is not just a very convincing observation of synchronisation, but also very nice in the sense that it employs a dynamical systems technique which we discussed in Section 3.5, namely the Poincaré surface of section. The stroboscopic recording of ϕ_{air} is an intersection of the space spanned by the phase ϕ_{air} and the unspecified variables representing the heart dynamics by a Poincaré surface of section located at fixed values for the heart dynamics. During synchronisation, the Poincaré map has a stable periodic orbit which is identified by the above mentioned histograms.

Further reading

The currently most complete treatment of synchronisation was already mentioned above: Pikovsky *et al.* (2001). Recent original work on this topic can be found in

Pikovsky (1996) and Pecora *et al.* (1995). Other works on generalised synchronisation are Rulkov *et al.* (1995) and Parlitz (1996). An early work on time series aspects is Schiff *et al.* (1996).

Exercises

14.1 Write your own program to compute $\hat{I}_{xy}$ given by Eq. (14.8). Generate a time series of the logistic equation $x_{n+1} = 1 - 2x_n$. Choose $x = y$ and compute the time lagged interdependence, and compare it to the autocorrelation function. You should observe that $\hat{I}_{x,y}(\tau) = 1$ for a range of positive τ (since $x_{n+\tau}$ is a deterministic function of x_n). This can be detected to be so up to about its 10th iterate when using 10 000 data points and requiring about 10–50 neighbours). The autocorrelation function, instead, is zero for all lags $\tau \neq 0$.

Chapter 15

Chaos control

Regarding applications, chaos control is surely one of the most exciting outcomes of the theory of dynamical systems. [See Ott & Spano (1995) for a nontechnical account.] There exists an impressive list of experiments where chaos control has been applied successfully. Examples include laser systems, chemical and mechanical systems, a magneto-elastic ribbon, and several others. Additionally, there are claims that the same mechanism also works in the control of biological systems such as the heart or the brain.[1] After the pioneering work of Ott, Grebogi & Yorke (1990), often referred to as the "OGY method", a number of modifications have been proposed. We want to focus here on the original method and only briefly review some modifications which can simplify experimental realisations. We give only a few remarks here on the time series aspects of chaos control technology. For further practical hints, including experimental details, the reader is asked to consult the rich original literature. (See "Further reading" below.)

In most technical environments chaos is an undesired state of the system which one would like to suppress. Think, for instance, of a laser which performs satisfactorily at some constant output power. To increase the power the pumping rate is raised. Suddenly, due to some unexpected bifurcation, the increased output starts to fluctuate in a chaotic fashion. Even if the average of the chaotic output is larger than the highest stable steady output, such a chaotic output is probably not desired. Chaos control can help to re-establish at least a regularly oscillating output at a higher rate, with judiciously applied minimal perturbations. The idea of all the control methods discussed in this chapter is to identify unstable periodic orbits inside the chaotic attractor and to stabilise them by slightly modifying some control parameters.

---

[1] We are not aware of any attempts to control the stock market as yet, though.

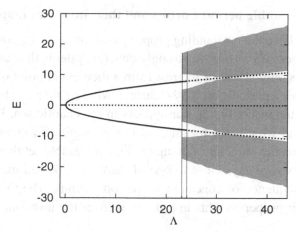

Figure 15.1 Bifurcation diagram for the Lorenz–Haken laser equations. The possible solutions in a suitable section, E, as a function of the pumping rate $\Lambda = r - 1$. The output power is related to E^2. In the range between the vertical lines at $\Lambda = 23.06$ and 23.74, both the chaotic attractor (grey) and the fixed points are stable (coexistence of attractors). When reducing Λ below 23.06, the chaotic attractor collides with the basins of attraction of the stable fixed points and thus turns into a *chaotic repeller*, until it disappears around $\Lambda \approx 20$. In this regime of Λ one can observe chaotic transients. The first bifurcation at $\Lambda = 0$ is a pitchfork bifurcation (in the laser system, $\Lambda < 0$ is unphysical).

Example 15.1 (Lorenz-like laser). Precisely the above scenario can be found in a far infrared (FIR) ring laser. The experimental set-up can be designed such that it can be modelled quite well by the Lorenz–Haken equations [Hübner *et al.* (1989) and (1993); see also Exercise 3.2 and Appendix B.1]. A typical chaotic signal from this system is shown in Fig. B.1. The pump power Λ enters the parameter r of the Lorenz equations, $r = \Lambda + 1$. In Fig. 15.1 we show the transition from regular motion to chaos as a function of Λ for the model equations. At $\Lambda = 23.74$ the steady states become unstable and disappear in the chaotic attractor.[2] Control in this case could consist of a stabilisation of one of the two unstable fixed points in the chaotic regime in order to establish a constant output power. More easily, one can stabilise a periodic orbit, such that the output oscillates regularly. Chaos control in laser systems has been successful in a number of cases, see e.g. Roy *et al.* (1992) and Gills *et al.* (1992). □

---

[2] There is a complication in the Lorenz system: the transition to chaos occurs via an inverse (also called sub-critical) Hopf bifurcation and thus for smaller Λ a chaotic attractor already coexists with the stable fixed points.

15.1 Unstable periodic orbits and their invariant manifolds

Chaotic systems have the outstanding property that the attractor possesses a skeleton of *unstable periodic orbits*. In strongly chaotic systems they are even dense.[3] Superficially, this only seems of interest from a theoretical point of view. If a trajectory approaches a periodic orbit by chance, it will remain close for a while, until it is repelled by the instability and disappears in the chaotic sea. But where does it disappear to and what is the meaning of it approaching by chance? If periodic orbits are dense on the attractor (although as an enumerable set they have measure zero), a trajectory is always close to one of them. So one can model the chaotic dynamics by a sequence of concatenated periodic orbits. They will then appear according to their proper weights in order to recover the invariant measure of the chaotic process.

The orbits we are speaking about are of the *saddle* type: they possess a stable manifold and an unstable one. Trajectories are attracted towards them along the first direction and are repelled along the second (how to find these manifolds is described in Section 15.1.2). Since a trajectory is repelled the faster the less stable an orbit is, it is natural to assume that the contribution of each orbit is proportional to its stability.[4]

15.1.1 Locating periodic orbits

For a systematic approach of the control problem, the first step is to find the location of the periodic orbits from the experimental time series. When we look at a typical chaotic time series, this seems to be a hopeless task. Nevertheless, when you remember what we said about periodic orbits and chaotic attractors, you can imagine that it is not only the periodic orbits that can be used to characterise chaos: conversely, a chaotic trajectory reveals information about the periodic skeleton. Let us note that the number $n(p)$ of different possible periodic orbits increases exponentially with the period p. The exponent is given by the *topo-*

---

[3] More precisely, in *Axiom A systems* [Smale (1967)]. For *non-hyperbolic systems* this is not known, but is generally assumed.

[4] Let us give you some flavour of the theoretical relevance of periodic orbits. In Section 11.2 we introduced the *linear instability* which describes the growth of small perturbations and is fully determined by the dynamics in tangent space. For a periodic orbit p of a map the instability is the largest eigenvalue of the product of Jacobians along the orbit, the logarithm of which is the largest Lyapunov exponent λ_p. In this way one can approximate the invariant natural measure of a chaotic attractor by the sum of Dirac δ-distributions located at all the unstable periodic orbits and with amplitudes given by $e^{-\lambda_p}$. Reordering this sum allows one to approximate the measure by a *periodic orbit expansion* and to compute averages such as the Lyapunov exponent. Since this sum also includes orbits which are just a repetition of shorter ones, one can interpret it as a kind of *grand canonical partition function* and develop parallels to the formalism of statistical physics (the number of particles in state p is then the number of repetitions of a certain orbit). This is one of the reasons why this theoretical approach is called *thermodynamic formalism*. The concept is very charming: not only can chaos be explained by the properties of ordered motion, but, moreover, by a highly developed formalism from a completely different field. The interested reader will find a review in Artuso *et al.* (1990).

logical entropy h_0: $n(p) \propto e^{h_0 p}$ for $p \to \infty$. For example, for the logistic map $x_{n+1} = 1 - 2x_n^2$ the topological entropy is $\ln 2$ and there are 2^p different orbits of length p.

The most direct signature of periodic orbits in a chaotic time series are close returns in a reconstructed state space [Auerbach *et al.* (1987), Gunaratne *et al.* (1989)]. The closer a chaotic trajectory happens to approach a periodic orbit the longer it remains in its neighbourhood. If the orbit has a short period and is not too unstable, one observes a few periods of such almost periodic motion, which is reflected by a series of close returns. To give an impression, Auerbach *et al.* (1987) were able to locate all periodic orbits of length ≤ 10 in the Hénon attractor with a time series of length 200 000. The best experimental example was given by Flepp *et al.* (1991) for a time series of the Zürich NMR laser system (operating at different parameters than those underlying the data set in Appendix B.2). Unfortunately, in many interesting situations such a scheme does not work satisfactorily. In unfavourable cases one may be able to identify the fixed points, but for higher periods this method fails in practice. The problem is, as usual, the lack of data. In order to observe a return to a sphere of radius ϵ for an orbit of period p and a Lyapunov exponent λ, the first approach has to be closer than $\epsilon e^{-\lambda p}$. If this is smaller than the average inter-point distance in the reconstructed state space, it is unlikely that one can find such an ϵ-return inside the data set.

Example 15.2 (Close returns for the NMR laser data). In Fig. 15.2 we visualise the problem using the NMR laser data, Appendix B.2. In a time delay embedding space unstable fixed points have to lie on the diagonal. We therefore show the

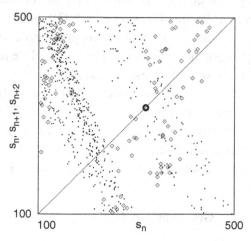

Figure 15.2 Visualisation of the method of close returns. The unstable fixed point of the NMR laser attractor lies somewhere inside the small circle on the diagonal. It should lie on the intersection of the graphs of s_n (diagonal line), s_{n+1} (dots), and s_{n+2} (diamonds) *versus* s_n.

relevant part of the upper right quadrant of Fig. 3.6 (s_{n+1} *versus* s_n), together with s_{n+2} *versus* s_n. The fixed point can only be on the diagonal in regions where the two representations of the attractor intersect each other. The circle shows where we located it using another method (see below).   □

The method of close returns is model free and is therefore not based on any assumptions about the dynamical system (apart from determinism). Unfortunately, this is also its weakness: the lack of data is its limitation. For more powerful methods we have to extract knowledge about the dynamics from the data in order to be able to interpolate towards the precise positions of the fixed points.

In the following we will restrict our discussion to scalar time series, since an extension to vector valued series is not straightforward. A periodic orbit of length p is a sequence of p values $s_1^*, \ldots, s_p^*$ (or a cyclic permutation of them), such that the p different m-tuples form the delay vectors of the points of the orbit in the delay embedding space. The deterministic equation $s_{n+1} = F(\mathbf{s}_n)$ describes, as usual, the dynamics in the delay embedding space. In order to be a periodic orbit, the values s_k^* above have to fulfil

$$F(\mathbf{s}_k^*) - s_{k+1}^* = 0, \qquad k = 1, \ldots, p, \tag{15.1}$$

where we again use a periodic continuation of the sequence s_k^*, so that, for example, $\mathbf{s}_1^* = (s_1^*, s_p^*, \ldots, s_{p-m+2}^*)$. This set of p nonlinear equations in p variables is what we would have to solve if we had complete knowledge of the function F.

In fact, in the best conceivable situation we possess a good global model for the dynamics in the minimal delay embedding space. In Chapter 12 we discussed this possibility thoroughly. The question is how to find *all* solutions of Eq. (15.1). If you consult the *Numerical Recipes* [Press *et al.* (1992)], you will read the very discouraging statement that "there never will be any good, general methods" (p.372). The arguments in support of this are quite convincing, and we recommend that you consult this book if you want to do something very elaborate.

Here we restrict ourselves to a brief presentation of a method which works if one has a good initial guess for the approximate location of the solution, called the Newton–Raphson method [Press *et al.* (1992)].[5] Let us read Eq. (15.1) as a p-dimensional vector equation and call its left hand side $\mathbf{G}$. We are looking for a p-dimensional vector $\mathbf{s}^*$ with $\mathbf{G}(\mathbf{s}^*) = \mathbf{s}^*$. In linear approximation, the equation reads

$$\mathbf{G}(\mathbf{s}^* + \delta\mathbf{s}^*) \approx \mathbf{G}(\mathbf{s}^*) + \mathbf{J}\delta\mathbf{s}^*. \tag{15.2}$$

---

[5] For a period orbit expansion in the spirit of Artuso *et al.* (1990), this would not be good enough: we need to be sure that we have found *all* existing orbits up to some period.

If the trial vector $\mathbf{s}^*$ yields the values $\mathbf{G}(\mathbf{s}^*) \neq 0$, we can solve Eq. (15.2) for the correction $\delta\mathbf{s}^*$ by inverting $\mathbf{J}\delta\mathbf{s}^* = -\mathbf{G}(\mathbf{s}^*)$, and the new trial vector is then $\mathbf{s}^* + \delta\mathbf{s}^*$. The matrix $\mathbf{J}$ is the Jacobian of $\mathbf{G}$, and its elements are simply the derivatives $(\mathbf{G})_{kl} = \partial(F(s_k^*) - s_{k+1}^*)/\partial s_l^*$. If one starts close enough to a solution of Eq. (15.1), an iteration of this scheme converges quite quickly. If the initial values are less reasonable, it might be necessary to reduce the weight of the correction $\delta\mathbf{s}^*$ in the first steps.

You might suspect that knowledge of the full nonlinear function F is not necessary to employ this scheme, since it only enters when we determine how far the trial values violate determinism. In fact, local linear approximations of the dynamics are completely sufficient. The elements of the Jacobian $\mathbf{J}$ can be taken from a local linear fit of the dynamics at each delay vector $\mathbf{s}_k^*$, obtained from the information about the time evolution of a small neighbourhood of $\mathbf{s}_k^*$. In Section 12.3.2 we derived local maps $s_{k+1} = a_0 + \sum_{l=1}^m a_l s_{k-l+1}$ for points $\mathbf{s}_k$ from a small neighbourhood of $\mathbf{s}_k^*$, which can easily be rewritten for the evolution of $\delta\mathbf{s}_k^* = \mathbf{s}_k - \mathbf{s}_k^*$. The values $F(\mathbf{s}_k^*)$ needed to compute the new trial values are just the constants in the latter expression. Thus the set of p linear equations for $\delta\mathbf{s}^*$ reads:

$$\sum_{l=1}^m a_l^{(k)} \delta s_{k-l+1}^* - \delta s_{k+1}^* = -\sum_{l=1}^m a_l^{(k)} s_{k-l+1}^* - a_0^{(k)} + s_{k+1}^*, \tag{15.3}$$

with periodic "boundary" conditions $\delta s_{k+p}^* = \delta s^*{}_k$. Again, one can iterate the procedure and find convergence. Let us recall that the size of the neighbourhood (both its diameter and the minimal number of points), in order to obtain the local linear dynamics, is a relevant parameter and, unfortunately, local linear fits are not a good tangent approximation of the global nonlinear dynamics on *all* parts of the attractor. Moreover, during the process of convergence, the local neighbourhoods change, and together with them, so do the local maps in a non-continuous way. Therefore, the condition $F(\mathbf{s}_k^*) = s_{k+1}^*$ cannot usually be fulfilled with infinite precision, since the method often traps itself in a situation where it oscillates between two almost periodic solutions with slightly different neighbourhoods. In summary, the periodic orbits are not as reliable as those obtained from a reasonable global model, since the location of the orbits of higher period may depend quite sensitively on the parameters of the maps.

Example 15.3 (Period-two orbits of the Hénon map). The period-two orbits of the Hénon map $x_{n+1} = 1 - ax_n^2 + bx_{n-1}$ are given by $x_\pm = (1 \pm \sqrt{-3 + 4a/(1-b)^2})(1-b)/2a$. In lowest order, this expression is proportional to any deviation δa from its true value, so that a 5% error in the determination of a leads to an equivalent error in the location of the periodic orbit.

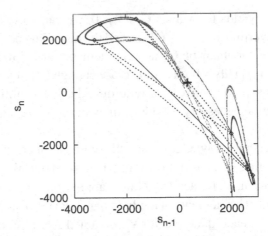

Figure 15.3 Period-one (cross), -two (solid line), and -four (dashed lines) orbits of the NMR laser dynamics. The results of the Newton–Raphson method using local linear fits of the dynamics are scattered within the area of the diamonds, whereas the orbits obtained from a global nonlinear model are unique and in optimal agreement with the local linear results. The only visible difference concerns the position of the fixed point (cross). Its coordinates turn out to be $s = 313.2$ in the global model and $s = 349 \pm 3$ in the local linear approach.

Another inconvenience comes along with the Newton–Raphson method. We finally end up with a huge collection of periodic orbits, but it is a nontrivial task to single out the different primary orbits, i.e. those which are not just repetitions of shorter ones. Even more difficult, we have to reduce this set to a set of orbits where each orbit occurs only once. Many different initial conditions of the Newton–Raphson method will lead to the same orbit.

Example 15.4 (Periodic orbits of the NMR laser data). For a demonstration of the Newton–Raphson method we compute the orbits up to period $p = 4$ of the NMR laser data, Appendix B.2. We use both local linear fits and a global polynomial of seventh order in three dimensional delay coordinates. Initial points in both cases are returns of the scalar time series during $p + 2$ successive time steps into intervals of 5% of the dynamical range of the data, i.e. $|s_{n+p} - s_n| < 335$ for $n = n_0, \ldots, n_0 + p + 1$. See Fig. 15.3 for the results. Using either of the methods, we are able to locate periodic orbits up to period 14 without problems. We do not find primary orbits with odd periods apart from period 13. This does not say that these orbits do not exist, but that due to the strong anticorrelation of successive measurements there are no sufficiently close returns. □

The numerical solution of the fixed point equations by a Newton–Raphson method is very efficient, when the starting points for the procedure are already close to a fixed point. The following method due to So *et al.* (1996), (1997) can yield rather precise estimates for those, and can even replace the Newton–Raphson method if one does not require the highest possible precision. It exploits essentially the same information used before, but in a different way. In So *et al.* (1996) a local transformation is introduced which is applied to the invariant measure on the attractor. After transformation, the measure has singularities at the unstable periodic points belonging to orbits of a selected period.

This periodic orbit transform can be most easily derived for a one dimensional linear map $x_{n+1} = ax_n + b$ with the fixed point $x_f = b/(1 - a)$ $(a > 1)$. One wishes to find a map from an arbitrary element x_k of the time series onto the fixed point, which means that we need to estimate a and b. For these estimates we make use of the fact that in our time series the values of x_{k+1} and x_{k+2} are related to x_k by iterating the linear map, hence

$$\hat{x}_k = \left(x_{k+1} - x_k \frac{x_{k+2} - x_{k+1}}{x_{k+1} - x_k} \right) \bigg/ \left(1 - \frac{x_{k+2} - x_{k+1}}{x_{k+1} - x_k} \right) . \tag{15.4}$$

For a nonlinear relation between x_{k+1} and x_k, this transform maps points onto the fixed points of the x_k-dependent local linear maps. For all x_k in the vicinity of the fixed point where we can linearise the dynamics, the transform identifies exactly this unique linear map and hence maps all these points onto the fixed point. For points far away from the fixed point, the parameters a and b of the linearisation contain the nonlinearities and hence depend on x_k. Hence points far away from the fixed point are simply mapped around but typically do not generate singularities. However, this transformation conserves singularities which are already present in the invariant measure, whose images hence have to be distinguished from the newly created singularities. Exactly this complication arises, e.g., in one dimensional maps such as the logistic equation $x_{n+1} = 1 - ax_n^2$. In So *et al.* (1996) a remedy of this problem is suggested, which involves a kind of randomisation in the transform which is active only far away from the fixed point. A simple alternative consists in applying the transformation Eq. (15.4) only to those points x_k which fulfil a close return condition $|x_{k+1} - x_k| < \epsilon$. The generalisation of the transform method to period orbits of period p, to higher dimensional systems, and the inclusion of second order effects is described in detail in So *et al.* (1997).

The periodic orbit transform (which is essentially a single Newton–Raphson step with a crude estimate of the derivative) maps many individual points into the vicinity of the orbit of the system, which is identified by the singularities of this distribution. The Newton–Raphson method discussed before iterates a single candidate until convergence, where the redundant information stored in the time series is exploited by the fits of the local linear dynamics. We suggest to use the periodic orbit transform for finding initial values of the Newton–Raphson method, in

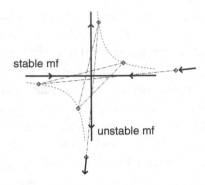

Figure 15.4 A schematic view of a fixed point and its stable and unstable manifold. The symbols represent the path of a chaotic trajectory in its vicinity.

order to save computation time and to avoid the problems related to the interpretation of the points found by the Newton–Raphson method.

15.1.2 Stable/unstable manifolds from data

Assume that we know the location of an orbit that we want to stabilise. When the system moves towards this point but misses it by some small amount, a tiny perturbation will suffice to push it onto the orbit, whereas it disappears in the chaotic sea otherwise. How can we find the correct perturbation? We have to know about the structure in tangent space around the fixed point.

A hyperbolic fixed point has a *stable* and an *unstable manifold*. A trajectory approaches it along the stable one and is repelled from it along the unstable one. The continuous curve underlying these steps is a hyperbola. As a complication, in a Poincaré map, one may additionally observe that the trajectory alternates from one branch of the manifold to the opposite one. This is shown schematically in Fig. 15.4. We can approximate these manifolds from the observed data by looking at all segments of the time series which form the close returns, and interpolating between them linearly.

More systematically, we can again make either a global (nonlinear) or locally linear fit to the dynamics and estimate the invariant manifolds from these fits. For the locally linear fit described in Section 12.3.2 we can form the product of the matrices **A** [cf. Eq. (12.7)]. The eigenvectors of this product yield the direction of the stable and unstable manifolds.

Example 15.5 (NMR laser data). In Fig. 15.5 we show the scenario depicted in Fig. 15.4 for experimental data (NMR laser, Appendix B.2). The full black circle is the unstable fixed point, and we show sequences of close returns by the symbols connected by the dashed lines. The trajectories enter the window from the left.

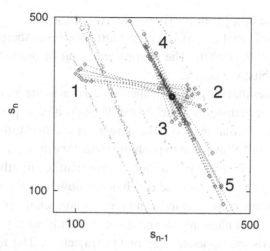

Figure 15.5 Part of the NMR laser attractor around the unstable fixed point and close returns. The numbers indicate in which sequence a typical trajectory visits this neighbourhood. The diagram thus reveals the structure of the stable and unstable manifold.

The first iteration leads to the cloud of points to the right of s^* and indicates the direction of the stable manifold. The next iterate is already almost on the unstable manifold, and some of the points lie above, some below the fixed point. They leave the window after one more iteration, which maps them onto the opposite side of the fixed point. □

Stable and unstable manifolds of periodic orbits of period $p > 1$ can be found in exactly the same way. For example, for period $p = 2$, one focuses on the second iterate of the map, and considers the two fixed points of the desired orbit separately.

In a recent paper by Pierson & Moss (1995) the numerical evidence for the stable and unstable manifolds was used as an additional criterion for a sequence of points close to the diagonal to represent a true fixed point. As they show, such signatures can also appear accidentally in sequences of random numbers, but with much less probability. The purpose of their investigation is to establish a new criterion to distinguish determinism from noise rather than to identify the orbits.

15.2 OGY-control and derivates

When we carry out chaos control by tiny perturbations we first have to wait. Due to ergodicity on the attractor, a chaotic trajectory will approach every periodic orbit arbitrarily close if we wait long enough. The individual transient times t_{trans} trajectories need to pass from an arbitrary point on the attractor to within a neighbourhood of size ϵ of a desired orbit are distributed exponentially, $P(t_{\text{trans}}) \propto \exp(t_{\text{trans}}/\langle t_{\text{trans}} \rangle)$, and the average time $\langle t_{\text{trans}} \rangle$ follows a power law in ϵ, $\langle t_{\text{trans}} \rangle \propto \epsilon^{-\gamma}$, where γ assumes a value close to the fractal dimension of the attractor [Grassberger (1983), Halsey *et al.* (1986), Ott *et al.* (1990)]. The waiting time can be dramatically reduced by the *targeting* of orbits; see below.

Once one observes that the chaotic trajectory arrives in the vicinity of the desired point, the control mechanism is switched on. If one is able to perturb the trajectory directly, the above information about the tangent space structure in the neighbourhood of the target orbit allows one to push the trajectory onto the stable manifold of the fixed point. Very often it is much more convenient to slightly change a control parameter of the system. In this case one has to know how the fixed point and its manifolds move in the reconstructed state space under changes of this parameter. Control then consists of changing the parameter slightly such that the stable manifold moves towards the opposite side of the trajectory. The trajectory feels the repulsion due to the instability towards the opposite direction and will therefore be pushed towards the original position of the stable manifold. Turning the parameter back to its original value, the trajectory moves towards the fixed point. Since one is working in a small area in phase space, it is sufficient to know only all functions on the level of linear approximation around the desired orbit.

Following Ott *et al.* (1990) again, we imagine an experiment which represents a continuous time dynamical system, a flow. The data are taken with a relatively high sampling rate, such that we can safely reduce them to data representing a Poincaré map. In a delay embedding we want to denote these map-like data by $\mathbf{x}_n$. Now let us assume that we identified a fixed point $\mathbf{x}^*$ using the methods of the last section. In order to develop the control algorithm, all dependencies are considered in linear order only. So we assume that under perturbation of a control parameter, $\alpha \mapsto \alpha + \Delta\alpha$, $\Delta\alpha \ll 1$, $F \mapsto \bar{F}$, the fixed point moves to a point $\bar{\mathbf{x}}^* = \mathbf{x}^* + \mathbf{g}\Delta\alpha$. Moreover, the directions of the stable and unstable manifolds, $\mathbf{e}_s$ and $\mathbf{e}_u$, as well as the corresponding eigenvalues λ_s and λ_u, are assumed to remain essentially unchanged under the small perturbation. Since $\mathbf{e}_u$ and $\mathbf{e}_s$ usually (as in Example 15.5) are not orthogonal, in the following we need their dual vectors $\tilde{\mathbf{e}}_u$ and $\tilde{\mathbf{e}}_s$, defined by $\tilde{\mathbf{e}}_u \cdot \mathbf{e}_u = \tilde{\mathbf{e}}_s \cdot \mathbf{e}_s = 1$ and $\tilde{\mathbf{e}}_u \cdot \mathbf{e}_s = \tilde{\mathbf{e}}_s \cdot \mathbf{e}_u = 0$. While the difference between the image of a point $\mathbf{x}_n$ and the fixed point $\mathbf{x}^*$ without perturbation would be

$$F(\mathbf{x}_n) - \mathbf{x}^* = \lambda_u[(\mathbf{x}_n - \mathbf{x}^*) \cdot \tilde{\mathbf{e}}_u]\,\mathbf{e}_u + \lambda_s[(\mathbf{x}_n - \mathbf{x}^*) \cdot \tilde{\mathbf{e}}_s]\,\mathbf{e}_s\,, \qquad (15.5)$$

we find under the perturbation $\Delta\alpha$

$$\bar{F}(\mathbf{x}_n) - \bar{\mathbf{x}}^* = \lambda_u[(\mathbf{x}_n - \bar{\mathbf{x}}^*) \cdot \tilde{\mathbf{e}}_u]\,\mathbf{e}_u + \lambda_s[(\mathbf{x}_n - \bar{\mathbf{x}}^*) \cdot \tilde{\mathbf{e}}_s]\,\mathbf{e}_s\,. \qquad (15.6)$$

Now we want to map $\mathbf{x}_n$ by the modified dynamics $\bar{F}$ onto the stable manifold of $\mathbf{x}^*$ of F. We need to determine the perturbation $\Delta\alpha$ for which $(\bar{F}(\mathbf{x}_n) - \mathbf{x}^*) \cdot \tilde{\mathbf{e}}_u = 0$. Using $\bar{\mathbf{x}}^* = \mathbf{x}^* + \mathbf{g}\Delta\alpha$ we find for the perturbation $\Delta\alpha$

$$\Delta\alpha = \frac{\lambda_u}{1 - \lambda_u} \frac{(\mathbf{x}_n - \mathbf{x}^*) \cdot \tilde{\mathbf{e}}_u}{\mathbf{g} \cdot \tilde{\mathbf{e}}_u}. \tag{15.7}$$

After application of the perturbation $\Delta\alpha$, the trajectory lies much closer to the stable manifold than before. Generally, it will not be exactly on the stable manifold, since in the above treatment we ignored higher than linear terms in both the parameter dependence and in the deviation from the fixed point. Moreover, in real world situations, the trajectory will most probably be subject to noise and will therefore feel additional perturbations. So one has to compute a new parameter perturbation $\Delta\alpha_{n+1}$ from the next observation $\mathbf{x}_{n+1}$, and so on. For a complete derivation of these ideas and a discussion of what to do for more than two-dimensional Poincaré maps see the paper by Romeiras *et al.* (1992).

In order to avoid the long transient times, one can apply a method called *targeting* [Shinbrot *et al.* (1990)]. One either needs a global model of the dynamics or enough experimental observations to form local approximations. Sensitive dependence on initial conditions allows one to dramatically change the time that a trajectory needs to pass from a point $\mathbf{x}_{\text{from}}$ into the ϵ-neighbourhood of a point $\mathbf{x}_{\text{to}}$. Simply take a reasonable neighbourhood of $\mathbf{x}_{\text{from}}$ and iterate it forward in time. Due to the positive Lyapunov exponents it will be stretched and eventually (say, after n iterations) have a linear extension of order 1. Also, take an ϵ-neighbourhood of $\mathbf{x}_{\text{to}}$ and iterate it backward in time. After some iterations it will also form a line segment which represents a part of the stable manifold of $\mathbf{x}_{\text{to}}$, and after the mth backward iteration it will intersect the nth image of the $\mathbf{x}_{\text{from}}$-neighbourhood somewhere. The nth pre-image of this intersection, call it $\mathbf{x}$, is a neighbour of $\mathbf{x}_{\text{from}}$ which will be mapped in only $n + m$ iterations into the neighbourhood of $\mathbf{x}_{\text{to}}$. The difference $\mathbf{x} - \mathbf{x}_{\text{from}}$ is the small ($< \epsilon$) perturbation to be applied to $\mathbf{x}_{\text{from}}$. Some refinements of this procedure allow us to also cope with noise and with imperfect models of the dynamics. Again one can substitute the direct manipulation of $\mathbf{x}_{\text{from}}$ by a small temporary change of a control parameter of the system.

Example 15.6 (Magneto-elastic ribbon). Ditto *et al.* (1990) were the first to apply the OGY method in an experimental set-up. The elasticity (Young's modulus) of a magneto-elastic ribbon can be changed by a small magnetic field. An oscillating magnetic field periodically modulates its elasticity and thus the energy stored in the material. If such a ribbon is mounted in the gravitational field of the earth, the phase space of its movement is three dimensional and chaotic motion is possible. Due to the periodic forcing, the best Poincaré surface of section is given by a stroboscopic view of the system. Hence the control method can be applied in a two-dimensional reconstructed state space. Ditto *et al.* (1990) could localise different unstable periodic orbits and stabilise them.   □

Modified versions of the OGY method for control and targeting have been applied successfully in several experimental situations. Most of them are clearly low dimensional, such as lasers [Roy *et al.* (1992), Gills *et al.* (1992)] or electric circuits [Hunt (1991)]. Others are only effectively low dimensional, such as the thermal convection experiment of Singer *et al.* (1991). The most spectacular, but also controversial, work concerns the control of the cardiac tissue of a rabbit, where chaos could be suppressed [Garfinkel *et al.* (1992)].[6]

The difficulty of the original OGY method clearly lies in the detailed information that one needs about the periodic orbits and their stable and unstable manifolds and in the computational effort involved in each single control step. On the other hand, it allows for direct and precise control. From the observation of a system over some time, one can extract the periodic orbits and decide which one wants to stabilise. One can switch on the control algorithm for the desired orbit and the signal will almost immediately follow the prescribed path. By switching from one target orbit to another, one can in principle force the system to adapt to different external requirements.

15.3 Variants of OGY-control

In most of the experiments cited above, simplified versions of OGY-control were used. The most direct simplification lies in a reduction of dimensionality. If the attractor in the Poincaré surface of section is close to one dimensional, one can approximate it by a single valued function, $x_{n+1} = F^{(\alpha)}(x_n)$. In this case there is no need to identify stable or unstable manifolds. The effect due to a parameter perturbation $\Delta\alpha$ is simply $\Delta\alpha\,\partial F/\partial\alpha|_{x_n}$. The gradient of F with respect to α has to be estimated experimentally.

Another variant is called *occasional proportional feedback* and was used in the experiments by Hunt (1991) and Roy *et al.* (1992). No periodic orbits are sought, and the parameter perturbation to achieve control is allowed to be fairly large. Control is activated every time the observable (in a Poincaré surface of section) enters a prescribed region. The control parameter of the experiment is changed proportionally to the difference between the observable and the value in the centre

---

[6] The criticism is that in such a system the presence of low-dimensional deterministic dynamics is highly questionable and thus it is not clear how the control actually works; doubtlessly, the authors were able to drive the system into a periodic state. But the state was not the desired one and the perturbations were not particularly small.

of the interval. This only leads to occasional control steps. In particular, when the resulting orbit is of period p, control is activated at most every pth step. By changing the centre and the width of the window and the amplitude of the feedback, one can search for all kinds of periodic orbit. However, due to the fact that the parameter perturbation may be quite large, the orbits found in this way do not necessarily exist in the unperturbed system. This method is very simple and does not require any knowledge about the system equations. In particular, it can be implemented in hardware without digital computing. On the other hand, it is a somewhat trial-and-error method and the resulting orbit cannot be chosen beforehand.

15.4 Delayed feedback

The control step of the OGY-control protocol needs some time. Some, although simple, numerical computation has to be performed, and its output has to be translated into the perturbation of the control parameter. Even with fast equipment this may take about 10^{-5} s, which is too long for certain experiments, where internal time scales are shorter. In such cases another, and more recent, class of control methods can be of interest, since it makes use of a simple delayed feedback. We briefly report on the ideas proposed by Pyragas (1992) (see also Pyragas & Tamaševičius) (1993), but similar suggestions have also been made by other authors.

Unlike in OGY-control, the idea is here that the time evolution of a variable of the system can be influenced by an additional, time dependent force $G(t)$. It will depend on the experimental situation how easily this can be achieved. The amplitude of $G(t)$ is then adjusted proportional to the difference between the present signal and its value a well-defined time before,

$$G(t) = g(s(t) - s(t - p)). \tag{15.8}$$

In a periodic solution of period p, this perturbation vanishes identically. Therefore, this is also a control method by tiny perturbations. The advantage with respect to OGY and related methods is that the delayed feedback method is extremely simple. It does not rely on a Poincaré surface of section and therefore acts on the system not only sporadically but continuously over time. More importantly, one does not need any specific *a priori* knowledge about the periodic orbits and their locations on the attractor. One just scans a reasonable range of values for p and tries different feedback amplitudes g for optimal control. Again, ergodicity on the attractor helps since the trajectory will find the period p orbit by itself, if it exists.

Meanwhile, also the time delay feedback control has been subject to intense theoretical studies. Hence, it is well understood, why certain periodic orbits cannot be stabilised by the standard procedure, and how to improve the control scheme by introducing multiple delays or even an additional destabilising term. Instead of listing the numerous original literature (which, by the way, is not directly related to time series and therefore outside the scope of this book) we refer to the *Handbook of Chaos Control* [Schuster (1999)].

15.5   Tracking

A relevant application of control both in experiments and in numerical investigations of model equations consists of tracking. Very often it is not at all easy to find the interesting periodic orbits in the chaotic regime of the dynamics. However, many unstable periodic orbits have stable counterparts in other regimes of some control parameter. For instance, in the logistic equation, all the periodic orbits of the Feigenbaum (period doubling) bifurcation scenario persist at higher values of the control parameter and just change their stability properties at some bifurcation point. Therefore, both experimentally and numerically, it can be very convenient to start in a regime of stability of a certain orbit, and then just force the system to stay on this particular orbit when tuning the parameter to the desired value.

Beyond the bifurcation point at which the desired orbit loses stability, one has to employ the control. But as a complication, when changing the parameter further, the position of the orbit changes in an *a priori* unknown way, so that the information entering Eqs. (15.6) and (15.7) is lacking. Fortunately, since one needs only local information, it is possible to acquire it from the behaviour of the trajectory in the vicinity of the desired periodic orbit.

For simplicity we again speak of fixed points only. Assume that $\mathbf{x}^*$ is the current position of the fixed point and we shift the control parameter a small step towards its goal value. The attempt to stabilise $\mathbf{x}^*$ instead of the true new fixed point $\mathbf{x}^{*\prime}$ will require control steps which tend to have the same direction. Since $\mathbf{x}^*$ has shifted, the trajectory will try to escape from it in a systematic way and has to be pushed back by each control step. From this nonzero average correction one can derive the true position of the fixed point. Thus the control parameter may be shifted at the next tiny step, and so on. In summary, one can follow the path of the fixed point in parameter space.

In applications, this means that one can follow periodic orbits into ranges of the system parameter where they are extremely *unstable*. A typical chaotic trajectory would almost never approach such an orbit closely. As a consequence, they are difficult to locate and one might have to wait for a long time until the trajectory comes close enough to initialise the control algorithm. Tracking offers the solution of this problem. For a recent experiment see In *et al.* (1995).

15.6 Related aspects

The control techniques we discussed in this chapter all try to stabilise some particular kind of motion which is embedded, but usually hidden, within a chaotic attractor. A different control problem is how to influence a chaotic system with multiple attractors so that it switches from one attractor to the other. As a practical example one might think of a diseased heart as a dynamical system which allows for two qualitatively different attractors. One is the normal cardiac rhythm, the other is a state of fibrillation, a life threatening arrhythmia. An interesting question is how to defibrillate the heart with the smallest invasion. Current clinical practice is to use a single but strong electric pulse which renders all cardiac tissue refractory, thus interrupting the re-entry arrhythmia.

To our knowledge, the problem of attractor switching has not yet been studied very systematically. Thus we only give references to single related works that we are aware of: Tsonis & Elsner (1990), Jackson (1990) and (1991), Nagai & Lai (1995), and Xu & Bishop (1994).

Although the control of chaotic systems is a rather new field, control theory in general is a well-studied subject with good coverage in the literature. [See e.g. Wiener (1961), Healey (1967), Berkovitz (1974), or Burghes & Graham (1980).] The difference between classical control theory and chaos control is mainly the following. Classically, one wants to stabilise a system in a desired state against external influences which would otherwise sooner or later drive the system out of that state. Without control, the system would show irregular or even unbounded behaviour due to external influences. Chaotic systems without control also show irregular motion, but not due to external noise, rather due to the intrinsic nonlinearity of the system. Much of the nonlinear dynamical structure can be studied while the system is running in the chaotic regime. This knowledge then leads to particular powerful ways of directing the system towards the desired state and then stabilising it there.

Further reading

Obviously, time series analysis and control are – although related – different topics. Thus we could only very briefly sketch the time series aspects involved. If you are interested in control, you should go back to the original articles, some of which are reprinted in Ott *et al.* (1994). Apart from the works already cited in the text, you will find interesting material in Cheng & Dong (1993), Dressler & Nitsche (1992), Auerbach *et al.* (1992) and Shinbrot *et al.* (1993), for example. The most complete discussion representing the state of the art in the year 1998 is given by the *Handbook of Chaos Control* edited by H. G. Schuster (1999).

We were very brief in the discussion of the relevance of periodic orbits. In fact, there exists a well-developed theory relating periodic orbits to the invariant measure of a dynamical system and all its further invariants. This is well presented by Artuso *et al.* (1990) and in the book by Beck & Schlögl (1993).

Exercises

15.1   Write your own program to locate periodic orbits in a time series. Implement both the method of close returns and (advanced) the Newton–Raphson method. Apply it to a series from the Hénon map ($a = 1.4$, $b = 0.3$; see Exercise 3.1) and (advanced) check your findings for completeness by reproducing the method of Biham & Wenzel (1989).

15.2   Reconstruct the stable and unstable manifolds of the unstable fixed point and the period-two orbit of the Hénon map, following Section 15.1.2. Determine the multipliers λ_u and λ_s from the data.

15.3   You know $\mathbf{x}^*$ and the tangent space structure from the preceeding exercises. Determine $\mathbf{x}^*$ for a slightly different value of the parameter a, $a' = a + 0.1$. This yields $\mathbf{g}$. Now apply the OGY-control, i.e., write a program which iterates the Hénon map and additionally perturbs the parameter a as soon as the trajectory approaches the unstable fixed point. Convince yourself that you can stabilise $\mathbf{x}^*$. Compute the average size of the perturbation as a function of some additional perturbation of the dynamics by noise.

15.4   The unstable fixed point of the Hénon map, together with its local stability properties, can be computed analytically. Compare the exact values with your numerical results. Find out what happens if you use slightly wrong values for some of the quantities $\mathbf{x}^*$, $\mathbf{g}$, $\mathbf{e}_u$, $\mathbf{e}_s$, λ_u, λ_s.

Appendix A
Using the TISEAN programs

In this chapter we will discuss programs in the TISEAN software package that correspond to algorithms described in various sections throughout this book. We will skip over more standard and utility routines which you will find documented in most software packages for statistics and data analysis. Rather, we will give some background information on essential nonlinear methods which can rarely be found otherwise, and we will give hints to the usage of the TISEAN programs and to certain choices of parameters.

The TISEAN package has grown out of our efforts to publicise the use of nonlinear time series methods. Some of the first available programs were based on the code that was printed in the first edition of this book. Many more have been added and some of the initial ones have been superseded by superior implementations. The TISEAN package has been written by Rainer Hegger and the authors of this book and is publicly available via the Internet from http://www.mpipks-dresden.mpg.de/~tisean

Our common aim was to spare the user the effort of coding sophisticated numerical software in order to just try and analyse their data. There is no way, however, to spare you the effort of understanding the methods. Therefore, still none of the programs can be used as a black box routine. Programs that would implement all necessary precautions (such as tests for stationarity, estimates of the minimal required number of points, etc.) would in many cases refuse to attempt, say, a dimension estimate. But, even then, we suspect that such a library of nonlinear time series analysis tools would rather promote the misuse of nonlinear concepts than provide a deeper understanding of complex signals.

The recommended way of using each of the programs discussed below is first to read the corresponding section of the book and understand what the routine does. Then one would test it on an example where something about the solution is already known. You can, for example, reduce the noise that you added artificially on a deterministic data set or determine the dimension of artificial data for which there is an accepted value in the literature. See also the exercises given in each chapter. Once acquainted with the behaviour of the algorithm, one can venture to use it on real data, provided the data set seems a reasonable candidate.

Specific information on the programs is correct to our knowledge for TISEAN 2.1. Details in the calling sequence and options may change, although some stability has been reached by now. The TISEAN programs are distributed to be used from a command prompt, for example a shell. On operating systems that permit forking of subprocesses, such as all varieties of UNIX, it is often convenient to use pipelines to connect the output of one program with the input of another. We often use this opportunity within a plotting program. The TISEAN package comes with its own documentation which lists all the supported flags and options and tells you what the input and output are. A review paper, Hegger *et al.* (1999), gives a survey with many examples of what you can do with the TISEAN programs.

In the next section we will give some general information which is relevant to almost all of the routines. Whereas the knowledge about neighbour search is more educational, the reading of the commonly used options is mandatory for the application of the programs. The Sections A.2–7 contain routines matching the topics of the Chapters 2–7 of the first part of the book, where the advanced topics are already included, and A.8–10 are concerned with the "leftovers" of the second part. The level of treatment will range from just mentioning the names of the routines and linking them to equations in the text, to rather elaborate discussions of the algorithms, when we believe that the subject deserves it.

A.1   Information relevant to most of the routines

A.1.1   *Efficient neighbour searching*

Finding nearest neighbours in m-dimensional space is the one ubiquitous computationally demanding task encountered in the context of phase space based time series analysis. As long as only small sets (say $N < 1000$ points) are evaluated, neighbours can be found in a straightforward way by computing the $N^2/2$ distances between all pairs of points. However, modern experiments are able to provide much larger amounts of data. With increasing data sets, efficient handling becomes more important.

Neighbour searching and related problems of computational geometry have been studied extensively in computing science, with a rich literature covering both theoretical and practical issues. For an introduction see Knuth (1973) or Preparata & Shamos (1985). In particular, the tree-like data structures are studied in Bentley (1980), and the bucket (or box) based methods in Devroye (1986). For a tutorial about neighbour searching in the context of time series analysis, see Schreiber (1995).

Although considerable expertise is required to find and implement an optimal algorithm, a substantial factor in efficiency can be gained with relatively little effort. The use of any intelligent algorithm can result in reducing CPU time (or increasing the maximal amount of data which can be handled with reasonable resources) by orders of magnitude, compared to which the differences among these methods and the gain through refinements of an existing algorithm are rather marginal.

How to find nearest neighbours in m-dimensional space? The textbook answer is to use multidimensional trees. From the point of view of computing theory they have the advantage that they can be proved to have an asymptotic number of operations proportional to $N \ln N$ for a set of N points, which is the best possible performance for sets of arbitrary distribution. We refer the reader to the literature cited above for general reference and to Bingham & Kot (1989) for an implementation in the context of time series analysis.

Although there is not much to say against tree methods, we prefer a box-assisted approach for its simplicity. Moreover, for the rather low-dimensional sets we usually encounter, these are generally faster. If the points are not too clustered, the operation count will only be $\propto N$ for N points. A simple box assisted method which is worth the effort even for sets of only moderate size has been heuristically developed in the context of time series analysis [Grassberger (1990); see Theiler (1987) for a similar approach].

Consider a set of N vectors $\mathbf{x}_n$ in m dimensions, for simplicity rescaled to fall into the unit cube. For each $\mathbf{x}_n$ we want to determine all neighbours closer than ϵ in the max norm. We could also say that we want to determine the set of indices

$$\mathcal{U}_n(\epsilon) = \{n' : \|\mathbf{x}_{n'} - \mathbf{x}_n\| < \epsilon\}. \tag{A.1}$$

This is preferable for technical reasons since we do not want to move the whole vectors around, in particular not if they are obtained as delay coordinates from a scalar time series.

The idea of box-assisted methods is the following. Divide the phase space into a grid of boxes of side length ϵ. Then each point falls into one of these boxes. All its neighbours closer than ϵ have to lie in either the same box or one of the adjacent boxes. Thus in two dimensions only nine boxes have to be searched for possible neighbours, i.e. for $\epsilon < \frac{1}{3}$ we have to compute fewer distances than in the simple approach (see Fig. A.1). Note that the same reasoning holds if we use other norms than the maximum distance, since neighbourhoods of radius ϵ in the Euclidean or any other L^q norms are always contained in the ϵ-box.

The technical problem is how to put the points into the boxes without wasting memory. How to provide the right amount of memory for each box without knowing

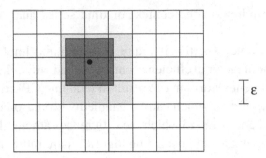

Figure A.1 All neighbours closer than ϵ to the reference point will be found within the nine shaded boxes if the grid spacing is chosen to be ϵ.

beforehand how many points it will contain. The conceptually simplest solution is to obtain this information by first filling a histogram which tells how many points fall into which box. With this information one can divide an array of N elements so that each box is assigned a contiguous section of memory just large enough to hold all the points in the box. For each box, one has to store the position where this section starts. The end of the section is given by the starting position for the next box. Once the indices of all the points are stored in the section corresponding to the appropriate box, scanning a box simply means scanning its section of the array.

The method works most efficiently if the expectation value of the number of points per box is $O(1)$. For small box size ϵ this can be achieved by providing roughly as many boxes as there are points. Thus we need $2N$ storage locations in addition to the data.

Sometimes we need a *minimal number* of neighbours rather than a neighbourhood of fixed radius. This is the case for the nonlinear noise reduction algorithm. In this case we form neighbourhoods of fixed radius but mark the points which do not have enough neighbours within this radius. These points are visited in a next sweep, where slightly larger neighbourhoods are formed. Again, we mark the points which cannot be served yet and repeat the procedure until all are satisfied. This is the most efficient approach if we don't have to process the points in their temporal order.

So far, we have only mentioned two dimensional grids of boxes. Of course, the technique described generalises easily to higher dimensions. However, we will need much more storage to store a decent cover of the point set. Instead, one can use fast neighbour search in two dimensions and use the resulting points as *candidates* for neighbours in higher dimensions, which then have to be confirmed. The same strategy is useful if we cannot cover the whole point set with a grid of given resolution. Then we simply wrap around at the edges and find candidates for real neighbours again.

The method described is used in many of the TISEAN programs. It is most efficient if the points are restricted to a lower dimensional subset, like an attractor. For high dimensional data, the savings are not as dramatic and other methods may be better.

A.1.2 Re-occurring command options

A.1.2.2 The help option

Every program lists the mandatory and the optional flags together with a brief description of its purpose when it is started together with the flag -h and quits hereafter, regardless of how complete the command is otherwise.

A.1.2.2 Input data

Input data are typically given as one or more columns inside a data file (ASCII format: integer, float, exponential or other. Note: decimal point, not comma!). Some routines also take input data from stdin, the command line. This can be useful when piping programs (see above). Every line in the input data is interpreted as one time step.

In virtually every routine, the number of points (length of the time series) to be read from the input file can be restricted by the option -l#, where # stands for a positive integer. Reasonable results cannot be expected for data sets smaller than several hundreds to several thousands of points, depending on the specific application and on the spatial dimension. The default value is always the whole file length. Initial lines of the data file can be skipped by the -x# (exclude) option. Sometimes, one likes to include comments on the origin of the data as a kind of header to a data file; in such cases the -x# will avoid that these will be taken as data (default: -x0). Since sometimes data files contain several columns (e.g., the first column might be the time), the -c# option can be used to select the proper column to be read. In those routines which are already written to work on multivariate data, as many columns have to be specified as input variables are to be read. The default is to start from the first column.

A.1.2.2 Embedding space

All those routines which are in the focus of this book work in a state space or an embedding space. Currently, almost all routines are being extended to the usage of multivariate data. Then, the option -m#1,#2 specifies the dimensions of the reconstructed space. The first argument fixes the dimensionality of the input data, wheres the second argument determines the temporal embedding dimension m.

Often, the default is -m1, 2, meaning that a scalar time series is to be embedded in two dimensions. Alternatively, e.g., -m2, 5 means that bivariate data are to be read and a two by five dimensional mixed embedding is to be used: a bivariate delay vector with five lags in total. As you are aware of, the time delay embedding requires a time lag. This is usually specified by -d#. When a default is given, this is -d1 (assuming map-like data). In several routines, no defaults are predefined for these essential parameters, in order to make sure that the user first thinks about reasonable values before starting the program.

A.1.2.2 Defining neighbourhoods

All those programs which require some locality in space work with neighbourhoods. Of course, first the space in which they live has to be set by the options of the last paragraph. Next, either the minimal number of neighbours has to be given by -k#, or the neighbourhood diameter has to be fixed. Sometimes, it makes sense to define a minimal, sometimes also a maximal neighbourhood diameter, together with the minimal number of neighbours. The proper neighbourhoods constructed may differ in the different programs. When input data are multivariate, a neighbourhood requires a specific metric. The TISEAN routines offer the possibility either to work with the unrescaled data, or to first rescale the input data to unit variance of each component separately, and then work with isotropic norms. As we discussed in Section 9.5, neither choice is guaranteed to be optimal, but they are general enough to be practicable. In certain cases, the user will have to rescale the data components as required by the situation.

A.1.2.2 Output data

Many routines can print the output to the screen (more precisely, to stdout). This is very useful when running the program inside a plot-program such as gnuplot, where the result can be presented visually without intermediate storage in a file. This makes life very easy if parameters are to be optimised interactively by visual inspection. Usually, the -o option without argument forces the output to be written to a standard output name, which typically is the name of the input file plus some extension which encodes the program by which it was created. Alternatively, an output name can be specified by -o fname, where fname is a character string.

A.2 Second-order statistics and linear models

TISEAN was not designed to present the most sophisticated implementation of linear tools. There are specialised toolboxes and statistics packages available with which we do not intend to compete. The linear tools in TISEAN are just for completeness, since it can be very convenient when one can compute, e.g., the power spectrum within the same environment as the mutual information.

Simple linear filters are `wiener` as defined by Eq. (2.11), `low121` which is a simple non-causal low-pass filter $\hat{x}_n = \frac{1}{4}(x_{n-1} + 2x_n + x_{n+1})$ (it can be iterated k times by `-Ik`), and `notch` implementing a notch filter, which can eliminate very efficiently a single frequency component such as 50 Hz AC power contaminations (see Example 2.6). For the numerical estimation of derivatives from differences, the Savitzky–Golay filter `sav_gol` is optimal (see Section 9.3.1).

The autocorrelation function can be estimated by `corr` in the time domain (the direct implementation of Eq. (2.5)) and by `autocorr` via a Fourier transform. For the power spectrum we offer `spectrum` and `memspec`. The first is obtained as the squared modulus of the Fourier transform, $|\tilde{s}_k|^2$ as defined in Eq. (2.8). The routine allows the suppression of fluctuations by decreasing the frequency resolution, The second contains a fit of an AR(p) model to the data and computes the spectrum of this AR model through the characteristic polynomial, as described after Eq. (12.2) on page 235. This has the advantage that the resulting spectrum is rather smooth and contains at most half as many peaks as the order p of the model. Additionally, there are particular routines for spike train data, `spikeauto` and `spikespec`, which compute the autocorrelation function and the power spectrum, repectively.

Finally, for linear global modelling, the twin routines `ar-model` and `ar-run` have been created. `ar-model` fits an AR model to the data, after subtracting their mean. Hence, it determines the coefficients of Eq. (2.12) by the solution of the minimisation of the one-step prediction error, Eq. (2.13) (see also Section 12.1.1). Additionally, it computes the residuals which are defined to be the differences between the deterministic part of the model and the next observation. In an ideal case, these residuals should be white noise. Having computed these residuals, they can be analysed in order to verify the validity of the model. The routine `ar-model` either can be used itself to produce a typical trajectory, or one can employ `ar-run`, which is also suitable when starting right away from some AR-coefficients. Notice that `ar-model` does not contain additional constraints which enforce the resulting model to be stable.

A.3 Phase space tools

Here we have collected a set of programs which correspond to Chapter 3, i.e., simple utilities for the visualisation of data and of structure inside the data. The most important program is `delay`. It takes a scalar time series and writes delay vectors. Notice that these vectors are always used in an overlapping way, i.e., the vector following $\mathbf{s}_n = (s_n, s_{n-\tau}, \ldots, s_{n-(m-1)\tau})$ is the vector $\mathbf{s}_{n+1} = (s_{n+1}, s_{n-\tau+1} \ldots)$. Even if this appears to be redundant at first sight, there are many situations where the down-sampling to temporal distances of τ or even of $(m-1)\tau$ eliminates valuable information. Since delay is often only used for visualisation (all TISEAN programs working in a time delay embedding space contain the embedding procedure themselves), the default value for the embedding dimension is two. All the two dimensional delay plots you find in this book have been created by this routine. Filtered delay embeddings as described in Section 9.3 can be created by svd and pc.

As you remember, a good estimate of the time lag τ can be obtained by either computing the autocorrelation function or the time delayed mutual information. The latter can be determined by mutual, Eq. (9.1). The program false_nearest is an implementation of Eq. (3.8) to compute the percentage of false neighbours.

In Section 3.5 the conversion of flow-like data into map-like data was recommended because of the enhancement of chaotic properties. For this purpose, poincare performs a Poincaré surface of section, giving as result the (interpolated) values of those components of the delay vectors which are not fixed by the surface of section, together with the time intervals in between successive intersections. In poincare, the surface of section is a hyperplane parallel to the axis of the coordinate system. By default, it passes through the mean value of the data, but can be shifted with respect to this by -a#. As you also remember, a Poincaré map is generated by considering only intersection points with one of the two possible directions of intersection. Since the resulting maps look geometrically different (although theory tells us that they are topologically conjugate to each other), the flag -C1 can be used to switch the orientation. The recipe of how to use these degrees of freedom is to optimise the signal to noise ratio in the resulting map. This is achieved by a large variation of the positions of points inside the Poincaré plane (i.e., do not cut the attractor where it is narrow), and by an almost perpendicular intersection of the trajectory with the plane (otherwise, interpolation errors get large). Alternatively, extrema determines all maxima or all minima of the time series, which is a Poincaré surface of section at $\dot{s} = 0$.

Recurrence plots as described in Section 3.6 are created by recurr. In order to prevent the output file from being unacceptably big, one can reduce the output by plotting only x percent of all points to be plotted, -%x.

The *space time separation plot* was introduced by Provenzale *et al.* (1992) as a means of detecting non-stationarity and temporal correlations in a time series. We discuss it in Section 6.5. It is a valuable tool to find an estimate of the time window $n_{\min}$ of Eq. (6.3) in the dimension estimates, by which we exclude pairs of points which are temporally correlated. The program stp produces a set of contour lines in the time–distance plane as shown in Fig. 6.8.

A.4 Prediction and modelling

In this section we collect all prediction and modelling programs we can offer. The corresponding models are discussed either in Chapter 4 or in Chapter 12. Most often, we do not make real predictions in the sense that we determine the most probable value for a yet unobserved measurement. Instead, the outcome to be predicted is already given. In order to avoid spuriously positive results, one has to respect causality. Points far in the future of a value to be predicted can be safely accepted as uncorrelated and hence can be included into the data base. A pitfall is given by predictions which go several sampling intervals into the future: all delay vectors whose images are in the future of the actual point have to be excluded from the neighbourhood, it is not enough to exclude the future delay vectors themselves. A flag taking care of causality is contained in most of the following programs.

A.4.1 Locally constant predictor

The simplest prediction scheme is given by Eq. (4.7) (see also Section 12.3). The easiest algorithmic implementation works with neighbourhoods of a fixed radius, but often does not give the best results. Such an approach is available in the TISEAN package as the program `predict`. There is a more sophisticated program as well, called `zeroth`, which allows one to specify a minimal radius and a minimal number of neighbours at the same time. This guarantees that no predictions are made which are based on one or two neighbours only and it solves the problem of what to do with empty neighbourhoods. Also, multivariate and mixed embeddings can be used as well as standard delay coordinates. As we have discussed in Section 12.7.2, such a locally constant predictor can also be used if the data represent a scalar Markov chain, hence it is fully justified to employ this program for stochastic data. The program `nstep` can be run with the option `-0` in order to iterate the locally constant model, similarly to what is described about the locally linear model below. The stationarity test based on cross-predictions between different segments of the data set as it is described in Section 4.4 is performed by `nstat_z`.

A.4.2 Locally linear prediction

This is the local version of AR models as discussed on page 240. The coefficients again are determined by the minimisation of the one-step prediction error. The program `onestep` performs this fit for the specified subset of points in the time series and reports the average rms forecast error as well as it lists all the residuals.

Both outputs yield insight into the predictive power of such a model. Of course, this power sensitively depends on the proper choice of the embedding parameters, which can thus be optimised. The program `nstep` generates iterated predictions, hence this is in the spirit of modelling: starting from the last point of the time series, the local model for the current endpoint is generated. Its prediction is then combined with the known past to a new delay vector, for which a new local model is built, and so on. If the model generates stable dynamics (there is nothing which guarantees that this is the case), then an arbitrarily long trajectory can be produced, which is noise free. In Example 12.2 (page 241) we show an attractor thus created. Of course, once an iteration creates a point which is far away from the original time series data, the procedure stops since no neighbours can be found and no local model can be constructed.

As we discussed before, an AR model is the global counterpart of the local linear models. The program `ll-ar` computes the average forecast errors as a function of the radius of the neighbourhood size, i.e., it makes, for all specified points of the time series, a local linear fit, for a whole sequence of radii. This yields insight into how much nonlinearity the deterministic part of the dynamics contains. For flow data, it is advisable to make s-step ahead predictions with $s > 1$, since otherwise the strong correlation of successive measurements hides all nonlinear structure. For this purpose, the parameter `-s#` can be used.

A.4.3 *Global nonlinear models*

The programs `rbf` and `polynom` determine global nonlinear models, namely a fit by a radial basis function network, Eq. (12.9), or by a multivariate polynomial, see Section 12.4.1. In both programs, the coefficients to be determined enter linearly in the model, and hence the minimisation of the one-step prediction errors eventually leads to the inversion of a matrix. In order to guarantee stability even if very many coefficients are to be determined, the SVD pseudo inverse is computed. The rbf fit uses Gaussian basis functions, with centres randomly distributed according to the invariant measure on the attractor. The initially chosen centres are then subject to some repelling force which distributes them more uniformly in the embedding space. Modelling by a multivariate polynomial contains the problem that the number of terms (and of coefficients to be determined) explodes when the highest power of the polynomial becomes large. Hence, we offer additionally different routines with different ways to eliminate seemingly irrelevant terms from the initially chosen polynomial. So a reduced and hopefully more robust model can be created. Details are provided in the manual pages of the TISEAN package.

A.5 Lyapunov exponents

The maximal Lyapunov exponent can be estimated by studying the growth of distances in the delay embedding space. The program `lyap_k` is an implementation of Eq. (5.2). Alternatively, `lyap_r` uses a different definition of neighbourhoods and of the distances, as proposed in Rosenstein *et al.* (1993). This yields sometimes a better scaling behaviour for flow data, but always reduces the scaling range, which can be anyway quite small for noisy map-like data. The routines suppress the worst statistical fluctuations by excluding reference points which have no more than `nfmin` neighbours closer than ϵ. We usually set `nfmin=10`. If you are concerned about the possibly resulting bias against sparsely populated areas, simply set `nfmin` equal to 1. In order to save computation time, our routine truncates the outer sum of Eq. (5.2) as soon as `ncmin` reference points are found with an acceptable number of neighbours. In most cases, `ncmin=500` should yield reasonable results. Increase this parameter if your data seem to be intermittent or if you have a fast computer.

The computation of the full Lyapunov spectrum through local linear fits, which yield the local Jacobians, is performed by `lyap_spec`. See Section 11.2.1 for what the algorithm is supposed to do. Always keep in mind that local linear fits can be terribly bad approximations to the true nonlinear dynamics. If such a bad fit occurs once in a while on a given data set, this may be negligible when, e.g., the average forecast error is computed, but it might produce very wrong Lyapunov exponents, since they are determined from the product of the fitted Jacobians. Hence, a single mis-estimation can have a long-range effect. As a rule of thumb, when `nstep` run with the same parameters as `lyap_spec` produces data which nicely represent the attractor of the original time series, then you can more or less trust in your Lyapunov exponents. If the `nstep`-attractor is very different, forget about the Lyapunov spectrum or try different embedding parameters.

A.6 Dimensions and entropies

A.6.1 The correlation sum

In Chapter 6 we discuss the Grassberger–Procaccia correlation dimension which can be obtained in the proper limit from the correlation sum (or integral). TISEAN contains the program `d2` which implements a fast algorithm to estimate the correlation sum for a sequence of length scales. The input data can be either scalar or multivariate. By default, the output is written to four different files, which are named by the input file name plus an extension. The file `input.c2` contains the numerical *estimate* of the correlation sum Eq. (6.3) as a function of ϵ for the specified range of the embedding dimension. As we discuss below in more detail, we do not compute the full double sum of Eq. (6.3). In input.d2, the corresponding local slopes as estimates of the dimensions $D_2(m, \epsilon)$ are listed. If these contain too many fluctuations, the auxiliary routine c2d can be applied to input.c2, with different smoothing options. The file input.stat contains statistical information about how many pairs of points are found when how many reference points have been treated. We only look at it when the program is running (which can take several hours if you provide really huge data sets with high dimensional attractors) and we want to see what it is doing. Finally, the file input.h2 contains entropy estimates to be described below.

Similarly to previous implementations [Theiler (1987), Grassberger (1990), Grassberger *et al.* (1991) and Schreiber (1995)], our code for the correlation integral uses a box-assisted neighbour search method. In those implementations (and, in fact, all that we are aware of), the range of length scales considered was split into two parts. For $\epsilon < \epsilon_0$ fast neighbour search was applied and for $\epsilon > \epsilon_0$, only part of the data was used in order to save computation time where statistics was not an issue. It turns out, however, that one can obtain the full correlation sum $C(\epsilon)$ with least computational effort if one collects only as many pairs as is necessary for *each* length scale. We fix a minimal required number of close pairs that we consider is sufficient to avoid statistical fluctuations, then we work our way through all the desired length scales. For each scale we accumulate more and more data points until this minimal number of pairs is reached. Of course, we need the precise number of potential pairs for the proper normalisation. Since we consider one value of ϵ at a time there is no histogram and no logarithms have to be computed. We just count neighbours. To ensure that the value of $C(\epsilon)$ for the large length scales is not based on too few consecutive reference points, we can specify their minimal number by the parameter ncmin. This value should be large enough to avoid the reference points all being correlated (and at least a few multiples of the major oscillation period). If you suspect that your data are *intermittent*, set ncmin to N, the total length of the time series (which makes the computation slower), and restrict the computation to the small length scales (which saves a lot of time).

A.6.2 *Information dimension, fixed mass algorithm*

The program c1 computes the curves which, when they exhibit a power law behaviour, yield an estimate of the information dimension by the fixed mass approach, Section 11.3.2. For a preselected fraction of points to be found inside a neighbourhood ("mass"), it computes the properly averaged neighbourhood size. The output of c1, which is written to a file whose name is the input file name extended by _c1, contains as two columns the average radius and the "mass". Although the "mass" is the independent quantity here, this is to conform with the output of the programs for the correlation dimension.

A logarithmic range of masses between $1/N$ and 1 is realised by varying the number of required neighbours k as well as the amount of data used, n. For a given mass k/n, n is chosen as small as possible as long as k is not smaller than the value specified by -K.

Since this is computationally quite expensive, it is usually reasonable to restrict the statistics to a smaller number of reference points than the time series offers, by specifying -n. This number of points is selected at random from all N potential reference points.

It is possible to use multivariate data, also with mixed embeddings. Contrary to the convention, the embedding dimension here specifies the total number of phase space coordinates, not just the number of time lags as usual. The number of components of the time series to be considered can only be given by explicit enumeration with the option -c.

Note: you will probably use the auxiliary programs c2d or c2t to process the output further, i.e., to extract numbers of the local slopes as dimension estimates.

A.6.3 Entropies

The correlation sum immediately also yields estimates of the KS-like order two entropy, the correlation entropy (see Section 11.4), as the logarithm of the ratios of the correlation sum for two successive embedding dimensions m and $m+1$ for fixed ϵ. Exactly this output is written by d2 into a separate file which by default has the extension .h2. The Figures 11.5 and 11.4 are examples for this output.

In principle, also the output of c1, which is the fixed mass estimate of the generalised correlation sum for $q = 1$, can be turned into entropy estimates, namely for the Kolmogorov–Sinai entropy. Right now there is no utility available which does the housekeeping automatically.

If you are interested in the computation of entropies of order $q \neq 1, 2$, we can offer boxcount. This is the implementation of the very first idea of partitioning the embedding space by hypercubes. As we pointed out in Section 11.4, this procedure suffers from severe finite sample effects and has the additional drawback of being quite memory consuming. On modern computers, the latter is not usually a problem any more. The finite sample effects can be partly compensated by corrections such as those proposed by Grassberger (1986) or by a sub-sampling as in Kantz &

Schürmann (1996), which are not included in our program. However, knowing the problem, you can try to identify a scaling range nonetheless.

A.7 Surrogate data and test statistics

In Chapter 7 we introduce the concept of surrogate data. The idea was to create a whole ensemble of different realisations of a null hypothesis, and to apply statistical tests to reject the null for our given data set. Hence, the task is split into algorithmically two different issues: the creation of the surrogate data sets, and the statistical tests.

The main program for surrogate data is surrogates. Without particular flags, is generates an ensemble of time series which are iterative amplitude adjusted Fourier transform surrogates, i.e., which share with the original data both the power spectrum (or, which is the same, the autocorrelation function), and the distribution of the data, see Section 7.1.2. Of course, there is always a small mismatch. By default, at the end of the iterative procedure, the surrogates have exactly the correct distribution. By -s, instead data with exactly the same amplitudes of the Fourier transform are created. Optionally, this program can be also used to create a random shuffle of the input data (-i0: same distribution, but fully uncorrelated), or to create phase randomised surrogates without adjustment of their distribution (-i0 -s). For each surrogate data set thus created a new output file is created. In order to make use of fast Fourier transform algorithms, the input data are truncated to the largest subinterval which can be factorised by 2, 3, and 5. Since a large discrepancy between the first and the last value of the data set introduces a bias in the Fourier transform which is performed under the hypothesis of a periodic continuation of the signal (i.e., no windowing!), for really sound results one should use the routine endtoend in order to select a subsequence of the data where initial and final values match in an optimal way.

If you have reasons to think that the amplitude adjusted Fourier transform surrogates do not represent the null hypothesis in a reasonable way, then the set of programs called randomize supplies many more sophisticated versions. We refer the interested reader to the manual pages of the TISEAN package and the articles by Schreiber & Schmidt (1996) and (2000).

As test statistics, you can in principle employ any of the time series routines described here. Of course, it should somehow be a test for nonlinearity, since by construction the linear properties of surrogates and original data should be identical. Two statistics have turned out to be of particular usefulness: the test for time inversion invariance, timerev, and the prediction error by locally constant predictions, predict, as described above.

A.8 Noise reduction

Noise reduction is the most delicate operation which the TISEAN package offers, because you are going to manipulate your data on which you base your further analysis. Doing the wrong thing can change your data properties tremendously and can lead to wrong conclusions afterwards. With these words of warning, you should apply these routines only if you have fully understood the ideas behind as explained in Section 4.5 and Chapter 10 (of course, this applies to linear filters as well; without care, you can easily remove the interesting part of the signal also by a linear filter, no sophisticated `project` is needed for that).

The programs `nrlazy` and `lazy` are two slightly different implementations of the simple scheme described in Section 4.5, i.e., they replace a delay vector by the mean of the vectors found in its neighbourhood. Eq. (4.11) describes the correction of exactly the central component of each delay vector. This is in fact done by `nrlazy`. However, experience showed that for highly sampled flow data, this is not always optimal, so that `lazy` is a version where all components of every delay vector are to be replaced by the mean of the corresponding component on its neighbourhood. Since every single time series element is element of m different delay vectors, this gives m typically conflicting corrections. `lazy` simply takes the arithmetic mean of all of these, so that the set of corrected delay vectors is converted back into a corrected time series at the end.

The full local projection scheme which is derived in Chapter 10 is implemented by the program `project`. Apart from the dimension of the embedding space, one needs to think about the dimensionality of the manifold onto which one wants to project. There is no time lag to be set, since this routine will always use a lag of unity. This is reasonable, since for noise reduction, one wants to exploit the maximal redundancy in the data. However, the value of m here should not be chosen as twice the attractor dimension, but it should be the embedding window. As an example, assume you have flow data from the Lorenz system, so that $m = 3$ and a time lag of $\tau = 10$ seems reasonable for a time delay embedding. The embedding window then covers $(m - 1)\tau + 1$ sample intervals, hence use here -m21. In this example, -q2, since the attractor is locally close to two dimensional. The program also requires a value for the minimal number of neighbours and the minimal radius of the neighbourhood size. As discussed in Section 10.3.2, the neighbourhood diameter has to be larger than the noise level. What you can achieve as a minimal number of neighbours depends on the length of the data set and its dimensionality, -k30 is a reasonable trial value to begin with when the attractor is expected to be low dimensional. It makes sense to iterate the procedure a few times, but be aware of the fact that it does not necessarily converge, but instead the resulting data may be distorted when you continue to iterate. Therefore, the output is not just the cleaned time series, but also the difference between input data and output data, which is the "noise" which has been subtracted. It is indispensable that you study the properties of this noise in order to verify that it is in fact noise that you have removed. More than for any other program of TISEAN we strongly recommend here that you collect some experience with `project` on numerically generated data whose properties are similar to your experimental data.

Our implementation `project` uses a fast neighbour searching method. As said, we demand a minimal number of neighbours for each reference point. Thus we search for neighbours closer than some value ϵ but we repeat the search for larger values ϵ for those points which have too few neighbours. For this purpose we keep a list of points which are yet uncorrected due to the lack of neighbours. We increase ϵ until this list is empty. The program implements the curvature correction discussed in Section 10.3.3. A very similar routine is given by `ghkss`, which does accept a nontrivial time lag τ, and which allows the user to activate a nontrivial metric for the projection.

A.9 Finding unstable periodic orbits

For controlling chaos, but also for other purposes, it might be relevant to identify locations of unstable periodic orbits embedded in the attractor. TISEAN supplies the program `upo` for this purpose. Given a trial point, the program creates a local approximation of the dynamics in the vicinity of the trial point $\mathbf{s}$ and at each of its $p - 1$ images. Let us call this sequence of p local functions $g_k(\mathbf{s})$, $k = 1, \ldots, p$. If there is a period p orbit in the neighbourhood, then there should exist a sequence of p scalar values x_k, such that $x_{k+1} = g_k(x_k, x_{k-1}, \ldots, x_{k-m+1})$. Here, $x_k = x_{k\pm p}$ is to be periodically continued. The algorithm searches for this point by a Levenberg–Marquardt method, in each step renewing the estimate of the local functions g_k in the neighbourhood of the actual trial values, until convergence. The stability of the orbits found is estimated from the product of the local Jacobians of the functions g_k.

A.10 Multivariate data

As we mentioned initially, most of the programs are written to work also with multivariate data and with mixed multivariate-time-delay embeddings. However, there are some programs which work *only* with multivariate data, since they compute statistical quantities which do not have a proper meaning for scalar time series.

Such programs are xcor, which computes the cross-correlation function given by Eq. (14.1) and xzero computing the cross-prediction error as mentioned in Section 14.1 and in Section 4.4. The interdependence and the transfer entropy did not yet make it into TISEAN, but are clearly on the list for the next release. Finally, there is a program called xc2, the cross-correlation sum. It helps to understand up to which scales two different invariant distributions are indistinguishable from each other, see Kantz (1994b). A more known but less sensitive test for this purpose is the Kolmogorov–Smirnov test (for an algorithm, see Press *et al.* (1992)).

Appendix B

Description of the experimental data sets

Throughout the text we have tried to illustrate all relevant issues by the help of experimental data sets, some of them appearing in several different contexts. In order to avoid repeats and to concentrate on the actual topic we did not describe the data and the systems they come from in any detail in the examples given in the text. This leeway we want to make up in this appendix, together with a list of all places where each set is referred to.

B.1 Lorenz-like chaos in an NH₃ laser

This data set was created at the PTB Braunschweig in Germany in an experiment run by U. Hübner, N. B. Abraham, C. O. Weiss and collaborators (1993). Within the time series competition organised in 1992 by N. A. Gershenfeld and A. Weigend at the Santa Fe Institute it served as one of the sample series and is available on the SFI server by anonymous FTP to `sfi.santafe.edu`.

A paradigmatic mathematical model for low dimensional chaos is the Lorenz system, Lorenz (1969), describing the convective motion of a fluid heated from below in a Rayleigh–Benard cell. Haken (1975) showed that under certain conditions a laser can be described by exactly the same equations, only the variables and constants have different physical meaning. The experiment in Braunschweig was designed to fulfil the conditions of being describable by the Lorenz–Haken equations as closely as possible.

The time series is a record of the output power of the laser, consisting of 10 000 data items. Part of it is shown in Fig. B.1. Similarly to the Lorenz model, the system exhibits regular oscillations with slowly increasing amplitude. When a critical value is reached, an instability occurs and the oscillation starts with a new small amplitude, the phase information is lost and its magnitude appears unpredictable. The sampling rate of the data is such that for each oscillation about eight measurements are recorded. The data are digitised by an 8-bit analogue/digital converter, such that

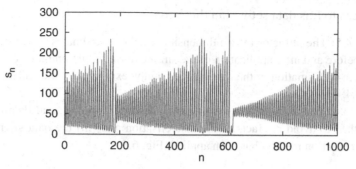

Figure B.1 1000 data points of the Lorenz-like laser chaos, connected by lines.

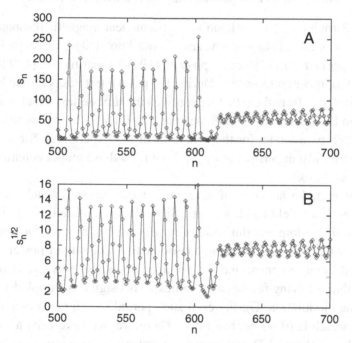

Figure B.2 The effect of taking the square root of the laser data. Panel A: original data. Panel B: their square root.

the discretisation error has an amplitude of $\frac{1}{512}$ of the signal's range. About 32 of the breakdowns are covered by the total observation time.

For several applications it is convenient to use the square root of this data instead of the data itself. From the physical point of view this is reasonable, since the observable, the output power, is the square of a more fundamental physical quantity, the amplitude of the electromagnetic field in the laser cavity. For the data analysis this transformation is helpful since it linearises the peaks. The effect is shown in Fig. B.2.

References to this data set within this book:

Example 4.5: The attractor in two dimensional delay coordinates is shown in Fig. 4.6 both before and after application of the simple noise reduction method.

Example 5.4: Estimation of the maximal Lyapunov exponent by the real space method, Fig. 5.4.

Example 6.4: The correlation sum is computed for (i) the original data (ii) the data smoothed to avoid artefacts of the discretisation and (iii) the data after the simple noise reduction method has been applied, Fig. 6.6.

B.2   Chaos in a periodically modulated NMR laser

At the ETH Zürich in Professor Brun's group a nuclear magnetic resonance (NMR) laser has been operated for many years. To our knowledge, the experiment goes back to Bösiger *et al.* (1977) [see Flepp *et al.* (1991), Finardi *et al.* (1992) and Badii *et al.* (1994) for recent references]. During the past year it has been run by L. Flepp and J. Simonet. It is from them that we received the time series, which was analysed previously in Kantz *et al.* (1993). Flepp *et al.* (1991) have been able to establish an accurate Bloch-type model for the dynamics of the experiment. Since a model is unavailable for most experiments we have not used these model equations for data analysis in this book.

The lasing particles are spins of Al atoms in a ruby crystal, which are positioned in a radio frequency field inside a cavity. Since the laser system primarily has only two degrees of freedom and thus cannot exhibit chaotic motion, the *quality factor* of the resonant structure is modulated periodically. The output power of the laser is decoupled by an antenna, and the voltage across this antenna is recorded as the signal (thus allowing for negative values). This signal is sampled 15 times per period of the modulation. During each such period the signal shows one regular oscillation, which is of no interest to us. Therefore, we have used a stroboscopic view (together with an SVD optimisation in order to reduce noise implicitly), thus obtaining 38 000 map-like data items. Although a part of the time series is plotted in Fig. 1.1, in Fig. B.3 we show a longer segment in a different representation: we have joined these map data by lines. Sometimes this facilitates the recognition of structure, since it makes the time ordering more obvious.

Material on this data set can be found in the following examples:

Example 1.1: Average prediction error for more step forecasts, Figs. 1.1 and 1.2. In Fig. 1.3 the attractor is shown in a two dimensional delay coordinate space after PCA optimisation of the signal. See Example 9.4.

Example 3.6: The attractor in a two dimensional delay coordinate space, Fig. 3.6.

Example 4.1: Predictions made with the simple algorithm, Fig. 4.1.

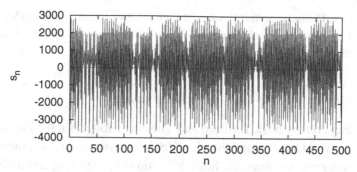

Figure B.3 Stroboscopic view of the NMR laser data.

Example 4.4: Noise reduction on the NMR laser data, Fig. 4.5, after they have been artificially contaminated with additional noise.

Example 5.1: Initially close states separate over the course of time, Fig. 5.1.

Example 5.2: The exponential divergence of nearby trajectories is demonstrated in Fig. 5.2.

Example 5.3: The maximal Lyapunov exponent obtained with the real space method, Fig. 5.3.

Example 6.1: The self-similarity of the attractor is demonstrated with successive enlargements of the phase portrait (noise reduced data, Fig. 6.1).

Example 6.2: Correlation sum and its slope for various embedding dimensions, Figs. 6.3 and 6.4.

Example 6.6: Space time separation plot, Fig. 6.8.

Example 9.3: Time delay embedding of the flow data with different time lags, Fig. 9.2.

Example 9.4: PCA optimisation of the stroboscopic view of the data, Figs. 9.3 and 9.4.

Example 10.2: Before noise reduction, the self-similarity of the attractor is broken by the noise, Fig. 10.1.

Example 10.3: We give details of the nonlinear noise reduction on this data set. The results are demonstrated by another enlargement of the phase portrait, Fig. 10.4, and by the computation of the maximal Lyapunov exponent by the real space method, Fig. 10.5.

Example 10.5: The typical behaviour of the correlation sum around the noise level is moved down two orders of magnitude in length scale by the nonlinear filtering, Fig. 10.6.

Example 11.5: Computation of the Lyapunov spectrum by various methods, Table 11.1.

Example 11.7: Information dimension estimated by the fixed-mass method, Fig. 11.2.

Example 11.13: Estimation of the order-two entropy, Fig. 11.5.

Example 11.14: Estimation of the Kolmogorov–Sinai entropy, Fig. 11.6.

Example 12.8: Modelling the dynamics: table of prediction errors with various methods (Table 12.1) and a view of a synthetic attractor, obtained through iteration of a global model, Fig. 12.5.

Example 15.2: The unstable fixed point on the attractor, located by close returns of the trajectories, Fig. 15.2.

Example 15.4: Orbits of period 1–4 found by the Newton method, Fig. 15.3.
Example 15.5: The fixed point on the attractor together with its stable and unstable
manifolds, Fig. 15.5.

B.3   Vibrating string

In an experiment run by Nick Tufillaro and Timothy Molteno at the Physics Department, University of Otago, Dunedin, New Zealand, a nonmagnetic tungsten wire is
mounted in a permanent magnetic field. Vibrations in the string are maintained by
an alternating electric current through the string. Within a small parameter range,
the amplitude envelope exhibits chaotic vibrations. The transverse amplitude displacement is recorded once every driving cycle at a fixed phase of the alternating
current. The interesting dynamics in the amplitudes are rather slow compared to
the driving frequency. Thus the data set is oversampled at more than 100 samples
per mean cycle and we treat it like flow data, although this is not really justified by
the underlying process. Data from this experiment have been analysed previously,
Brown et al. (1994), Tufillaro et al. (1995), etc.

We use data from two different parameter settings. An intermittent data set appears in Examples 1.3 and 8.3, Figs. 8.1, 8.2, and 8.4. For a slightly stronger forcing
the output is more chaotic. Such data are used in Example 11.6: return map of the
data in a Poincaré surface of section and fit of their dynamics, Fig. 11.1.

B.4   Taylor–Couette flow

The motion of a viscous fluid between two coaxial cylinders is called Taylor–
Couette flow. The experimental device consists of a rotating inner cylinder while
the bottom and top plates, as well as the outer cylinder, are at rest. The cylinder
length can be varied continuously and the aspect ratio can be used as a geometrical
control parameter. To provide boundary conditions without cylindrical symmetry,
the top plate of the apparatus can be inclined by a small amount. This inclination
is a second control parameter of the system. See Buzug et al. (1990) for more
details of the set-up. The local velocity component of the flow field is measured by
a laser-Doppler-anemometer and subsequently converted to 12-bit numbers.

Rotational Taylor–Couette flow is a hydrodynamical system which shows many
scenarios of low dimensional chaos [see also Brandstäter et al. (1983) and
Brandstäter & Swinney (1987)]. Depending on the boundary conditions one can
find period doubling, intermittency [Pfister et al. (1992)], homoclinic orbits [Pfister
et al. (1991)], breakup of 2-tori [Buzug et al. (1992)], as well as stable 3-tori. The
time series we used here is measured in the chaotic regime at a Reynolds number
$Re = 705$. The sampling time is 20 ms and the total number of data points is 32 768.

The same data set has been analysed previously in Kantz *et al.* (1993). We used the data in the following examples:

Example 6.3: Slopes of the correlation sum, Fig. 6.5.

Example 6.5: Spurious correlation sum when temporal neighbours are not taken care of, Fig. 6.7.

Example 6.7: Space time separation plot revealing temporal correlations, Fig. 6.9.

Example 6.8: The noise level present in the data is determined using the correlation sum, Fig. 6.10.

Example 7.6: A statistical test for determinism, employing Takens's estimator of the correlation dimension, Fig. 7.5.

B.5 Multichannel physiological data

This data set was provided by A. Goldberger from Beth Israel Hospital in Boston for the Santa Fe Institute time series competition in 1991/92 [see Weigend & Gershenfeld (1993)]. The quite long, non-stationary, multichannel data set contains the instantaneous heart rate, air flow, and blood oxygen concentration, recorded twice a second for one night. We used the multichannel data briefly in Example 14.5. Further, we selected two samples of 2048, respectively, 4096, successive measurements (about 16, respectively, 32, min) of the air flow through the nose of the human subject. The shorter segment starts with the 12 750th measurement of the original series, the longer one with the 305th. See Rigney *et al.* (1993) for more information on the data.

We used these data in the following examples:

Example 1.2: A plot of the shorter data segment that we use, Fig. 1.4, and in a two dimensional delay embedding, Fig. 1.5.

Example 4.2: Linear and nonlinear predictions on the shorter example, Fig. 4.2.

Example 4.3: The full length of the breath rate channel is investigated with respect to non-stationarity, Fig. 4.3.

Example 7.4: Measured data and a random surrogate with identical spectrum and distribution, Fig. 7.3.

Example 8.1: It is argued, that the air flow data represent a nonlinear limit cycle rather than damped and noise driven linear oscillations.

Example 10.7: Nonlinear noise reduction on th air flow data.

B.6 Heart rate during atrial fibrillation

This data set is based on record 08219 from the MIT-BIH Arrhythmia Database, Moody & Mark (1991). Inter-beat intervals have been extracted automatically from

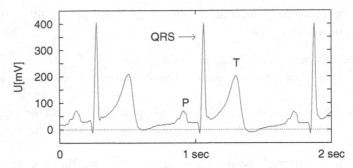

Figure B.4 Two full cycles of a human electrocardiogram. For one cycle, we identified the P-wave, corresponding to the depolarisation of the *atrium*, the QRS complex of the depolarisation of the larger *ventricles*, and the T-wave during the repolarisation phase. [See Goldberger (1977) for an introduction to the terminology and electrocardiography in general.] Note that although the signal appears to be periodic at first sight, the lengths of the two beats shown differ by 20 ms. The inter-beat intervals can show considerable fluctuations.

an electrocardiogram of a human patient undergoing several periods of atrial fibrillation, a potentially life-threatening arrhythmia of the heart. For the time series together with typical surrogates see Fig. 7.2. Part of the data is used in Example 7.7 (symbolic encoding). We also concatenated disjoint sections of fibrillation in order to obtain a more stationary time series. All we can say about the resulting data set is that we can reject the null hypothesis that it contains identically distributed independent random numbers. See Example 7.2. We have also used these data to show the lack of invariance properties of certain pattern entropies, Example 7.8.

B.7 Human electrocardiogram (ECG)

The ECG is a common clinical tool which aids the cardiologist in the diagnosis of cardiac disease. The particular recordings of healthy subjects that we used were made by Petr Saparin at Saratov State University, Saparin (1995). Figure B.4 shows two seconds from a longer recording. The sampling rate is 200 Hz. More about ECGs can be found in Example 2.5, Fig. 2.4, Example 2.6, Fig. 2.5, and Example 3.5, Fig. 3.4.

To illustrate the foetal ECG subtraction, Example 10.8, we first superposed two of the above normal ECGs to mimic an abdominal ECG recording during pregnancy. As a real example, we used a foetal ECG taken by John. F. Hofmeister, University of Colorado, Denver (see also Example 1.4). The foetus is about 16 weeks old. Electrodes were placed on the mother's abdomen and cervix. The use of cervical electrodes is described in Hofmeister *et al.* (1994). Single-channel ECGs taken with abdominal electrodes have very similar properties.

B.8 Phonation data

These data are measurements from a pathological male voice suffering from acute *laryngitis*. It has been analysed in Herzel & Reuter (1996). The data record comprises about five seconds and contains 100 000 points. The spectrogram, Fig. 2.3, shows *biphonation* and the occurrence of sub-harmonics of one-half or one-third of the fundamental pitch. We used it in Examples 2.2 and 2.4 (running variance of the data and spectrogram).

B.9 Postural control data

Bertenthal *et al.* (1996) performed several experiments with humans in order to understand the human posture control system. One such investigation concerns upright standing. A person is asked to stand still on a *force plate*, a device for measuring the location of the centre of the pressure under the feet. We analyse two data sets, one of the location in the x-direction, one of the y-direction. They contain 15 000 data each, covering 300 seconds. A segment of the time series is shown in Fig. B.5. Tests for nonlinearity using surrogate data (see Section 7.1) do not show any hint of a deterministic nature for this signal. Instead, the analysis described in Example 6.10 reveals some very interesting power law behaviour consistent with an anomalous diffusion process. The corresponding stochastic model, fractional Brownian motion, possesses an infinite memory and is thus non-Markovian. Translated to the postural control, this reflects that the control system remembers the desired upright position even over long stretches of time.

B.10 Autonomous CO$_2$ laser with feedback

A prerequisite of chaos is a three dimensional phase space. A standard single mode CO$_2$ laser is essentially a two dimensional system. Arecchi *et al.* (1986) designed an experiment where a feedback loop introduces an additional variable. The output intensity acts on an intra-cavity modulator and thus causes output dependent losses. The time scales of this process are adjusted to be comparable to the scales of the system without feedback, so that the dynamics are essentially governed by three active degrees of freedom (note that there is no delay line and phase space remains finite dimensional!). Depending on the values of several control parameters, the laser either produces constant output, or it oscillates periodically or even chaotically.

Our time series (Fig. B.6) consists of 128 000 data of the intensity of the laser output (voltages), sampled at a frequency of 2 MHz, in the chaotic regime. This amounts to about 54 measurements per oscillation. The dynamics live on a ribbon-like attractor (Fig. B.7), cycling with roughly a frequency of 40 kHz. The attractor dimension is extremely close to 2. In the Poincaré map, we can determine the

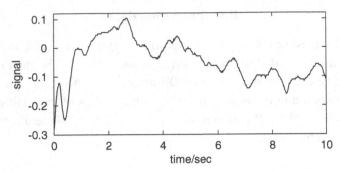

Figure B.5 Twenty seconds out of a longer recording of the side-to-side position of the centre of pressure.

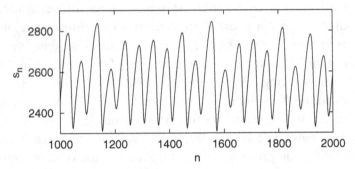

Figure B.6 A segment of length 2000 of the time series of laser output intensities.

Lyapunov exponent to $\lambda_+ \approx 0.38$ (both with a one dimensional fit of the dynamics of the graph in Fig. 9.7 and with the real space method). Divided by the average number of measurements between successive intersections with the Poincaré surface of section, we find the exponent of the flow data: $\lambda_{flow} \approx 0.007$. More details about the data and detailed model equations can be found in Varone *et al.* (1995).

We refer to these data in the following examples:

Example 2.3: Spectrum of the full time series (flow data), Fig. 2.2.
Example 9.6: Poincaré map and embedding of the Poincaré recurrence times, Fig. 9.7.
Example 9.7: Artificially enhanced maxima interpreted as a spike train, embedding of the inter-spike intervals, Fig. 9.8.
Example 12.2: Iterated local linear fits of the dynamics, Fig. 12.1.

B.11 Nonlinear electric resonance circuit

An electric resonance circuit is often depreciatorily called an *analogue computer*. In some sense this is right, since a resonance circuit is a very direct physical realisation of a solver for the initial value problem of a corresponding ordinary

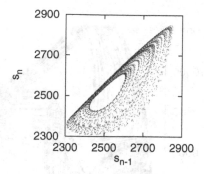

Figure B.7 The attractor reconstructed from the data of Fig. B.6 with a delay of $\tau = 4$.

(non-autonomous) differential equation. In other words, the correspondence between physical system and mathematical model is typically close to perfect. Nonetheless, it is a real experiment with many complications such as lack of constancy of system parameters (e.g., temperature dependence of resistors, or the instability of the power supply), such as the coupling to external noise sources, and of course such as measurement noise. Hence, it is an ideal experiment if one wants to test how the signatures of deterministic chaos survive such experimental deficiencies. Our data stem from a serial resonance circuit, a linear arrangement of Ohmian resistance, inductance, and capacitance, driven by a sinusoidal voltage. It is designed and operated by M. Diestelhorst at the Physics Department of the Martin Luther University in Halle (Germany). The nonlinearity is introduced through a *ferroelectric* material inside the capacitance. This means that below the critical temperature of $T_c \approx 322K$ the material possesses a spontaneous electric polarisation, which renders the characteristics of the capacitance nonlinear. The proper functional form of the nonlinearity is unknown, and a rough but reasonable model equation is

$$L\ddot{Q} + R\dot{Q} - \alpha Q + \gamma Q^3 + U_0 \cos \omega t = 0,$$

where the first term represents the inductance of the coil, the second the Ohmian resistances, the third and fourth the nonlinear capacitance, and the last one the external sinusoidal driving voltage. The quantity Q is the charge on the capacitance. This equation is the well known damped and driven Duffing oscillator. As mentioned, the cubic nonlinearity is an approximation of the unknown true nonlinearity inside the capacitance, so that one time series issue here was to extract from the data a better functional approximation of the nonlinearity. Time series data are measurements of a quantity q which is proportional to the charge on the nonlinear capacitance, through the voltage drop across an additional small linear capacitance. Before recording, these data are converted by a 12 bit A/D converter

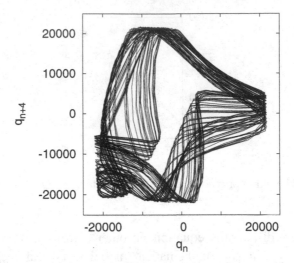

Figure B.8 Two dimensional time delay embedding with time lag $\tau = 4$ of 2000 data from the electric resonance circuit, connected by lines.

and are sampled with 100 kHz. The driving frequency ω is 3930 Hz. Figure B.8 shows a two dimensional time delay embedding of a part of the data set we use in this book. A detailed discussion of the time series analysis of these data, which represent the dynamics on a slightly more than two dimensional chaotic attractor, can be found in Hegger *et al.* (1998). We refer to these data in Example 3.7, in Example 12.4, Example 12.7 and in Example 13.6.

B.12 Frequency doubling solid state laser

In the Department of Energy and Semiconductor Physics of the University of Old-enburg, a multi-mode laser with nonlinear mode coupling is operated. The optically active medium of this solid state laser is a **n**eodymium **d**oped yttrium **a**luminium **g**arnet (Nd:YAG) which is pumped by a laser diode and produces laser light in the infrared. A frequency doubling KTP crystal is placed inside the laser cavity, converting part of the light into green laser light. As it was first observed by Baer (1986), this introduces a nonlinear coupling of the infrared laser modes and thus causes a dynamical instability. A very similar experiment was set up by R. Roy and collaborators, who also contributed to the development of rate equations for this system [Bracikowski & Roy (1991)]. Such a model is composed of two variables, the intensity and the gain, for every active laser mode. As one possible experimental observable the intensity of the green laser light can be measured. Alternatively, the total power (green plus infrared) can be split into the two directions of polarisa-tion. Each of these directions represents a subset of the laser modes. These two

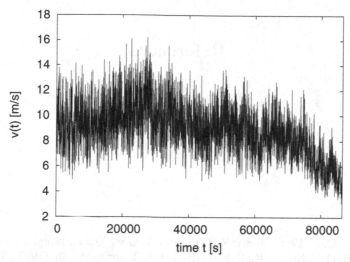

Figure B.9 One day (= 86400 s) of wind speed data. The wind speed decreases towards the end of the day.

sum intensities can be recorded simultaneously. The data of the period doubling bifurcation shown in Example 8.5 represent one of these sum intensities and stem most likely from a situation where four modes (two in each polarisation direction) are active. The comparison of numerical simulations of the model equations and experimental observations give hints that for suitable parameter values, chaos on three to five dimensional attractors should be present. The experimental time series data for such regimes clearly show strongly aperiodic behaviour, but noise on the data makes it impossible to identify a finite attractor dimension. See also Gills *et al.* (1992) for a study of the chaotic dynamics.

B.13 Surface wind velocities

Wind speed data measured close to the surface of the earth represent atmospheric turbulence in the boundary layer. Their study is of high relevance for the safe and efficient operation of wind energy turbines. The research centre in Risø operates a measurement site on the western shore of Denmark, in Lammefjord. We use a data set recorded with 8 Hz by cup anemometers. These anemometers are sensitive only to the wind speed perpendicular to its axis, which here is the x–y-plane, and measure the absolute speed. The down-sampled time series of one of these anemometers is shown in Fig. B.9. We refer to these data in Section 12.7.7 as an example for the prediction of stochastic processes and in Example 14.2 for the illustration of cross-correlation functions.

References

Abarbanel, H. D. I. (1996). *Analysis of Observed Chaotic Data.* Springer, New York.

Abarbanel, H. D. I., Brown, R., Sidorowich, J. J. & Tsimring, L. Sh. (1993). The analysis of observed chaotic data in physical systems. *Rev. Mod. Phys.,* **65**, 1331.

Akaike, H. (1974). A new look at the statistical model identification. *IEEE Trans. Automat. Contr.,* **AC-19**, 716.

Albano, A. M., Muench, J., Schwartz, C., Mees, A. I. & Rapp, P. E. (1988). Singular value decomposition and the Grassberger–Procaccia algorithm. *Phys. Rev. A,* **38**, 3017. Reprinted in [Ott *et al.* (1994)].

Alexander, J. C., Yorke, J. A., You, Z. & Kan, I. (1992). Riddled basins. *Intl. J. of Bifurcation and Chaos,* **2**, 795.

Allen, M. R. & Smith, L. A. (1996). Monte Carlo SSA: detecting irregular oscillations in the presence of coloured noise. *J. Climate,* **9**, 3373.

Alligood, K. T., Sauer, T. D. & Yorke, J. A. (1997). *Chaos: an introduction to dynamical systems.* Springer, Heidelberg.

Arecchi, F. T., Gadomski, W. & Meucci, R. (1986). Generation of chaotic dynamics by feedback on a laser. *Phys. Rev. A,* **34**, 1617.

Arnhold, J., Grassberger, P., Lehnertz, K. & Elger, C. E. (1999). A robust method for detecting interdependence: application to intracranially recorded EEG. *Physica D,* **134**, 419.

Artuso, R., Aurell, E. & Cvitanović, P. (1990). Recycling of strange sets, I and II. *Nonlinearity,* **3**, 325, 361.

Auerbach, D., Cvitanović, P., Eckmann, J. P., Gunaratne, G. & Procaccia, I. (1987). Exploring chaotic motion through periodic orbits. *Phys. Rev. Lett.,* **58**, 2387.

Auerbach, D., Grebogi, C., Ott, E. & Yorke, J. A. (1992). Controlling chaos in high dimensional systems. *Phys. Rev. Lett.,* **69**, 3479.

Babloyantz, A. (1989). Some remarks on nonlinear data analysis of physiological time series. In Abraham, N. *et al.*, editors, *Quantitative Measures of Dynamical Complexity in Nonlinear Systems*, Plenum Press, New York.

Badii, R. (1989). Conservation laws and thermodynamic formalism for dissipative dynamical systems. *Revista del Nuovo Cimento,* **12**, 3.

Badii, R. & Politi, A. (1985). Statistical description of chaotic attractors. *J. Stat. Phys.,* **40**, 725.

(1986). On the fractal dimension of filtered chaotic signals. In [Mayer-Kress (1986)].

(1999). *Complexity.* Cambridge University Press.

Badii, R., Broggi, G., Derighetti, B., Ravani, M., Ciliberto, S., Politi, A. & Rubio, M. A. (1988). Dimension increase in filtered chaotic signals. *Phys. Rev. Lett.,* **60**, 979.

Badii, R., Brun, E., Finardi, M., Flepp, L., Holzner, R., Parisi, J., Reyl, C. & Simonet, J. (1994). Progress in the analysis of experimental chaos through periodic orbits. *Rev. Mod. Phys.,* **66**, 1389.

Baer, J. (1986). J. Opt. Soc. Am. **B 3**, 1175.

Baier, G. & Klein, M. (1990). Maximum hyperchaos in generalized Hénon maps. *Phys. Lett. A,* **151**, 281.

Barahona, M. & Poon, C.-S. (1996). Detection of nonlinear dynamics in short, noisy time series. *Nature,* **381**, 215.

Bauer, M., Heng, H. & Martiensen, W. (1993). Characterization of spatiotemporal chaos from time series. *Phys. Rev. Lett.* **71**, 521.

Beck, C. & Schlögl, F. (1993). *Thermodynamics of Chaotic Systems.* Cambridge University Press.

Bélair, J., Glass, L., an der Heiden, U. & Milton, J. (1995). *Dynamical Disease,* AIP Press, New York.

Benedicks, M. (1994). New developments in the ergodic theory of nonlinear dynamical systems. *Phil. Trans. Roy. Soc.,* **A 346**, 145.

Benettin, G., Galgani, L. & Strelcyn, J. M. (1976). Kolmogorov entropy and numerical experiments. *Phys. Rev. A,* **14**, 2338.

Benettin, G., Galgani, L., Giorgilli, A. & Strelcyn, J. M. (1978). Lyapunov characteristic exponents for smooth dynamical systems and for Hamiltonian systems: a method for computing all of them. *C. R. Academie Sci. Paris A,* **206**, 431.

Ben-Mizrachi, A., Procaccia, I., Rosenberg, N. & Schmidt, A. (1985). Real and apparant divergencies in low-frequency spectra of nonlinear dynamical systems. *Phys. Rev. A,* **31**, 1830.

Bentley, J. L. (1980). Multidimensional divide-and-conquer. *Communications of the ACM,* **23**, 214.

Bergé, P., Dubois, M., Manneville, M. & Pomeau, Y. (1980). Intermittency in Rayleigh–Bénard convection. *J. Phys. (Paris) Lett.,* **41**, L 341.

Bergé, P., Pomeau, Y. & Vidal, C. (1986). *Order Within Chaos: Towards a deterministic approach to turbulence.* Wiley, New York.

Berkovitz, L. D. (1974). *Optimal Control Theory.* Springer, New York.

Bertenthal, B. I., Boker, S. M. & Rose, J. L. (1996). Dynamical analyses of postural development. *J. Sport & Exercise Psychology,* **17**, 8.

Biham, O. & Wenzel, W. (1989). Characterization of unstable periodic orbits in chaotic attractors and repellers. *Phys. Rev. Lett.,* **63**, 819.

Bingham, S. & Kot, M. (1989). Multidimensional trees, range searching, and a correlation dimension algorithm of reduced complexity. *Phys. Lett. A,* **140**, 327.

Bösiger, P., Brun E. & Meier, D. (1977). Solid state nuclear spin–flip maser pumped by dynamic nuclear polarization. *Phys. Rev. Lett.,* **38**, 602.

Bowen, R. & Ruelle, D. (1975). The ergodic theory of axiom-A flows. *Inv. Math.,* **19**, 181.

Box, G. E. P. & Jenkins, G. M. (1976). *Time Series Analysis.* Holden-Day, San Francisco.

Bracikowski, C. & Roy, R. (1991). *Chaos,* **1**, 49.

Brandstäter, A. & Swinney, H. L. (1987). Strange attractors in weakly turbulent Couette–Taylor flow. *Phys. Rev. A,* **35**, 2207. Reprinted in [Ott *et al.* (1994)].

Brandstäter, A., Swift, J., Swinney, H. L. & Wolf, A. (1983). Low-dimensional chaos in a hydrodynamic system. *Phys. Rev. Lett.,* **51**, 1442.

Brock, W. A., Dechert, W. D., Scheinkman, J. A. & LeBaron, B. (1988). *A Test for Independence Based on the Correlation Dimension*. University of Wisconsin Press, Madison.

Brockwell, P. & Davis, R. (1987). *Time Series: Theory and methods*. Springer series in Statistics, Springer, New York.

Broomhead, D. & King, G. P. (1986). Extracting qualitative dynamics from experimental data. *Physica D,* **20**, 217. Reprinted in [Ott *et al.* (1994)].

Broomhead, D. & Lowe, D. (1988). Multivariable function interpolation and adaptive networks. *Complex Systems,* **2**, 321.

Broomhead, D., Jones, R. & King, G. P. (1987). Topological dimension and local coordinates from time series data. *Jour. Phys. A,* **20**, L563.

Broomhead, D., Huke, J. P. & Muldoon, M. R. (1992). Linear filters and nonlinear systems. *J. Roy. Stat. Soc. B,* **54**, 373.

Brown, R., Rulkov, N. F. & Tufillaro, N. B. (1994). Synchronization of chaotic systems: The effects of additive noise and drift in the dynamics of the driving. *Phys. Rev. E,* **50**, 4488.

Bünner, M. J., Ciofini, M., Giaquinta, A., Hegger, R., Kantz, H., Meucci, R., Politi, A. (2000a). Reconstruction of systems with delayed feedback: (I) Theory. *European Physical Journal D,* **10**, 165.

(2000b). Reconstruction of systems with delayed feedback: (II) Application. *European Physical Journal D,* **10**, 177.

Burghes, D. & Graham, A. (1980). *Introduction to Control Theory Including Optimal Control Theory.* Ellis Harwood, Chichester.

Buzug, Th. (1994). *Analyse Chaotischer Systeme.* B.I. Wissenschaftsverlag, Mannheim.

Buzug, Th. & Pfister, G. (1992). Comparison of algorithms calculating optimal parameters for delay time coordinates. *Physica D,* **58**, 127.

Buzug, Th., Reimers, T. & Pfister, G. (1990). Dimensions and Lyapunov spectra from measured time series of Taylor–Couette flow. In Busse, F. H. & Kramer, L., editors, *Nonlinear Evolution of Spatio-Temporal Structures in Dissipative Continuous Systems,* Plenum, New York.

(1990a). Optimal reconstruction of strange attractors from purely geometrical arguments. *Europhys. Lett.,* **13**, 605.

Buzug, Th., v. Stamm, J. & Pfister, G. (1992). Fractal dimensions of strange attractors obtained from the Taylor–Couette experiment. *Physica A,* **191**, 559.

Casdagli, M. (1989). Nonlinear prediction of chaotic time series. *Physica D,* **35**, 335. Reprinted in [Ott *et al.* (1994)].

(1991). Chaos and deterministic versus stochastic nonlinear modeling. *J. Roy. Stat. Soc. B,* **54**, 303.

Casdagli, M. C. (1997). Recurrence plots revisited. *Physica D,* **108**, 12.

Casdagli, M. & Eubank, S. (1992). *Nonlinear Modeling and Forecasting.* Santa Fe Institute Studies in the Science of Complexity, Proc. Vol. XII, Addison-Wesley, Reading, MA.

Casdagli, M., Des Jardins, D., Eubank, S., Farmer, J. D., Gibson, J., Hunter, N. & Theiler, J. (1991). Nonlinear modeling of chaotic time series: theory and applications. In Kim, J. H. & Stringer, J., editors, *Applied Chaos,* Wiley, New York.

Casdagli, M., Eubank, S., Farmer, J. D. & Gibson, J. (1991a). State space reconstruction in the presence of noise. *Physica D,* **51**, 52.

Cawley, R. & Hsu, G. H. (1992). Local-geometric-projection method for noise reduction in chaotic maps and flows. *Phys. Rev. A,* **46**, 3057.

Cencini, M., Falcioni, M., Kantz, H., Olbrich E. & Vulpiani, A. (2000). Chaos or Noise – Difficulties of a Distinction. *Phys. Rev. E,* **62**, 427.

Čenys, A. (1993). Lyapunov spectrum of the maps generating identical attractors. *Europhys. Lett.*, **21**, 407.

Chaitin, G. J. (1987). *Algorithmic Information Theory*. Cambridge University Press.

Cheng, G. & Dong, X. (1993). From chaos to order – perspectives and methodologies in controlling chaotic nonlinear dynamical systems. *Int. J. Bifurcation and Chaos*, **3**, 1363.

Cladis, P. E. & Palffy-Muhoray, P. (1994). *Spatio-Temporal Patterns in Nonequilibrium Complex Systems*. SFI Studies in the Sciences of Complexity, Addison–Wesley, Reading.

Collins, S. & De Luca, C. J. (1995). Upright, correlated random walk: A statistical– biomechanics approach to the human postural control system. *Chaos*, **5**, 57.

Cross, M. C. & Hohenberg, P. C. (1993). Pattern formation outside of equilibrium. *Rev. Mod. Phys.*, **65**, 851.

Crutchfield, J. P. & Kaneko, K. (1988). Are attractors relevant to turbulence? *Phys. Rev. Lett.*, **60**, 2715.

Crutchfield, J. P. & McNamara, B. S. (1987). Equations of motion from a data series. *Complex Systems*, **1**, 417.

Crutchfield, J. P. & Young, K. (1989). Inferring statistical complexity. *Phys. Rev. Lett.*, **63**, 105.

Citing the webbook as a whole:

Cvitanović, P., Artuso, R., Mainieri, R., Tanner, G. & Vattay, G. (2001). *Classical and Quantum Chaos.*, www.nbi.dk/ChaosBook/, Niels Bohr Institute, Copenhagen.

Darbyshire, A. G. & Broomhead, D. S. (1996). Robust estimation of tangent maps and Lyapunov spectra. *Physica D*, **89**, 287.

Davies, M. E. (1992). Noise reduction by gradient descent. *Int. J. Bifurcation and Chaos*, **3**, 113.

(1994). Noise reduction schemes for chaotic time series. *Physica D*, **79**, 174.

Devroye, L. (1986). *Lecture Notes on Bucket Algorithms*. Progress in Computer Science no. 6, Birkhäuser, Boston.

Diks, C. (1996). Estimating invariants of noisy attractors. *Phys. Rev. E*, **53**, 4263.

(1999). *Nonlinear Time Series Analysis*. World Scientific, Singapore.

Ding, M., Grebogi, C., Ott, E., Sauer, T. & Yorke, J. A. (1993). Plateau onset for correlation dimension: When does it occur?. *Phys. Rev. Lett.*, **70**, 3872. Reprinted in [Ott *et al.* (1994)].

Ditto, W. L., Rauseo, S. N. & Spano, M. L. (1990). Experimental control of chaos. *Phys. Rev. Lett.*, **65**, 3211. Reprinted in [Ott *et al.* (1994)].

Drazin, P. G. & King, G. P. (1992). *Proceedings of the IUTAM Symposium and NATO Advanced Research Workshop on Interpretation of Time Series for Nonlinear Mechanical Systems, Physica D*, **58**, 1.

Dressler, U. & Nitsche, G. (1992). Controlling chaos using time delay coordinates. *Phys. Rev. Lett.*, **68**, 1.

Eckmann, J. P. & Ruelle, D. (1985). Ergodic theory of chaos and strange attractors. *Rev. Mod. Phys.*, **57**, 617.

Eckmann, J. P., Oliffson Kamphorst, S., Ruelle, D. & Ciliberto, S. (1986). Lyapunov exponents from a time series. *Phys. Rev. A*, **34**, 4971. Reprinted in [Ott *et al.* (1994)].

Eckmann, J. P., Oliffson Kamphorst, S. & Ruelle, D. (1987). Recurrence plots of dynamical systems. *Europhys. Lett.*, **4**, 973.

Efron, B. (1982). *The Jackknife, the Bootstrap and Other Resampling Plans*. SIAM, Philadelphia, PA.

Essex, C., Lookman, T. & Nerenberg, M. A. H. (1987). The climate attractor over short time scales. *Nature*, **326**, 64.

Falconer, K. J. (1985). *The Geometry of Fractal Sets.* Cambridge University Press.

Farmer, J. D. & Sidorowich, J. J. (1987). Predicting chaotic time series. *Phys. Rev. Lett.,* **59**, 845. Reprinted in [Ott *et al.* (1994)].

(1988). Exploiting chaos to predict the future and reduce noise. In Lee, Y. C., editor, *Evolution, Learning and Cognition,* World Scientific, Singapore.

(1991). Optimal shadowing and noise reduction. *Physica D,* **47**, 373.

Farmer, J. D., Ott, E. & Yorke, J. A. (1983). The dimension of chaotic attractors. *Physica D,* **7**, 153.

Finardi, M., Flepp, L., Parisi, J., Holzner, R., Badii, R. & Brun, E. (1992). Topological and metric analysis of heteroclinic crises in laser chaos. *Phys. Rev. Lett.,* **68**, 2989.

Fisher, R. A. (1925). Theory of statistical estimation. *Trans. Cambridge Philos. Soc.,* **22**, 700.

Flepp, L., Holzner, R., Brun, E., Finardi, M. & Badii, R. (1991). Model identification by periodic-orbit analysis for NMR-laser chaos. *Phys. Rev. Lett.,* **67**, 2244. Reprinted in [Ott *et al.* (1994)].

Fraedrich, K. (1986). Estimating the dimensions of weather and climate attractors. *J. Atmos. Sci.,* **43**, 419.

Fraser, A. M. & Swinney, H. L. (1986). Independent coordinates for strange attractors from mutual information. *Phys. Rev. A,* **33**, 1134.

Friedrich R. & Peinke, J. (1997). Description of a turbulent cascade by a Fokker–Planck equation. *Phys. Rev. Lett.,* **78**, 863.

Froehling, H., Crutchfield, J. P., Farmer, J. D., Packard, N. & Shaw, R. (1981). On determining the dimension of chaotic flows. *Physica D,* **3**, 606.

Galka, A. (2000). *Topics in Nonlinear Time Series Analysis.* World Scientific, Singapore.

Gardiner, C. W. (1997). *Handbook of stochastic methods: for physics, chemistry and the natural sciences.* 2nd edition, Springer series in synergetics; Springer, Berlin.

Garfinkel, A., Spano, M., Ditto, W. L. & Weiss, J. (1992). Controlling cardiac chaos. *Science,* **257**, 1230.

Gaspard, P. & Wang, X.-J. (1993). Noise, chaos and (ϵ, τ)-entropy per unit time. *Phys. Rev.* **235** 291

Geist, K., Parlitz, U. & Lauterborn, W. (1990). Comparison of different methods for computing Lyapunov exponents. *Prog. Theo. Phys.,* **83**, 875.

Gershenfeld, N. A. (1992). Dimension measurements on high-dimensional systems. *Physica D,* **55**, 135.

Gershenfeld, N. A. & Weigend, A. S. (1993). The future of time series. In [Weigend & Gershenfeld (1993)].

Gibson, J. F., Farmer, J. D., Casdagli, M. & Eubank, S. (1992). An analytic approach to practical state space reconstruction. *Physica D,* **57**, 1.

Gills, Z., Iwata, C., Roy, C., Schwartz, I. B. & Triandaf, I. (1992). Tracking unstable steady states: Extending the stability regime of a multimode laser system. *Phys. Rev. Lett.,* **69**, 3169. Reprinted in [Ott *et al.* (1994)].

Giona, M., Lentini, F. & Cimagalli, V. (1991). Functional reconstruction and local prediction of chaotic time series. *Phys. Rev. A,* **44**, 3496.

Giovanini, F. & Politi, A. (1991). Homoclinic tangencies, generating partitions, and curvature of invariant manifolds. *Jour. Phys. A,* **24**, 1837.

Glendinning, P. (1994). *Stability, Instability and Chaos: An introduction to the theory of nonlinear differential equations.* Cambridge University Press.

Goldberger, A. L. (1977). *Clinical Electrocardiography,* Mosby, St Louis.

Gollub, J. P. & Swinney, H. L. (1975). Onset of turbulence in a rotating fluid. *Phys. Rev. Lett.,* **35**, 927.

Grassberger, P. (1983). Generalized dimension of strange attractors. *Phys. Lett. A*, **97**, 227.
  (1985). Generalizations of the Hausdorff dimension of fractal measures. *Phys. Lett. A*, **107**, 101.
  (1986). Towards a quantitative theory of self-generated complexity. *Int. J. Theoret. Phys.*, **25**, 907.
  (1988). Finite sample corrections to entropy and dimension estimates. *Phys. Lett. A*, **128**, 369.
  (1989). Information content and predictability of lumped and distributed dynamical systems. *Physica Scripta*, **40**, 346.
  (1990). An optimized box-assisted algorithm for fractal dimensions. *Phys. Lett. A*, **148**, 63.
Grassberger, P. & Kantz, H. (1985). Generating partitions for the dissipative Hénon map. *Phys. Lett. A*, **113**, 235.
Grassberger, P. & Procaccia, I. (1983). Measuring the strangeness of strange attractors. *Physica D*, **9**, 189.
  (1983a). Characterization of strange attractors. *Phys. Rev. Lett.*, **50**, 346.
Grassberger, P., Badii, R. & Politi, A. (1988). Scaling laws for invariant measures on hyperbolic and nonhyperbolic attractors. *J. Stat. Phys.*, **51**, 135.
Grassberger, P., Schreiber, T. & Schaffrath, C. (1991). Non-linear time sequence analysis. *Int. J. Bifurcation and Chaos*, **1**, 521.
Grassberger, P., Hegger, R., Kantz, H., Schaffrath, C. & Schreiber, T. (1993). On noise reduction methods for chaotic data. *Chaos*, **3**, 127. Reprinted in [Ott *et al.* (1994)].
Grebogi, C., Ott, E. & Yorke, J. A. (1983). Fractal basin boundaries, long-lived chaotic transients, and unstable–unstable pair bifurcations. *Phys. Rev. Lett.*, **50**, 935.
  (1986). Critical exponents of chaotic transients in nonlinear dynamical systems. *Phys. Rev. Lett.*, **57**, 1284.
Grebogi, C., Ott, E., Romeiras, F. & Yorke, J. A. (1987). Critical exponents for crisis-induced intermittency. *Phys. Rev. A*, **36**, 5365.
Grebogi, C., Hammel, S. M., Yorke, J. A. & Sauer, T. (1990). Shadowing of physical trajectories in chaotic dynamics: Containment and refinement. *Phys. Rev. Lett.*, **65**, 1527.
Green, M. L. & Savit, R. (1991). Dependent variables in broadband continous time series. *Physica D*, **50**, 521.
Guckenheimer, J. & Buzyna, G. (1983). Dimension measurements for geostrophic turbulence. *Phys. Rev. Lett.*, **51**, 1438. Reprinted in [Ott *et al.* (1994)].
Guckenheimer, J. & Holmes, P. (1983). *Nonlinear Oscillations, Dynamical Systems, and Bifurcations of Vector Fields*, Springer, New York.
Gunaratne, G. H., Linsay, P. S. & Vinson, M. J. (1989). Chaos beyond onset: A comparison of theory and experiment. *Phys. Rev. Lett.*, **63**, 1.
Güttler, S., Kantz, H. & Olbrich, E. (2001). Reconstruction of the parameter spaces of dynamical systems. *Phys. Rev. E*, **63**, 056215.
Haken, H. (1975). Analogy between higher instabilities in fluids and lasers. *Phys. Lett. A*, **53**, 77.
Haken, H. (1977). *Synergetics*. Springer, Berlin.
  (1983). At least one Lyapunov exponent vanishes if the trajectory of an attractor does not contain a fixed point. *Phys. Lett. A*, **94**, 71.
Hale, J. K. (1977). *Theory of Functional Differential Equations*. Springer, New York.
Halsey, T. C., Jensen, M. H., Kadanoff, L. P., Procaccia, I. & Shraiman, B. I. (1986). Fractal measures and their singularities: The characterization of strange sets. *Phys. Rev. A*, **33**, 1141.

Hammel, S. M. (1990). A noise reduction method for chaotic systems. *Phys. Lett. A*, **148**, 421. Reprinted in [Ott *et al.* (1994)].

Hamming, R. W. (1983). *Digital Filters.*, 2nd edn. Prentice-Hall, Englewood Cliffs, NJ.

Hausdorff, F. (1918). Dimension und äuseres Mas. *Math. Annalen*, **79**, 157.

Healey, M. (1967). *Principles of Automatic Control.* van Nostrand, Princeton.

Hediger, T., Passamante, A. & Farrell, M. E. (1990). Characterizing attractors using local intrinsic dimensions calculated by singular-value decomposition and information-theoretic criteria. *Phys. Rev. A*, **41**, 5325.

Hegger, R & Kantz, H. (1997). Embedding of sequences of time intervals. *Europhys. Lett.*, **38**, 267.

    (1999). Improved false nearest neighbour method to detect determinism in time series data. *Phys. Rev. E*, **60**, 4970.

Hegger, R. & Schreiber, T. (1992). A noise reduction method for multivariate time series. *Phys. Lett. A*, **170**, 305.

Hegger, R., Kantz, H., Schmüser, F., Diestelhorst, M., Kapsch, R. P. & Beige, H. (1998). Dynamical properties of a ferroelectric capacitor observed through nonlinear time series analysis. *Chaos*, **8**, 727.

Hegger, R., Kantz, H. & Schreiber, T. (1999). Practical implementation of nonlinear time series methods: The TISEAN package., *Chaos*, **9**, 413.

    The TISEAN routines can be downloaded from
    http://www.mpipks-dresden.mpg.de/~tisean

Hegger, R., Kantz, H., Matassini, L. & Schreiber, T. (2000). Coping with non-stationarity by overembedding., *Phys. Rev. Lett.*, **84**, 4092.

Hénon, M. (1976). A two-dimensional mapping with a strange attractor. *Commun. Math. Phys.*, **50**, 69.

Herzel, H. & Reuter, R. (1996). Biphonation in voice signals. In Katz, R. A., Frison, T. W., Kadtke, J. B. & Bulsara, A. R., editors, *Non-Linear Dynamics and Full Spectrum Processes*, AIP Press, Woodbury.

Hofmeister, J. F., Slocumb, J. C., Kottmann, L. M., Picchiottino, J. B. & Ellis, D. G. (1994). A noninvasive method for recording the electrical activity of the human uterus *in vivo.*, *Biomed. Instr. Technol.*, **Sept.**, 391.

Hübner, U., Abraham, N. B. & Weiss, C. O. (1989). Dimensions and entropies of chaotic intensity pulsations in a single-mode far-infrared NH_3 laser. *Phys. Rev. A*, **40**, 6354.

Hübner, U., Weiss, C. O., Abraham, N. B. & Tang, D. (1993). Lorenz-like chaos in NH_3-FIR lasers. In [Weigend & Gershenfeld (1993)].

Hunt, E. R. (1991). Stabilizing high-period orbits in a chaotic system: the diode resonator. *Phys. Rev. Lett.*, **67**, 1953.

Ikeda, K. (1979). Multiple valued stationary state and its instability of the transmitted light by a ring cavity system. *Opt. Commun.*, **30**, 257.

In, V., Ditto, W. L. & Spano, M. L. (1995). Adaptive control and tracking of chaos in a magnetoelastic ribbon. *Phys. Rev. E*, **51**, R2689.

Isliker, H. & Kurths, J. (1993). A test for stationarity: Finding parts in a time series apt for correlation dimension estimates. *Int. J. Bifurcation and Chaos*, **3**, 1573.

Jackson, E. A. (1990). The entrainment and migration controls of multiple-attractor systems. *Phys. Lett. A*, **151**, 478. Reprinted in [Ott *et al.* (1994)].

    (1991). On the control of complex dynamical systems. *Physica D*, **50**, 341.

Jaeger, L. & Kantz, H. (1996). Unbiased reconstruction underlying a noisy chaotic time series. *Chaos*, **6**, 440.

Jenkins, G. M. & Watts, D. G. (1986). *Spectral analysis and its applications.* Holden–Day, San Francisco.

Just, W., Kantz, H., Ragwitz, M. & Schüser, F. (2003). Nonequilibrium physics meets time series analysis: measuring Probability currents from data. *Europhys. Lett., 62,* 28.

Kantelhardt, J. W., Koscielny-Bunde, E., Rego, H. H. A., Havlin, S. & Bunde, A. (2001). Detecting long-range correlations with detrended fluctuation analysis. *Physica A, 295,* 441.

Kantz, H. (1994a). A robust method to estimate the maximal Lyapunov exponent of a time series. *Phys. Lett. A, 185,* 77.

Kantz, H. (1994b). Quantifying the closeness of factal measures. *Phys. Rev., E, 49,* 5091.

Kantz, H. & Grassberger, P. (1985). Repellers, semi-atrractors and long lived chaotic transients. *Physica D, 17,* 75.

Kantz, H. & Jaeger, L. (1997). Improved cost functions for modelling based on noisy chaotic time series. *Physica D, 109,* 59.

Kantz, H. & Schürmann, T. (1996). Enlarged scaling ranges in entropy and dimension estimates. *Chaos, 6,* 167.

Kantz, H., Schreiber, T., Hoffmann, I., Buzug, T., Pfister, G., Flepp, L. G., Simonet, J., Badii, R. & Brun, E. (1993). Nonlinear noise reduction: a case study on experimental data. *Phys. Rev. E, 48,* 1529.

Kantz, H., Hegger, R. & Matassini, L. (2001). Noise reduction for human voice by local projections in reconstructed phase spaces. *IEEE Transactions on Circuits and Systems I, 48,* 1454.

Kaplan, D. T. & Glass, L. (1992). Direct test for determinism in a time series. *Phys. Rev. Lett., 68,* 427. Reprinted in [Ott *et al.* (1994)].

(1995). *Understanding Nonlinear Dynamics.* Springer, New York.

Kaplan, J. & Yorke, J. (1979). Chaotic behavior of multidimensional difference equations. In Peitgen, H. O. & Walther, H. O., editors, *Functional Differential Equations and Approximation of Fixed Points,* Springer, Heidelberg–New York.

Katok, A. & Hasselblatt, B. (1996). *Introduction to the Modern Theory of Dynamical Systems.* Cambridge University Press.

Kennel, M. B. (1997). Statistical test for dynamical nonstationarity in observed time-series data. *Phys. Rev. E, 56,* 316.

Kennel, M. B. & Isabelle, S. (1992). Method to distinguish possible chaos from colored noise and to determine embedding parameters. *Phys. Rev. A, 46,* 3111.

Kennel, M. B., Brown, R. & Abarbanel, H. D. I. (1992). Determining embedding dimension for phase-space reconstruction using a geometrical construction. *Phys. Rev. A, 45,* 3403. Reprinted in [Ott *et al.* (1994)].

Klafter, J. & Zumofen, G. (1993). Dynamically generated enhanced diffusion: the stationary state case. *Physica A, 196,* 102.

Knuth, D. (1973). *The Art of Computer Programming.* Vol. 3: *Sorting and Searching.* Addison-Wesley, Reading, MA.

Kolmogorov, A. N. (1958). A new invariant of transitive dynamical systems. *Dokl. Akad. Nauk. SSSR, 119,* 861.

Kostelich, E. J. & Schreiber, T. (1993). Noise reduction in chaotic time-series data: a survey of common methods. *Phys. Rev. E, 48,* 1752.

Kostelich, E. J. & Yorke, J. A. (1988). Noise reduction in dynamical systems. *Phys. Rev. A, 38,* 1649. Reprinted in [Ott *et al.* (1994)].

Kriso, S., Peinke, J., Friedrich, R. & Wagner, P. (2002). Reconstruction of dynamical equations for traffic flow. *Phys. Lett. A, 299,* 287.

Kruel, T. M., Eiswirth, M. & Schneider, F. W. (1993). Computation of Lyapunov spectra: Effect of interactive noise and application to a chemical oscillator. *Physica D, 63,* 117.

Kugiumtzis, D. (1996). State space reconstruction parameters in the analysis of chaotic time series – the role of the time window length. *Physica D*, **95**, 13.

Kullback, S. (1959). *Information Theory and Statistics*. Dover, New York.

Kurths, J. & Herzel, H. (1987). An attractor in a solar time series. *Physica D*, **25**, 165.

Layne, S. P., Mayer-Kress, G. & Holzfuss, J. (1986). Problems associated with dimensional analysis of EEG data. In [Mayer-Kress (1986)].

Ledrappier, F. & Young, L. S. (1985). The metric entropy of diffeomorphisms, Parts I and II. *Ann. Math.*, **122**, 509.

Liebert, W. & Schuster, H. G. (1989). Proper choice of the time delays for the analysis of chaotic time series. *Phys. Lett. A*, **142**, 107.

Liebert, W., Pawelzik, K. & Schuster, H. G. (1991). Optimal embeddings of chaotic attractors from topological considerations. *Europhys. Lett.*, **14**, 521.

Lorenz, E. N. (1963). Deterministic non-periodic flow. *J. Atmos. Sci.*, **20**, 130.

    (1969). Atmospheric predictability as revealed by naturally occurring analogues. *J. Atmos. Sci.*, **26**, 636.

Mackey, M. C. & Glass, L. (1977). Oscillations and chaos in physiological control systems. *Science*, **197**, 287.

Malinetskii, G. G., Potapov, A. B., Rakhmanov, A. I. & Rodichev, E. B. (1993). Limitations of delay reconstruction for chaotic systems with a broad spectrum. *Phys. Lett. A*, **179**, 15.

Mandelbrot, B. B. (1985). *The Fractal Geometry of Nature*. Freeman, San Francisco.

Manuca, R. & Savit, R. (1996). Stationarity and nonstationarity in time series analysis., *Physica D*, **99**, 134.

Marschinski, R. & Kantz H. (2002). Analysing the information flow between financial time series. An improved estimator for transfer entropy. EPJB30 275.

Mayer-Kress, G. (1986). *Dimensions and Entropies in Chaotic Systems*. Springer.

Mees, A. (editor) (2000). *Nonlinear Dynamics and Statistics*. Birkhäuser, Boston.

Milnor, J. (1985). On the concept of attractor. *Commun. Math. Phys.*, **99**, 177, *Commun. Math. Phys.*, **102**, 517.

Mitschke, F. (1990). Acausal filters for chaotic signals. *Phys. Rev. A*, **41**, 1169.

Mitschke, F., Möller, M. & Lange, W. (1988). Measuring filtered chaotic signals. *Phys. Rev. A*, **37**, 4518.

Möller, M., Lange, W., Mitschke, F., Abraham, N. B. & Hübner, U. (1989). Errors from digitizing and noise in estimating attractor dimensions. *Phys. Lett. A*, **138**, 176.

Moody, G. B. & Mark, R. (1991). MIT-BIH Polysomnographic Database. Biomedical Engineering Centre, MIT, Cambridge, MA.

Morfill, G. E. & Schmidt, G. (1994). Komplexitätsanalyse in der Kardiologie. *Phys. Blätter*, **50**, 156.

Nagai, Y. & Lai, Ying-Cheng (1995). Selection of a desirable chaotic phase using small feedback control. *Phys. Rev. E*, **51**, 3842.

Nicolis, C. & Nicolis, G. (1984). Is there a climatic attractor? *Nature*, **311**, 529.

Oltmans, H. & Verheijen, P. J. T. (1997). The influence of noise on power law scaling functions and an algorithm for dimension estimations. *Phys. Rev. E*, **56**, 1160.

Osborne, A. R. & Provenzale, A. (1989). Finite correlation dimension for stochastic systems with power-law spectra. *Physica D*, **35**, 357.

Osborne, A. R., Kirwan, A. D., Provenzale, A. & Bergamasco, L. (1986). A search for chaotic behavior in large and mesoscale motions in the Pacific Ocean. *Physica D*, **23**, 75.

Oseledec, V. I. (1968). A multiplicative ergodic theorem. Lyapunov characteristic numbers for dynamical systems. *Tran. Moscow Math. Soc.*, **19**, 197.

Ott, E. (1993). *Chaos in Dynamical Systems*. Cambridge University Press.

Ott, E. & Spano, M. (1995). Controlling chaos. *Physics Today,* **May 1995**, 34.

Ott, E., Grebogi, C. & Yorke, J. A. (1990). Controlling chaos. *Phys. Rev. Lett.,* **64**, 1196. Reprinted in [Ott *et al.* (1994)].

Ott, E., Sauer, T. & Yorke, J. A. (1994). *Coping with Chaos*. Wiley, New York.

Packard, N. H., Crutchfield, J. P., Farmer, J. D. & Shaw, R. S. (1980). Geometry from a time series. *Phys. Rev. Lett.,* **45**, 712. Reprinted in [Ott *et al.* (1994)].

 (1987). Anomalous scaling laws in multifractal objects. *Phys. Rep.,* **156**, 147.

Paluš, M. (1995). Testing for nonlinearity using redundancies: Quantitative and qualitative aspects. *Physica D,* **80**, 186.

Paparella, F., Provenzale, A., Smith, L. A., Taricco, C. & Vio, R. (1997). Local random analogue prediction of nonlinear processes. *Phys. Rev. A,* **235**, 233.

Parlitz, U. (1992). Identification of true and spurious Lyapunov exponents from time series. *Int. J. Bifurcation and Chaos,* **2**, 155. Reprinted in [Ott *et al.* (1994)].

 (1996). Estimating model parameters from time series by autosynchronization. *Phys. Rev. Lett.,* **76**, 1232.

Parlitz, U., Zöller, R., Holzfuss, J. & Lauterborn, W. (1994). Reconstructing physical variables and parameters from dynamical systems. *Int. J. Bifurcation and Chaos,* **4**, 1715.

Pawelzik, K. (1991). *Nichtlineare Dynamik und Hirnaktivität*. Verlag Harri Deutsch, Frankfurt (M).

Pecora, L. M. & Carroll, T. L. (1990). Synchronization in chaotic systems. *Phys. Rev. Lett.,* **64**, 821.

 (1991). Driving systems with chaotic signals. *Phys. Rev. A,* **44**, 2374.

Pecora, L. M., Carroll, T. L. & Heagy J. F. (1995). Statistics for mathematical properties of maps between time series embeddings. *Phys. Rev. E,* **52**, 3420.

Peng, C.-K., Buldyrev, S. V., Havlin, S., Simons, M., Stanley, H. E. & Goldberger, A. L. (1994). On the mosaic of organization of DNA sequences. *Phys. Rev. E,* **49**, 1691.

Pesin, Ya. B. (1977). Characteristic Lyapunov exponents and smooth ergodic theory. *Russ. Math. Surveys,* **32**, 55.

Pfister, G., Schulz, A. & Lensch, B. (1991). Bifurcations and a route to chaos of a one-vortex-state in Taylor–Couette flow. *Eur. J. Mech. B,* **10**, 247.

Pfister, G., Buzug, Th. & Enge, N. (1992). Characterization of experimental time series from Taylor–Couette flow. *Physica D,* **58**, 441.

Pierson, D. & Moss, F. (1995). Detecting periodic unstable points in noisy chaotic and limit cycle attractors with applications to biology. *Phys. Rev. Lett.,* **75**, 2124.

Pikovsky, A. (1986). Discrete-time dynamic noise filtering. *Sov. J. Commun. Technol. Electron.,* **31**, 81.

Pikovsky, A. & Grassberger, P. (1991). Symmetry breaking bifurcation for coupled chaotic attractors. *Jour. Phys. A,* **24**, 4587.

Pikovsky, A., Rosenblum, M. & Kurth, J. (2001). *Synchronization: a universal concept in nonlinear science*. Cambridge University Press, Cambridge, U. K.

Pomeau, Y. & Manneville, P. (1980). Intermittent transition to turbulence in dissipative dynamical systems. *Comm. Math. Phys.,* **74**, 189.

Preparata, F. P. & Shamos, M. I. (1985). *Computational Geometry, an Introduction*. Springer, New York.

Press, W. H., Flannery, B. P., Teukolsky, S. A. & Vetterling, W. T. (1992). *Numerical Recipes*. 2nd edn., Cambridge University Press.

Prichard, D. & Theiler, J. (1994). Generating surrogate data for time series with several simultaneously measured variables. *Phys. Rev. Lett.,* **73**, 951.

(1995). Generalized redundancies for time series analysis. *Physica D,* **84**, 476.

Priestley, M. B. (1981). *Spectral Analysis and Time Series.* Vols. I and II, Academic Press, London.

(1988). *Non-linear and Non-stationary Time Series Analysis.* Academic Press, London.

Provenzale, A., Smith, L. A., Vio, R. & Murante, G. (1992). Distinguishing between low-dimensional dynamics and randomness in measured time series. *Physica D,* **58**, 31.

Pyragas, K. (1992). Continuous control of chaos by self-controlling feedback. *Phys. Lett. A,* **170**, 421.

Pyragas, K. & Tamaševičius, A. (1993). Experimental control of chaos by delayed self-controlling feedback. *Phys. Lett. A,* **180**, 99.

Ragwitz, M. & Kantz, H. (2000). Detecting nonlinear structure and predicting turbulent gusts in surface wind velocities. *Europhys. Lett.,* **51**, 595.

(2001). Indispensible finite time corrections for Fokker–Planck equations from time series data. *Phys. Rev. Lett.,* **87**, 254501.

(2002). Markov models from data by simple nonlinear time series predictors in delay embedding spaces. *Phys. Rev. E,* **65**, 056201.

Renyi, A. (1971). *Probability Theory.* North Holland, Amsterdam.

Rieke, C., Sternickel, K., Andrzejak, R. G., Elger, C. E., David, P. & Lehnetz, K. (2002). Measuring nonstationarity by analyzing the loss of recurrence in dynamical systems. *Phys. Rev. Lett.,* **88**, 244102.

Rigney, D. R., Goldberger, A. L., Ocasio, W., Ichimaru, Y., Moody, G. B. & Mark, R. (1993). Multi-channel physiological data: Description and Analysis. In [Weigend & Gershenfeld (1993)].

Risken, H. (1989). *The Fokker–Planck equation: methods of solution and applications.* Springer, Berlin.

Rissanen, J. (1980). Consistent order estimates of autoregressive processes by shortest description of data. in Jacobs, O. *et al.*, editors, *Analysis and Optimisation of Stochastic Systems,* Academic, New York.

Romeiras, F., Grebogi, C., Ott, E. & Dayawansa, W. P. (1992). Controlling chaotic dynamical systems. *Physica D,* **58**, 165. Reprinted in [Ott *et al.* (1994)].

Rosenblum, M. G., Pikovsky, A. S. & Kurths, J. (1996). Phase synchronisation of chaotic attractors. *Phys. Rev. Lett.,* **76**, 1804.

Rosenstein, M. T., Collins, J. J. & De Luca, C. J. (1993). A practical method for calculating largest Lyapunov exponents from small data sets. *Physica D,* **65**, 117.

(1994). Reconstruction expansion as a geometry-based framework for choosing proper delay times. *Physica D,* **73**, 82.

Rössler, O. E. (1979). An equation for hyperchaos. *Phys. Lett. A,* **71**, 155.

Roy, R., Murphy, T. W., Maier, T. D., Gills, Z. & Hunt, E. R. (1992). Dynamical control of a chaotic laser: Experimental stabilization of a globally coupled system. *Phys. Rev. Lett.,* **68**, 1259.

Ruelle, D. (1978). *Thermodynamic Formalism.* Addison-Wesley, Reading.

(1989). *Chaotic Evolution and Strange Attractors: The statistical analysis of time series for deterministic nonlinear systems.* Cambridge University Press.

(1991). *Chance and Chaos.* Princeton University Press, Princeton, NJ.

Rulkov, N. F., Sushchick, M. M., Tsimring, L. S. & Abarbanel, H. D. I. (1995). Generalized synchronization of chaos in directionally coupled chaotic systems. *Phys. Rev. E,* **51**, 980.

Rushdie, S. (1990). *Haroun and the Sea of Stories.* Granta Books, London.

Salvino, L. W. & Cawley, R. (1994). Smoothness implies determinism: A method to detect it in time series. *Phys. Rev. Lett., 73*, 1091.

Sano, M. & Sawada, Y. (1985). Measurement of the Lyapunov spectrum from a chaotic time series. *Phys. Rev. Lett., 55*, 1082.

Saparin, P. (1995). Private communication.

Sauer, T. (1992). A noise reduction method for signals from nonlinear systems. *Physica D, 58*, 193.

(1993). Times series prediction using delay coordinate embedding. In [Weigend & Gershenfeld (1993)]. Reprinted in [Ott *et al.* (1994)].

(1994). Reconstruction of dynamical systems from interspike intervals. *Phys. Rev. Lett., 72*, 3811.

(1995). Interspike embedding of chaotic signals. *Chaos, 5*, 127.

Sauer, T. & Yorke, J. A. (1993). How many delay coordinates do you need? *Int. J. Bifurcation and Chaos, 3*, 737.

Sauer, T., Yorke, J. A. & Casdagli, M. (1991). Embedology. *J. Stat. Phys., 65*, 579.

Schäfer, C., Rosenblum, M. G., Kurths, J. & Abel, H.-H. (1998). Heartbeat synchronized with ventilation. *Nature, 392*, 239.

Schreiber, T. (1993). Extremely simple nonlinear noise reduction method. *Phys. Rev. E, 47*, 2401.

(1993a). Determination of the noise level of chaotic time series. *Phys. Rev. E, 48*, R13.

(1995). Efficient neighbour searching in nonlinear time series analysis. *Int. J. Bifurcation and Chaos, 5*, 349.

(1997). Detecting and analysing nonstationarity in a time series with nonlinear cross predictions. *Phys. Rev. Lett., 78*, 843.

(1999). Interdisciplinary application of nonlinear time series methods. *Phys. Rep., 308*, 2.

(2000). Measuring information transfer. *Phys. Rev. Lett., 85*, 461.

Schreiber, T. & Grassberger, P. (1991). A simple noise-reduction method for real data. *Phys. Lett. A, 160*, 411.

Schreiber, T. & Kantz, H. (1996). Observing and predicting chaotic signals: Is 2% noise too much? In Kravtsov, Y. & Kadtke, J., editors, *Predictability of Complex Dynamical Systems*. Springer, New York.

Schreiber, T. & Kaplan, D. T. (1996a). Nonlinear noise reduction for electrocardiograms. *Chaos, 6*, 87.

(1996b). Signal separation by nonlinear projections: The fetal electrocardiogram. *Phys. Rev. E, 53*, 4326.

Schreiber, T. & Schmitz, A. (1996). Improved surrogate data for nonlinearity tests. *Phys. Rev. Lett., 77*, 635.

(1997). Discrimination power of measures for nonlinearity in a time series. *Phys. Rev. E, 55*, 5443.

(2000). Surrogate time series. *Physica D, 142*, 346.

Schuster, H. G. (1988). *Deterministic Chaos: An introduction*. Physik Verlag, Weinheim.

Schuster, H. G., (editor) (1999). *Handbook of chaos control*. Wiley-VCH, Weinheim.

Shannon, C. E. & Weaver, W. (1949). *The Mathematical Theory of Information*. University of Illinois Press, Urbana, IL.

Shinbrot, T., Ott, E., Grebogi, C. & Yorke, J. A. (1990). Using chaos to direct trajectories to targets. *Phys. Rev. Lett., 46*, 3250.

Shinbrot, T., Grebogi, C., Ott, E. & Yorke, J. A. (1993). Using small perturbations to control chaos. *Nature, 363*, 411.

Sinai, A. G. (1959). On the concept of entropy of a dynamical system. *Dokl. Akad. Nauk. SSSR,* **124**, 768.

Singer, J., Wang, Y. Z. & Bau, H. H. (1991). Controlling a chaotic system. *Phys. Rev. Lett.,* **66**, 1123.

Smale, S. (1967). Differentiable dynamical systems. *Bull. Amer. Math. Soc.,* **73**, 747.

Smith, L. A. (1992). Identification and prediction of low-dimensional dynamics. *Physica D,* **58**, 50.

So, P., Ott, E., Schiff, S. J., Kaplan, D. T., Sauer, T. & Grebogi, C. (1996). Detecting unstable periodic orbits in chaotic experimental data. *Phys. Rev. Lett.,* **76**, 4705.

So, P., Ott, E., Sauer, T., Gluckman, B. C., Grebogi, C. & Schiff, S. J. (1997). Extracting unstable periodic orbits from chaotic time series data. *Phys. Rev. E,* **55**, 5398.

Stoop, R. & Meier, P. F. (1988). Evaluation of Lyapunov exponents and scaling functions from time series. *J. Opt. Soc. Am. B,* **5**, 1037.

Stoop, R. & Parisi, J. (1991). Calculation of Lyapunov exponents avoiding spurious elements. *Physica D,* **50**, 89.

Stoop, R., Peinke, J., Parisi, J., Röhricht, B. & Huebener, R. P. (1989). A p-Ge semiconductor experiment showing chaos and hyperchaos. *Physica D,* **35**, 425.

Subba Rao, T. & Gabr, M. M. (1984). *An Introduction to Bispectral Analysis and Bilinear Time Series Models.* Lecture Notes in Statistics Vol. 24, Springer, New York.

Sugihara, G. & May, R. (1990). Nonlinear forecasting as a way of distinguishing chaos from measurement error in time series. *Nature,* **344**, 734. Reprinted in [Ott *et al.* (1994)].

Takens, F. (1981). *Detecting Strange Attractors in Turbulence.* Lecture Notes in Math. Vol. 898, Springer, New York.

(1983). Invariants related to dimension and entropy. In Atas do 13º Coloquio Brasileiro de Matematica, Instituto de Matematica Pura e Aplicada, Rio de Janeiro.

(1985). On the numerical determination of the dimension of an attractor. In Braaksma, B. L. J., Broer, H. W. & Takens, F., editors, *Dynamical Systems and Bifurcations,* Lecture Notes in Math. Vol. 1125, Springer, Heidelberg–New York.

Theiler, J. (1986). Spurious dimension from correlation algorithms applied to limited time series data. *Phys. Rev. A,* **34**, 2427.

(1987). Efficient algorithm for estimating the correlation dimension from a set of discrete points. *Phys. Rev. A,* **36**, 4456.

(1988). Lacunarity in a best estimator of fractal dimension. *Phys. Lett. A,* **135**, 195.

(1990). Statistical precision in dimension estimators. *Phys. Rev. A,* **41**, 3038.

(1990a). Estimating fractal dimension. *J. Opt. Soc. Am. A,* **7**, 1055.

(1991). Some comments on the correlation dimension of $1/f^{\alpha}$ noise. *Phys. Lett. A,* **155**, 480.

Theiler, J. & Eubank, S. (1993). Don't bleach chaotic data. *Chaos,* **3**, 771.

Theiler, J. & Prichard, D. (1995). Constrained-realization Monte-Carlo method for hypothesis testing. *Physica D,* **94**, 221.

(1996). Using 'Surrogate Surrogate Data' to calibrate the actual rate of false positives in tests for nonlinearity in time series. Fields Institute Communications, to appear.

Theiler, J., Eubank, S., Longtin, A., Galdrikian, B. & Farmer, J. D. (1992). Testing for nonlinearity in time series: The method of surrogate data. *Physica D,* **58**, 77. Reprinted in [Ott *et al.* (1994)].

Theiler, J., Galdrikian, B., Longtin, A., Eubank, S. & Farmer, J. D. (1992a). Using surrogate data to detect nonlinearity in time series. In [Casdagli & Eubank (1992)].

Theiler, J., Linsay, P. S. & Rubin, D. M. (1993). Detecting nonlinearity in data with long coherence times. In [Weigend & Gershenfeld (1993)].

Timmer, J., Rust, H., Horbelt, W. & Voss, H. U. (2000). Parametric, nonparametric and parametric modelling of a chaotic circuit time series. *Phys. Lett. A,* **274**, 123.

Tong, H. (1983). *Threshold Models in Non-Linear Time Series Analysis.* Lecture Notes in Statistics Vol. 21, Springer, New York.

(1990). *Non-Linear Time Series Analysis.* Oxford University Press, Oxford.

Tong, H. (editor) (1993). *Dimension Estimation and Models.* World Scientific, Singapore.

(1995). *Chaos and Forecasting.* World Scientific, Singapore.

Torcini, A., Politi, A., Puccioni, G. P. & D'Alessandro, G. (1991). Fractal dimension of spatially extended systems. *Physica D,* **53**, 85.

Tsimring, L. S. (1993). Nested strange attractors in spatiotemporal chaotic systems. *Phys. Rev. E,* **48**, 3421.

Tsonis, A. A. & Elsner, J. B. (1988). The weather attractor over very short time scales. *Nature,* **333**, 545.

(1990). Multiple attractors, fractal basins and longterm climate dynamics. *Contrib. Atmos. Phys.,* **63**, 171.

Tufillaro, N. B., Abbott, T. & Reilly, J. (1992). *An Experimental Approach to Nonlinear Dynamics and Chaos.* Addison-Wesley, Redwood City, CA.

Tufillaro, N. B., Wyckhoff, P., Brown, R., Schreiber, T. & Molteno, T. (1995). Topological time series analysis of a string experiment and its synchronized model. *Phys. Rev. E,* **51**, 164.

van de Water, W. & Schram, P. (1988). Generalized dimensions from near-neighbor information. *Phys. Rev. A,* **37**, 3118.

van Kampen, N. G. (1992). *Stochastic Processes in Physics and Chemistry.* 2nd edn. North Holland, Amsterdam.

Varone, A., Politi, A. & Ciofini, M. (1995). CO_2 laser with feedback. *Phys. Rev. A,* **52**, 3176.

Vautard, R., Yiou, P. & Ghil, M. (1992). Singular-spectrum analysis: a toolkit for short, noisy chaotic signals. *Physica D,* **58**, 95.

Voss, H. V., Kolodner, P., Abel, M. & Kurths, J. (1999). Amplitude equations from spatio-temporal binary-fluid convection data. *Phys. Rev. Lett.,* **83**, 3422.

Wayland, R., Bromley, D., Pickett, D. & Passamante, A. (1993). *Phys. Rev. Lett.,* **70**, 580.

Weigend, A. S. & Gershenfeld, N. A. (1993). *Time Series Prediction: Forecasting the future and understanding the past.* Santa Fe Institute Studies in the Science of Complexity, Proc. Vol. XV, Addison-Wesley, Reading.

Weigend, A. S., Huberman, B. A. & Rumelhart, D. E. (1990). Predicting the future: A connectionist approach. *Int. J. Neural Syst.,* **1**, 193.

Whitney, H. (1936). Differentiable manifolds. *Ann. Math.,* **37**, 645.

Wiener, N. (1961). *Cybernetics.* MIT Press, Cambridge, MA.

Witt, A., Kurths, J. & Pikovsky, A. (1998). Testing stationarity in time series. *Phys. Rev. E,* **58**, 1800.

Wolf, A., Swift, J. B., Swinney, H. L. & Vastano, J. A. (1985). Determining Lyapunov exponents from a time series. *Physica D,* **16**, 285.

Xu, Daolin & Bishop, S. R. (1994). Steering dynamical trajectrories to target a desired state. *Chaos, Solitons and Fractals,* **4**, 1931.

Zaslavsky, G. M. (1993). Self-similar transport in incomplete chaos. *Phys. Rev. E,* **48**, 1683.

Index

Printed in the United States
By Bookmasters